Phosphates as Food Ingredients

Author:
R. H. ELLINGER
Kraft Foods
Chicago, Ill.

CRC Press
Taylor & Francis Group
Boca Raton London New York

CRC Press is an imprint of the
Taylor & Francis Group, an **informa** business

First published 1972 by CRC Press
Taylor & Francis Group
6000 Broken Sound Parkway NW, Suite 300
Boca Raton, FL 33487-2742

Reissued 2018 by CRC Press

A Library of Congress record exists under LC control number: 72192987

ISBN 13: 978-1-315-89641-0 (hbk)
ISBN 13: 978-1-351-07551-0 (ebk)

Visit the Taylor & Francis Web site at http://www.taylorandfrancis.com and the
CRC Press Web site at http://www.crcpress.com

CRC UNISCIENCE SERIES

Another information program recently initiated by CRC is the "Uniscience" Series. The function of "Uniscience" has been defined as the publication of highly authoritative scientific reference works dealing with those subjects for which there exists an urgent necessity for published information due to significant new developments.

To achieve this function, the "Uniscience" program utilizes the conventional reference book format, but with one, all-important difference: it has radically reduced the traditional lag between completion of the manuscript and publication of the printed book. Utilizing newly developed editorial procedures together with newly perfected computerized type-setting equipment, the time-span normally required from manuscript to printed book has been dramatically shortened to a matter of only a few months.

This very real "break-through" in production time makes it possible for the Uniscience Series to provide a new dimension in the distribution of scientific information.

B. J. Starkoff, President
THE CHEMICAL RUBBER CO.

THE AUTHOR

Rudolph H. Ellinger, Ph.D., is currently the Manager of Regulatory Compliance at Kraft Foods Company in Chicago, Illinois. Dr. Ellinger received his B.A. degree in chemistry in 1950 from Michigan State University. From Iowa State University, he received his M.S. degree in organic chemistry in 1953 and his Ph.D. in biochemistry from Iowa State University in 1954. Prior to joining Kraft, he was Manager of Food Products Development for Stauffer Chemical Company at their Eastern Research Center at Dobbs Ferry, New York. It was here he obtained his experience with phosphates as food additives.

TABLE OF CONTENTS

Chapter 1
Introduction . . . 1

Chapter 2
Nomenclature, Classification and Structure of Phosphates Used in Foods . . . 3
The Orthophosphates . . . 3
Orthophosphoric Acid . . . 3
Ammonium Orthophosphates . . . 3
Calcium Orthophosphates . . . 3
Anhydrous Monocalcium Phosphate . . . 6
Monocalcium Phosphate Monohydrate . . . 6
Dicalcium Phosphate Dihydrate . . . 6
Tricalcium Phosphate . . . 7
Ferric Orthophosphate . . . 7
Potassium Orthophosphates . . . 7
Sodium Orthophosphates . . . 7
The Pyrophosphates . . . 9
Pyrophosphoric Acid . . . 9
Calcium Pyrophosphates . . . 9
Potassium Pyrophosphate . . . 9
Sodium Pyrophosphates . . . 9
Ferric Pyrophosphate . . . 10
The Tripolyphosphates . . . 10
Tripolyphosphoric acid . . . 10
Potassium Tripolyphosphate . . . 10
Sodium Tripolyphosphate . . . 10
The Straight-Chain Polyphosphates . . . 10
Polyphosphoric Acids . . . 11
Potassium Polyphosphate . . . 11
Sodium Polyphosphates . . . 11
The Cyclic Metaphosphates . . . 14

Chapter 3
Some General Chemical Characteristics of Phosphates . . . 15

Chapter 4
Toxicology of the Food Phosphates . . . 19
Acute Toxicity . . . 19
Chronic Toxicity of Phosphates in Animal Diets . . . 19
Orthophosphates . . . 19
Pyrophosphates . . . 23
Polyphosphates . . . 23
Cyclic Polyphosphates . . . 23
Biochemical Aspects of Phosphate Toxicity . . . 24
Acceptable Daily Intake for Humans . . . 24
Summary . . . 25

Chapter 5
The Attitudes of Regulatory Agencies . . . 27
The U.S. Food and Drug Administration . . . 27
The Meat Inspection Division, USDA . . . 27

Chapter 6

The Functions and Applications of Phosphates in Food Systems 31
Some General Functions of Phosphates in Foods 31
Complexing Metal Ions 31
Complexing Organic Polyelectrolyte Food Constituents 32
Direct Chemical Reactions with Food Constituents 32
Buffering or pH Stabilization 32
Dispersion of Food Constituents 32
Emulsion Stabilization 32
Increasing Hydration and Water Binding 32
Mineral Supplementation 32
Acidification or Lowering pH 32
Alkalization or Raising pH 32
Prevention of Caking 32
Food Preservation 33
Phosphate Applications in Beverages 33
Applications in Alcoholic Beverages 33
Buffering 33
Bacteriological Control 33
Complexing Metal Ions 33
Applications in Carbonated Beverages 34
Complexing Metal Ions 34
Acidification 34
Improving Flow of Powders 34
Applications in Other Beverages 34
Acidification 34
Mineral Supplementation 35
Complexing Metal Ions 35
Phosphate Applications in Cereal Products 35
Applications as Leavening Acids 35
Characteristics of Leavening Systems 35
Characteristics of Commercial Leavening Acids 36
Other Potential Phosphate Leavening Acids 40
Leavening Acids in Baking Powders 40
Leavening Acids in Prepared Mixes 40
Leavening Acids in Refrigerated Doughs and Batters 44
Non-Leavening Applications in Cereal Products 46
Dough Conditioning 46
Inhibition of Enzyme Activity 46
Antioxidant Activity 46
Inhibition of Microbiological Growth 46
Mineral Enrichment 47
Decreasing Cereal Cooking Time 50
Miscellaneous Applications 50
Phosphate Applications in Dairy Products 51
Interaction of Phosphates with Milk Constituents 51
Interaction with Milk Calcium 51
Effect of Phosphates on Heat Stability of Milk 52
Interactions with Milk Proteins 52
Applications in Milk Beverages 53
Applications in High Butterfat Dairy Products 53
Butter 53

Buttermilk 54
Cream Products 54
Applications in Condensed and Evaporated Milks 54
Applications in Sterile Concentrated Milk 55
Applications in Milk Gels and Puddings 57
Applications in Nonfat Milk 60
Applications in Frozen Dairy Desserts 62
Applications in Whey, Lactose, and Lactalbumin 63
Applications in Imitation Dairy Products 65
Imitation Milk 66
Coffee Whiteners 66
Whipped Toppings 68
Imitation Sour Cream, Sour Cream Dressings, and Chip-Dips 68
Imitation Ice Cream or "Frozen Desserts" 68
Imitation Cheese, Cream Cheese and Cheese Spreads 69
Margarine 69
Applications in Cheese Products 69
Cottage Cheese 69
Natural Cheese 69
Pasteurized Process Cheese Products 72
Phosphate Applications in Egg Products 76
Applications in Shell Eggs 76
Applications in Whole Egg Products 77
Applications in Egg Whites 79
Phosphate Applications in Fats and Oils 80
Applications in Oil Extraction 80
Applications in Processing Fats and Oils 80
Alkali Refining 80
Acid Refining 80
Other Refining Applications 81
Bleaching 81
Rearrangement 82
Hydrogenation 82
Monoglyceride Preparation 82
Other Processing Applications 82
Applications in Fat Antioxidant Systems 82
Dry Fat Systems 82
Aqueous Fat Systems 83
Applications in Oil and Water Emulsions 85
Phosphate Applications in Fruit and Vegetable Products 85
Applications as Inhibitors of Microbiological Spoilage 85
Surface Applications 85
Applications in Juices 86
Applications as Antioxidants 86
Applications in Stabilizing Fruit and Vegetable Color 87
Applications in Obtaining Optimum Texture 89
Pectin Gels 89
Tomato Products 90
Potato Texture 91
Tenderization 91
Firming Tissues 93
Miscellaneous Applications 93

Phosphate Applications in Gums and Gels . . . 94
Applications in Agar Gel Systems . . . 94
Applications in Alginate Gel Systems . . . 94
Applications in Carageenan Gel Systems . . . 96
Applications in Other Gum Gel Systems . . . 96
Phosphate Applications in Meat Processing . . . 97
The Biochemistry of Phosphate Interactions with Meat Proteins . . . 98
General Meat Applications . . . 105
Color Preservation . . . 105
Increasing Tenderness . . . 105
Increasing Binding . . . 106
Increasing Moisture Retention . . . 106
Flavor Improvement . . . 107
Preventing Off-Flavors . . . 107
Preventing Microbiological Spoilage . . . 107
Applications in Fresh Whole Meats . . . 108
Color Preservation . . . 108
Increasing Tenderness . . . 108
Increasing Binding . . . 110
Increasing Moisture Retention . . . 110
Flavor Improvement . . . 111
Applications in Fresh Comminuted Meat Products . . . 111
Color Preservation . . . 111
Increasing Tenderness . . . 112
Increasing Binding . . . 113
Increasing Moisture Retention . . . 113
Applications in Cured Meat Products . . . 114
Color Preservation . . . 114
Increasing Tenderness . . . 117
Increasing Binding . . . 117
Increasing Moisture Retention . . . 118
Improving Fat Emulsification . . . 120
Flavor Improvement . . . 121
Preventing Microbiological Spoilage . . . 122
Miscellaneous uses . . . 122
Additional Patents in the Literature . . . 123
Phosphate Applications in Poultry Processing . . . 124
Color Preservation . . . 124
Increasing Tenderness . . . 124
Increasing Binding . . . 125
Increasing Moisture Retention . . . 126
Increasing Fat Emulsification . . . 127
Improving flavor . . . 127
Preventing Microbiological Spoilage . . . 130
Phosphate Applications in Seafood Processing . . . 131
The Biochemistry of Phosphate Interactions with Seafood Flesh . . . 131
Color Preservation . . . 133
Increasing Tenderness . . . 133
Increasing Binding . . . 134
Increasing Moisture Retention . . . 135
Improving Flavor . . . 143
Preventing Microbiological Spoilage . . . 145

Preventing Struvite Crystals 147
Phosphate Applications as Microbiological Inhibitors 147
Applications as Inhibitors of Bacterial Growth 148
General Inhibitory Effects 148
Effects of Metal Chelation 149
Applications as Inhibitors of Yeast and Fungal Growth 152
Applications as Inhibitors of Viruses 152
Applications as Inhibitors of Viruses 152
Phosphate Applications in Processing Food Protein 152
Applications in Protein Dispersion 154
Increasing Protein Water Holding Capacity and Gelling Properties 154
Improving Protein Whipping Properties 154
Applications as Protein Precipitants 155
Phosphate Applications in Starch Processing 157
Effects of Phosphates on Starch Properties 157
The Starch Phosphates 157
Starch Phosphate Monoesters 158
Starch Phosphate Diesters 159
Phosphate Applications in Sugar Processing 160
Clarification of the Sugar-Bearing Juice 160
Bleaching the Juice 161
Other Applications 161
Food Applications For Organic Phosphates 161
Applications as Antioxidants 161
Applications as Emulsifying Agents 162
Applications as Whipping Agents 162
Other Applications 162

Chapter 7
Future Trends 165

References 167

Index 187

Chapter 1

INTRODUCTION

The element phosphorus is essential to life. It is present in every organism, whether it is a single-celled or a multi-celled organism. It is essential to the functions that give life and allow life to continue. Phosphorus, as the phosphate anion, is found in the various types of nucleic acids vital to genetic reproduction and to restoration of injured tissues. It is involved in all of the energy-producing reactions in the life processes. The phosphate anion is an essential component of many enzyme systems and is usually involved in the reactions of such enzymes. It is involved in the synthesis and metabolism of carbohydrate, fat, and protein and the formation and structure of such tissues as brain, muscle, and skeleton.

Phosphorus is absorbed into all living organisms mainly in the form of the phosphate ion.[1,2] No living organism is capable of synthesizing the phosphate anion;[1] it must, therefore, be absorbed through the food supply. Although some natural compounds found in living tissues contain phosphorus linked directly to carbon and other elements,[3] by far the most frequent linkage in living systems is that of the phosphate ester linked to the compound through oxygen.

The vital role played by the phosphate ion in the synthesis of compounds essential to life is demonstrated by the evidence that it was involved in the origin of life.[4-8] Some evidence exists that the earliest living organisms used polyphosphate or pyrophosphate rather than adenosine triphosphate (ATP) as an energy carrier,[5] although it has also been demonstrated that ATP could have been synthesized under conditions that existed on the primitive earth,[8] making it available for the origin and replication of life.

The occurrence of the orthophosphate anion in living tissue has been known for a considerable period of time. However, the occurrence of polyphosphate in a living tissue was first discovered by Liebermann in 1888.[9] Since that time numerous examples of the occurrence of polyphosphates in microbiological, botanical, and animal tissues have been demonstrated, reviews of which have recently appeared in the literature.[9,10] There are recent reports of investigations that demonstrate the occurrence of polyphosphates and their physiological necessity in microbiological cells.[11-15] Evidence has also been reported for occurrence of polyphosphates and their metabolic role in higher forms of life.[16-20] Numerous tissues contain enzymes capable of synthesizing as well as hydrolyzing polyphosphates ranging in chain length from pyrophosphate, containing two phosphate units, to those of several thousand units in length, as well as for cyclic metaphosphates.[19-21] Recent reviews of this topic have been published by Harold[10] and Mattenheimer.[21]

An appreciation of the involvement of the phosphate and polyphosphate anions in life processes may be obtained by examining the chart of biological pathways, attached to the *Handbook of Biochemistry*.[22] This chart demonstrates the requirement for phosphates in almost every vital function of respiration and biological synthesis. Sanchelli[2] states that phosphorus is so vital to plant tissues that it can be transferred from one organ to another during short supplies of phosphorus in the soil. The plant will drain phosphorus in the form of phosphate from the older, mature leaves into the new, growing tissues to make optimum use of the available phosphate.

Phosphate is a natural constituent of almost every type of food that we eat. It is impossible to eat any food derived from a once-living organism without obtaining phosphate. Table 1 shows typical phosphorus levels of common food products as examples.

The orthophosphates and polyphosphates serve many useful functions as additives in food applications. Among these are functions in ion-exchange reactions, as buffers, in interactions with other polyelectrolytes, as sequestrants of unwanted metal ions, and in microbiological inhibition. These will be discussed in more detail under the individual food products.

There is an enormous quantity of literature, including numerous patents, on the functions and applications of phosphates in foods. No attempt has been made to review every publication, as such an exhaustive review would require a sizeable book. Anyone interested in further details on any topic covered in this treatise should consult the references listed at the end of the chapter. The bibliographies provided in the references for each topic should provide more thorough coverage. It is hoped that the references cited are those with the most useful information on the phosphate applications and their effects on foods.

There are numerous reviews on the use of phosphates in certain types of food products. These have been vary helpful in obtaining references to original papers that could be accumulated and their data evaluated for use. The reviews by Kiermeier and Mohler,[24] Ruf,[25] and Shettino[26] and the Symposium on the Condensed Phosphates in Foods, held in Mainz in 1957,[27] were especially helpful. Uses of phosphates in foods have also been reviewed by Barackman and Bell[28] and in Stauffer Chemical Company's *Food Industry Release* No. 1.[29]

TABLE 1

Phosphorus Content of the Edible Portion of Some Typical Foods

Type of food	Phosphorus content, mg/100g
Roots, tubers, and bulbs,	
Beets	41 ± 1.1
Carrots	40± 1.2
Potatoes	56 ±0.8
Onion	33 ±1.1
Edible leaves and buds	
Broccoli	72 ± 2.3
Cabbage	30 ± 0.6
Lettuce	28 ± 0.9
Fruits and juices	
Apples	11 ± 0.17
Oranges	21 ± 0.5
Orange juice	17 ± 0.4
Peaches	18 ± 0.55
Cereal products	
Corn, sweet	120 ± 1.8
Oats (oatmeal)	395 ± 14.9
Wheat flour	101 ± 2.1
White bread	97 ± 2.0
Animal products	
Milk	93 ± 0.3
Cheese (cheddar)	524 ± 1.4
Eggs	224 ± 1.4
Beef, lean	204 ± 2.5

Source: *Table based on data in Sherman, H. C., *Calcium and Phosphorus*, Columbia University Press, New York, 1947.

Chapter 2

NOMENCLATURE, CLASSIFICATION, AND STRUCTURE OF PHOSPHATES USED IN FOODS

Most food scientists find the nomenclature and classification of the phosphates, particularly the polyphosphates, extremely confusing. This is very obvious to anyone who studies the food-science literature dealing with phosphate additives. In many cases it is impossible to determine the exact nature of the compound used. It is, therefore, thought of benefit to include a brief discussion of the nomenclature, classification, and structure of the common food phosphates in this chapter. The phosphates will be discussed according to their classification by the number of phosphorus atoms in the phosphate molecule. An attempt will also be made to point out the discrepancies in the nomenclature of these compounds.

A complete discussion of the chemistry of the food phosphates is beyond the scope of this chapter; however, the literature contains numerous sources of this information. One of the best and most complete discussions of the chemistry of phosphorus compounds has been published by Van Wazer.[30]

Table 2 summarizes the common nomenclature, structure, formula, pH, and solubility characteristics of the phosphates in current usage as food additives.[24,25,30-32] Each of the individual groups, and the compounds in them, will be discussed in more detail in the following paragraphs of this section.

The Orthophosphates

Orthophosphoric Acid

Orthophosphoric acid (H_3PO_4), or more commonly called phosphoric acid, is a well-known acidifying agent in foods as well as a natural constituent of many fruits and their juices. It is commercially available as a viscous, colorless, syrupy liquid. It is a tribasic acid, as it has three replaceable hydrogens. Its first replaceable hydrogen reacts as a strong acid, the second as a weaker acid, and the third as a very weak acid. Its dissociation constants are as follows at 25°C:[32]

$$K_1 - 0.71 \times 10^{-2};$$
$$K_2 - 0.63 \times 10^{-4};$$
$$K_3 - 4.7 \times 10^{-13}.$$

A number of liquid forms of the acid, all designated as phosphoric acid, are available. These are usually described by the percent of H_3PO_4 or percent of P_2O_5. It should, however, be emphasized that most of these commercially available products are mixtures of orthophosphoric acid with the higher polymers, such as pyrophosphoric, tripolyphosphoric, and higher acids. The amount of the polymeric acids present in any single preparation can be estimated from the percent P_2O_5 in the product. Pyrophosphoric and polyphosphoric acids begin to appear as the P_2O_5 increases above 69%, as shown in Table 3.[33,34]

The composition of the phosphoric acid at any given level of P_2O_5 depends entirely on the H_2O/P_2O_5 ratio; this composition will be constant regardless of how the acid has been prepared or of its previous history. Tables such as Table 3, showing the composition of phosphoric acids containing various percentages of P_2O_5, are available in the literature[30,34] and from suppliers of the acids. These compositions have been determined by paper chromatography by a method developed by Huhti and Gartaganis[34] in 1956. Most tables available from suppliers of these acids are based on this work.

Several salts of orthophosphoric acid are used in foods. Structures and some characteristics of phosphoric acids and the phosphate salts used in foods are shown in Table 2. Only those inorganic phosphates of significance to food processing are shown.

Ammonium Orthophosphates

Two ammonium orthophosphate salts are used in a few food applications. There are *monoammonium phosphate*, $NH_4H_2PO_4$, and *diammonium phosphate*, $(NH_4)_2HPO_4$. Both exist as the anhydrous salts. Although triammonium phosphate is also known, it is unstable and not used in foods.

Calcium Orthophosphates

There are five common calcium salts of orthophosphoric acid used in foods; four of these are shown in Table 2. The monobasic calcium salt,

TABLE 2

Nomenclature, Structure, and Some Characteristics of Phosphate Acids and Salts Commonly Used in Foods[1]

Class of phosphate	Number P atoms per molecule	General structure[2]	Common names	Formulas[3]	pH[4]	Solubility[5]
Orthophosphates (monomer)	One	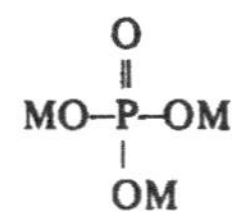	Phosphoric acid	H_3PO_4	2.0-2.2	High
			Monoammonium phosphate	$NH_4H_2PO_4$	4.5	28
			Diammonium phosphate	$(NH_4)_2HPO_4$	8.0	41
			Monocalcium phosphate	$Ca(H_2PO_4)_2$	4.5	[6]
			Dicalcium phosphate	$CaHPO_4$	7.5	insol
			Tricalcium phosphate	$Ca_5(PO_4)_3(OH)$[7]	7.2	insol
			Ferric orthophosphate	$FePO_4$[7]	3.8-4.4	insol
			Monopotassium phosphate	KH_2PO_4	4.4	20
			Dipotassium phosphate	K_2HPO_4	8.8	63
			Tripotassium phosphate	K_3PO_4	11.9	51
			Hemisodium phosphate	$NaH_2PO_4 \cdot H_3PO_4$	2.2	High
			Monosodium phosphate	NaH_2PO_4	4.4	48
			Disodium phosphate	Na_2HPO_4	8.8	11
			Trisodium phosphate	Na_3PO_4	11.8	13
			Sodium aluminum phosphate, acidic	$NaAl_3H_{14}(PO_4)_8$	2.4-2.5	Slight
			Sodium aluminum phosphate, acidic	$Na_3Al_2H_{15}(PO_4)_8$	2.6	Slight
			Sodium aluminum phosphate, alkaline	$Na_{15}Al_{2.8}(PO_4)_8$[7]	9.2-9.4	Slight
	Two (dimer)	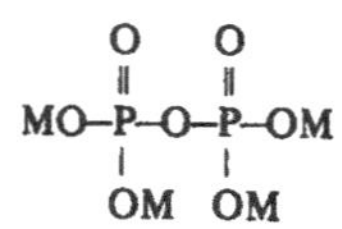	Pyrophosphoric acid	$H_4P_2O_7$	V. acid	High
			Calcium pyrophosphate	$Ca_2P_2O_7$	6.0	insol
			Tetrapotassium pyrophosphate	$K_4P_2O_7$	10.2	61
			Ferric pyrophosphate	$Fe_4(P_2O_7)_3$[7]		insol
			Sodium acid pyrophosphate	$Na_2H_2P_2O_7$	4.2	13
			Tetrasodium pyrophosphate	$Na_4P_2O_7$	10.2	6
			Sodium iron pyrophosphate	$Na_8Fe_4(P_2O_7)_5$[7]	7.0-7.7	Insol

Linear or straight-chain polyphosphates (polymer)	Three (trimer)	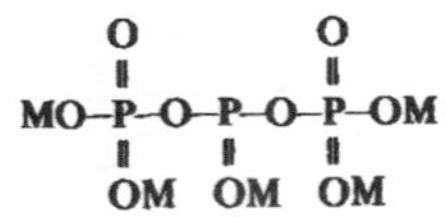	Tripolyphosphoric acid	$H_5P_3O_{10}$	V. acid	High
			Potassium tripoly-phosphate	$K_5P_3O_{10}$	9.8	65
			Sodium tripolyphosphate	$Na_5P_3O_{10}$	9.8	13+
			Polyphosphoric acids	$(HPO_3)_n$	V. acid	high
			Potassium metaphos-phates (Kurrol's salt)	$(KPO_3)_n$ n = 400-20,000	4-8	insol[8]
	Four to 10[5]	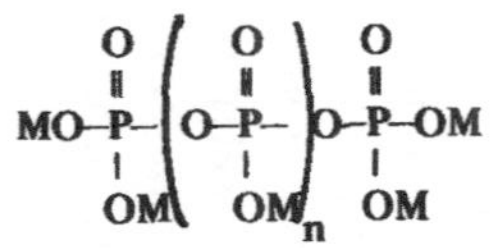 $M_{(n+2)}P_nO_{(3n+1)}$	Sodium tetrapolyphos-phate	$(NaPO_3)_n$ n = 4-10	7-8	High
			Sodium hexametaphos-phate (Graham's salt)	$(NaPO_3)_n$ n = 10-15	7.0	High
			Soluble sodium metaphos-phate (Graham's salt)	$(NaPO_3)_n$ n = 50-100	6.2	High
			Insoluble sodium meta-phosphate (Kurrol's salt)	$(NaPO_3)_n$ n = 100-500	5.5	insol[9]
Cyclic poly-phosphates	Three	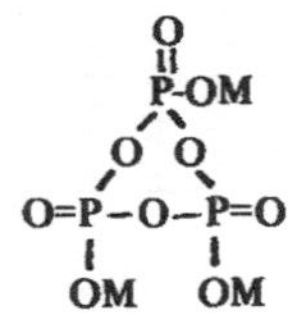	Sodium trimetaphosphate	$(NaPO_3)_3$	6.7	23
	Four	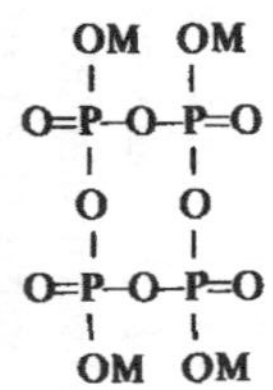	Sodium tetrametaphos-phate	$(NaPO_3)_4$	6.2	18

[1] The information in this table was compiled from References 24,30-32,35, and 36.
[2] The letter *M* in these structures can be hydrogen or a metal ion in the compounds used as food additives.
[3] Only the formulas for the anhydrous compounds are given. The hydrates important as food additives are described in subsequent sections.
[4] The pH of soluble phosphates was determined in a 1% solution, and in 10 to 50% slurries of insoluble compounds.
[5] Solubility values represent g/100g of saturated solutions at 25°C.
[6] Monocalcium phosphate dissolves incongruently, or disproportionates to form insoluble dicalcium phosphate and phosphoric acid. However, it is commonly called soluble calcium phosphate, because it appears to be very soluble.
[7] These formulas are the molecular ratios shown by analysis. The iron compounds also contain water of hydration.
[8] Some long-chain potassium polyphosphates are made soluble by the presence of sodium ions.
[9] Some long-chain sodium polyphosphates are made soluble by the presence of potassium ions.

TABLE 3

Composition of Polyphosphoric Acids

Composition, wt % P_2O_5	Ortho-	Pyro-	Tri-	Tetra-	Penta-	Hexa-	Hepta-	Octa-	Nona-	"Hypoly-"
68.80	100.00	Trace								
69.81	97.85	2.15								
70.62	95.22	4.78								
72.04	89.91	10.09								
72.44	87.28	12.72								
73.43	76.69	23.31								
74.26	67.78	29.54	2.67							
75.14	55.81	38.88	5.31							
75.97	48.93	41.76	8.23	1.08						
77.12	39.86	46.70	11.16	2.28						
78.02	26.91	49.30	16.85	5.33	1.60					
78.52	24.43	48.29	18.27	6.75	2.26					
79.45	16.73	43.29	22.09	10.69	4.48	1.92	0.80			
80.51	13.46	35.00	24.98	13.99	6.58	3.14	2.84			
81.60	8.06	27.01	22.28	16.99	11.00	5.78	3.72	2.31	1.55	1.28
82.57	5.10	19.91	16.43	16.01	12.64	8.89	6.41	4.11	3.51	6.99
83.48	4.95	16.94	15.82	15.91	12.46	9.71	6.77	5.04	2.99	9.42
84.20	3.63	10.60	11.63	13.05	12.17	9.75	8.19	5.92	4.91	20.16
84.95	2.32	6.97	7.74	11.00	10.45	9.62	8.62	7.85	6.03	29.41
86.26	1.54	2.97	3.31	5.16	5.32	5.54	3.51	3.30	3.30	66.03

Note: These figures are given to two decimal places for further computation purposes, but the precision may not be better than 1% total phosphorus in some cases.

Sources: Bell, R.N., *Ind. Eng. Chem*, 40, 1464, 1948 and Huhti, A.L. and Gartaganis, P.A., *Can. J. Chem.* 34, 785, 1956. Reproduced by permission of the National Research Council of Canada.

monocalcium phosphate, exists in the anhydrous and monohydrate forms.

Anhydrous Monocalcium Phosphate

(AMCP)* has the general formula of $Ca(H_2PO_4)_2$. The salt is very hygroscopic in its pure form. Since its major food application is as a leavening acid in dry, flour-based mixes, the hygroscopicity prevents use of the pure salt. Therefore, a coated AMCP has been developed for this application. The coating consists of a mixed potassium, aluminum, calcium, and magnesium metaphosphate. It is formed by heating AMCP containing minute quantities of potassium, aluminum, and magnesium to 220°C until the coating has been formed. This thin coating delays the solubilization of the AMCP for a short period of time after it comes in contact with water. It thus becomes a slower reacting leavening acid then the hydrated form, monocalcium phosphate monohydrate.

Monocalcium Phosphate Monohydrate

($MCP \cdot H_2O$), $Ca(H_2PO_4)_2 \cdot H_2O$, is a commonly used acidifying agent. It is commercially available in 94 or 95% pure crystalline form and is noted for rapid release of its acidity. It does not form a true solution; it partially disproportionates or is incongruently soluble in water, forming dicalcium phosphate and phosphoric acid. This reaction is believed to proceed according to the following equation:

$$Ca(H_2PO_4)_2 \cdot H_2O + xH_2O = CaHPO_4 + H_3PO_4 + (x + 1)H_2O$$

Dicalcium Phosphate Dihydrate

(DCP), $CaHPO_4$ also exists in an anhydrous and a hydrated form. The anhydrous salt is used to some extent as a polishing agent in toothpaste, but it is not used in foods. The common form of dicalcium phosphate used in foods is the dihydrate, $CaHPO_4 \cdot 2H_2O$. This acidic salt is relatively insoluble in water at temperatures under 140°F

*The abbreviations in parentheses following each name of a food phosphate will be used to designate that phosphate in all subsequent references to it in the following section.

(60°C). It is, therefore, very stable in cake and other flour-based mixes. It does not react during the batter mixing stage, again because of its insolubility. After the batter or dough has been placed in the oven and the temperature has reached approximately 140°F, the DCP begins to go into solution with disproportionation to form $Ca(H_2PO_4)_2$ and $Ca_5(PO_4)_3OH$, and only then reacts with bicarbonate. This delay in the release of its acidity until the product reaches higher temperatures is also useful in other food applications, as well as the baking applications already mentioned.

Tricalcium Phosphate

The final form of calcium orthophosphate commonly used in foods is a crystalline precipitate, commonly called tricalcium phosphate (TCP). However, the compound in common use has a highly variable composition. Its formula, based on x-ray patterns, is usually written as $Ca_5(OH)(PO_4)_3$, a hydroxyl apatite. This is an insoluble salt and is used in applications requiring such properties.

Ferric Orthophosphate

Ferric orthophosphate is the commercial name for an iron salt of orthophosphoric acid commonly used as an iron enrichment compound. Although it is a complex salt, it is usually given the formula $FePO_4 \cdot xH_2O$. This formula represents the molecular ratios determined by chemical analysis. Ferric orthophosphate is normally insoluble in water, but it is soluble to varying degrees, according to the method of manufacture, in dilute hydrochloric acid solutions, such as exist in the stomach.

Potassium Orthophosphates

While the potassium salts of orthophosphoric acid are used to some extent in foods, they are, in general, more costly to produce than sodium salts. As a result the sodium salts are used wherever possible. Potassium salts usually do not introduce the same level of astringent flavors as do the sodium orthophosphates; therefore, most of the applications for the potassium salts are those in which flavor is so important that the added cost is worthwhile. Three potassium salts or orthophosphoric acid are listed in Table 2. They are usually available and used as the anhydrous salts.

Sodium Orthophosphates

Several sodium salts of orthophosphoric acid are used in foods. Some are available as anhydrous as well as hydrated salts. If the various hydrates are included, ten different crystalline sodium orthophosphate salts are known; these are shown in Table 4. Any one of these anhydrous or hydrated salts can be used in food applications as long as it has been manufactured under conditions that produce food-grade ingredients.

Monosodium Orthophosphate (MSP) is a mild acidulant and is often used in food systems for that purpose. *Disodium* and *trisodium orthophosphate* (DSP and TSP) are alkaline salts. They are used to increase the pH of a food. Because the sodium orthophosphates are also excellent buffers, they are utilized to stabilize food systems at the pH that is optimum for the desired effect. Usually combinations of MSP and DSP are used in buffer systems. Some food applications require the orthophosphate anion for interactions with one or more ingredients in the food system. Because these interactions are often

TABLE 4

Sodium Ortho- and Polyphosphate Salts and Hydrates Used in Foods[1]

Chemical name and synonyms	Type of hydrate	Chemical formula
Orthophosphates		
Monosodium monophosphate		
Monosodium dihydrogen monophosphate[2]		
Sodium biphosphate	Anhydrous	NaH_2PO_4
Sodium dihydrogen phosphate	Monohydrate	$NaH_2PO_4 \cdot H_2O$
Acid sodium phosphate	Dihydrate	$NaH_2PO_4 \cdot 2H_2O$
Monosodium orthophosphate		
Monosodium phosphate[3]		
Primary sodium phosphate		
Sodium phosphate monobasic		

TABLE 4 (continued)

Chemical name and synonyms	Type of hydrate	Chemical formula
Disodium monophosphate		
Disodium monohydrogen monophosphate[2]		
Disodium orthophosphate	Anhydrous	Na_2HPO_4
Disodium hydrogen phosphate	Dihydrate	$Na_2HPO_4 \cdot 2H_2O$
Disodium phosphate[3]	Heptahydrate	$Na_2HPO_4 \cdot 7H_2O$
Secondary sodium phosphate	Dodecahydrate	$Na_2HPO_4 \cdot 12H_2O$
Sodium phosphate dibasic		
Trisodium monophosphate[2]		
Trisodium orthophosphate	Anhydrous	Na_3PO_4
Trisodium phosphate[3]	Monohydrate	$Na_3PO_4 \cdot H_2O$
Sodium phosphate tribasic	Dodecahydrate	$Na_3PO_4 \cdot 12H_2O$
Pyrophosphates		
Disodium diphosphate[2]		
Disodium dihydrogen diphosphate[2]		
Disodium dihydrogen pyrophosphate	Anhydrous	$Na_2H_2P_2O_7$
Disodium pyrophosphate		
Sodium acid pyrophosphate[3]		
Tetrasodium diphosphate[2]		
Tetrasodium pyrophosphate[3]	Anhydrous	$Na_4P_2O_7$
Sodium pyrophosphate	Decahydrate	$Na_4P_2O_7 \cdot 10H_2O$
Tripolyphosphates		
Pentasodium triphosphate[2]		
Sodium tripolyphosphate[3]	Anhydrous	$Na_5P_3O_{10}$
Tripolyphosphate		
Straight-Chain Polyphosphates		
Sodium polyphosphate[4]		
Sodium metaphosphate[5]		
Sodium hexametaphosphate[6]		
Glassy sodium phosphate	Anhydrous	$(NaPO_3)_n$
Sodium tetraphosphate[7]		
Graham's salt		
Sodium Kurrol's salt[8]		
Insoluble sodium metaphosphate[8]		
Cyclic Polyphosphates		
Sodium trimetaphosphate	Anhydrous	$(NaPO_3)_3$
Sodium tetrametaphosphate	Quadrahydrate	$(NaPO_3)_4 \cdot 4H_2O$

[1] Information accumulated from References, 30, 32, and 35 through 38, with permission.
[2] Name considered most correct.
[3] Common name for commercial product in the U. S.
[4] Straight-chain sodium polyphosphates would most accurately be named with the average chain-length or the Na_2O/P_2O_5 ratio added e.g., sodium polyphosphate, n = 10-15 or sodium polyphosphate, $Na_2O/P_2O_5 = 1.1$
[5] Common name for soluble sodium polyphosphates with n = 20-100.
[6] Common name for soluble sodium polyphosphates with n = 10-15.
[7] Common name for soluble sodium polyphosphates with n = 4-10.
[8] Common name for insoluble, long-chain sodium polyphosphates.

pH dependent, the necessary pH may dictate which orthophosphate salts or mixture must be used.

Double sodium phosphate salts are also known.[30,32] However, the only one of importance in food applications is *hemisodium orthophosphate.* This compound is considered to be the double salt of monosodium phosphate and phosphoric acid.[31,32,35] Brout[39] reports, however, that hemisodium phosphate is actually the double salt of monosodium phosphate and polyphosphoric acid. He has proposed the revised formula $(HPO_3)_n \cdot NaH_2PO_4 \cdot H_2O$, although no other evidence for this formula has been found.

Sodium aluminum phosphates (SALP) have recently been approved for use as food ingredients – as leavening acids and cheese emulsifiers. The two acidic salts commonly used in leavening systems have the formulas $NaH_{14}Al_3(PO_4)_8 \cdot 4H_2O$ and $Na_3H_{15}Al_2(PO_4)_8$. An *alkaline sodium aluminum phosphate* used in cheese processing has the formula $Na_{15}Al_2 \cdot 8(PO_4)_8$. These three salts are only slightly soluble in water at room temperature, but they increase in solubility as the temperature is increased. Therefore, they are used in applications requiring delayed release of acidity or alkalinity.

The Pyrophosphates

The pyrophosphates, also called diphosphates in some of the European literature, are the simplest of the condensed or polymeric phosphates. The pyrophosphate anion contains two phosphorus atoms linked through a shared oxygen atom, as follows:

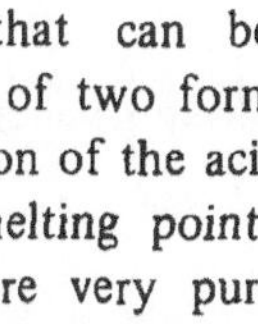

Pyrophosphoric Acid

Pyrophosphoric acid is the only polymeric phosphoric acid that can be crystallized. It crystallizes in one of two forms from a seeded, concentrated solution of the acid cooled to about 10°C below its melting point. Although these crystalline forms are very pure pyrophosphoric acid, if melted or dissolved in water, the pyrophosphoric acid immediately begins to revert to the equilibrium mixture of ortho- and polymeric phosphoric acids represented by the P_2O_5 content of the solution.[30,32] It is impossible to have a solution of any of the polymeric phosphoric acids that is not a mixture of two or more acids of different chain lengths (see Table 3).

Crystalline pyrophosphoric acid is very difficult to maintain in its pure crystalline state, as it is extremely hygroscopic. Therefore, unless the crystalline compound is well protected, it will rapidly absorb water from the atmosphere and become a solution of the equilibrium distribution of polyphosphoric acids indicated for its P_2O_5 content in Table 3.

Calcium Pyrophosphates

Although *calcium acid pyrophosphate*, $CaH_2P_2O_7$, has been proposed as a leavening acid, few, if any, calcium salts of pyrophosphoric acid are used in foods. The *dibasic calcium pyrophosphate*, $Ca_2P_2O_7$, has no applications in foods.

Potassium Pyrophosphate

Anhydrous *tetrapotassium pyrophosphate* (TKPP) is commercially available and is used to a limited extent in foods. Its applications are similar to the sodium salts, but, as in the orthophosphates, the costliness of the potassium salt limits its use.

Sodium Pyrophosphates

Two crystalline sodium salts of pyrophosphoric acid are used in foods. One is the acidic salt *sodium acid pyrophosphate* (SAPP), $Na_2H_2P_2O_7$, used in numerous acidification applications. It is possible to reduce the rate at which sodium acid pyrophosphate goes into solution by modifying its manufacturing processes and the use of certain additives that form slowly soluble coatings around each crystal. Such compounds have been especially useful as delayed reaction leavening acids.

The second crystalline sodium salt of pyrophosphoric acid is its "neutral" salt containing sodium atoms in place of all of the hydrogens, i.e., *tetrasodium pyrophosphate* (TSPP), $Na_4P_2O_7$. It is useful because of its alkalinity and its ability to complex or precipitate alkaline earth and heavy metal ions.

A complex sodium salt of pyrophosphoric acid also used in foods is the iron enrichment compound, *sodium iron (or ferric) pyrophosphate* (SIP), $Na_8Fe_4(P_2O_7)_5 \cdot xH_2O$ (formula derived from the molecular ratios obtained by chemical analysis). SIP is insoluble in water but is soluble in dilute hydrochloric acid, such as exists in the stomach.

Ferric Pyrophosphate

Ferric pyrophosphate, $Fe_4(P_2O_7)_3$, also used an an iron enrichment compound, has characteristics similar to SIP. It is insoluble in water, but it is soluble in dilute hydrochloric acid. This compound and SIP are used in food containing fats that may become rancid in contact with more soluble iron enrichment compounds.

The Tripolyphosphates

The tripolyphosphates, also called triphosphates in some of the European literature, are composed of a straight chain of three phosphorus atoms linked through shared oxygen atoms. The structure of the tripolyphosphate anion is as follows:

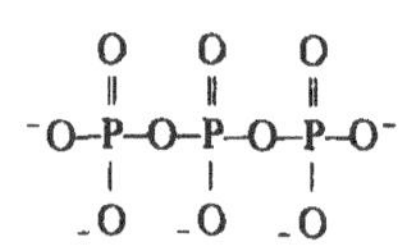

Tripolyphosphoric Acid

Tripolyphosphoric acid exists as one of the acids in the equilibrium mixture in strong phosphoric acids containing over 74% P_2O_5, as shown in Table 3. It increases to a level of approximately 25% at 80% P_2O_5 and then decreases as the P_2O_5 concentration increases and longer-chain polyphosphoric acids appear in greater quantities. Pure tripolyphosphoric acid cannot be crystallized from solution but can be prepared from its salts by special ion-exchange techniques. It has a very short life, however, as it immediately begins to form the equilibrium mixture of ortho- and polyphosphoric acids that would be expected for the P_2O_5 concentration in the solution involved.[30,32] If the P_2O_5 content is below 74%, the tripolyphosphoric acid solution hydrolyzes to a mixture of ortho- and pyrophosphoric acid. At levels of P_2O_5 above 74%, the equilibrium mixture formed will be that shown in Table 3 for the P_2O_5 concentration of the solution.[30,32]

Potassium Tripolyphosphate

There is only one commercial potassium salt of tripolyphosphoric acid – *pentapotassium pyrophosphate* (KTP), $K_5P_3O_{10}$, the fully neutralized salt. Since the potassium salt is more expensive than the sodium salt, it is used only in food applications that require its greater solubility or its less astringent flavor.

Sodium Tripolyphosphate

The only sodium salt of tripolyphosphoric acid used in food is the fully neutralized salt, commonly called *sodium tripolyphosphate* (STP), $Na_5P_3O_{10}$. Sodium tripolyphosphate is about 20% as soluble as its potassium counterpart, and it can produce astringent flavors at high levels; fortunately, high levels are usually unnecessary in food applications. There are two crystalline forms of STP – a high-temperature form known as Form-I and a low-temperature form known as Form-II.[30,32] The form utilized in food applications is of importance, since Form-I rapidly tends to produce the hexahydrate, which readily cakes or lumps when added to poorly agitated water during preparation of solutions. Form-II, in contrast, has greater immediate solubility in water and does not cake or form lumps, even if added to unagitated water. Thus, Form-II, the low-temperature crystalline type, is preferable for food applications where it is necessary for production personnel to prepare solutions of STP.

The Straight Chain Polyphosphates

Polyphosphates having more than three phosphorus atoms per chain are often designated as glassy, amorphous, or condensed phosphates. Their chains are composed of a number of phosphorus atoms linked together by shared oxygen atoms, as shown by the following general formula for the polyphosphate anion:

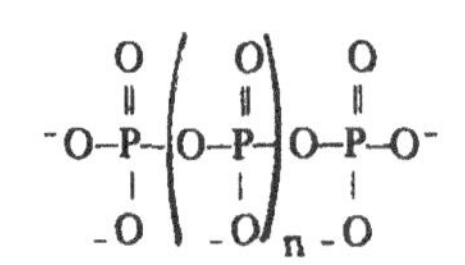

The salts of these acids form amorphous or glassy particles, not true crystals – hence the term *amorphous phosphates.*

Polyphosphoric Acids

As shown in Table 3, the concentrations of polyphosphoric acids containing four or more phosphorus atoms increase as the P_2O_5 content increases above 75.5%. Highly involved methods for preparing the pure acids of four to nine chain lengths through ion-exchange techniques have been described in the literature (see References 30 and 32-34), but only the equilibrium mixtures are commercially available. It is commercial practice to designate strong phosphoric acids that contain from 82 to 85% P_2O_5 as polyphosphoric acids. Although preparations containing up to 90% P_2O_5 have been reported in the literature, concentrations above 85% are not available without using special preparation methods.

The equilibrium distribution of polyphosphoric acids in any single preparation can be determined by analyzing its P_2O_5 content and by referring to tables such as Table 3. There is, however, a second analytical value of use in identifying the acid. This is the average chain length of the acid, or $\bar{n}$. It is obtained from titrations of the acid according to the following equation:[30,32,40]

$$\bar{n} = \frac{2\ (\text{ml of base to titrate to end point near pH 4.2})}{\text{ml of base to titrate from 4.2 end point to one near pH 9}}$$

The polyphosphoric acids exists as clear, colorless, viscous, hygroscopic liquids. When they are cooled in moisture-free sealed tubes, they gradually increase in viscosity and finally form rigid glasses but never crystallize. If they are cooled in containers exposed to the atmosphere, crystals of pyrophosphoric acid form at the surface as the acids absorb moisture. Polyphosphoric acids are soluble in water but quickly hydrolyze to lower molecular weight and shorter chain-length phosphoric acids as the proportion of water to P_2O_5 increases (or percent of P_2O_5 in the system decreases). Orthophosphoric acid may be prepared from the polyphosphoric acids if they are diluted with the necessary amount of water. Considerable heat is generated as the water is added. The polyphosphoric acids are not used in foods. Their salts, however, have many applications.

Potassium Polyphosphate

The one type of potassium salt of polyphosphoric acid used in foods is commonly designated *potassium metaphosphate* or *potassium polymetaphosphate*; a more accurate term would be *potassium polyphosphate*. It was first prepared just before the turn of the century by Kurrol and, therefore, often is called Kurrol's salt.[30,32] Potassium polyphosphate compositions of varying average chain lengths are easily prepared from monopotassium orthophosphate heated above 150°C. The preparation forms a solid that contains crystals and appears to have a fibrous structure much like asbestos. The average chain length of any particular preparation varies with the degree of dehydration accomplished by the heat treatment given the monopotassium phosphate from which it is prepared.[41,42]

Potassium polyphosphates are nearly insoluble in water, although there are reports that they can be solubilized by passing an aqueous suspension through a colloid mill.[32] They can be dissolved in a solution of a soluble ammonium or alkali metal salt other than one having the potassium cation. Thus, viscous solutions of potassium metaphosphates have been prepared by dissolving them in a solution of sodium chloride or of one of the soluble sodium phosphates.

The average chain lengths of these salts cannot be determined by Van Wazer's method.[40] Molecular weights for potassium metaphosphate have been determined by ultracentrifuge and viscosity techniques. These indicate that portions of the salts have very high molecular weights, ranging between 50,000 and 3 million; they also indicate that the average chain lengths range from 400 to 20,000 phosphorus atoms per chain.[32] Potassium polyphosphate is not a pure salt, but rather a mixture of numerous molecules with a wide distribution of chain lengths.[41,42] Several food applications for this salt will be discussed in later sections of this chapter.

Sodium Polyphosphates

The soluble sodium metaphosphates were first described by Graham in the early 1800's. Because of the method of preparation, Graham mistakenly expected to obtain a cyclic metaphosphate; the methods of analysis that he used could not distinguish between cyclic and straight-chain polyphosphates. As a result Graham attached the name *sodium metaphosphate*, and it has been impossible to obtain complete agreement to change this name to the more preferable designation, *sodium polyphosphate*. Several generic names have been applied interchangeably to the soluble, amorphous, glassy sodium phosphates –

sodium metaphosphate;
sodium hexametaphosphate;
sodium polyphosphate;
Graham's salt;
glassy sodium phosphate.

Two commercial names are also used frequently in the food-science literature to designate this group of compounds –

Calgon and
Quadrafos.

Calgon usually refers to polyphosphates having an average chain length of 10 to 20 phosphorus atoms, as determined by the titration method of Van Wazer.[40] Quadrafos usually refers to compounds having an average chain length of between four and eight phosphorus atoms. However, both compounds are composed of mixtures of polyphosphates of widely varying chain lengths.[41,42] Table 5 shows the chain-length distributions of sodium polyphosphates with $\bar{n}$ between 2.0 and 9.0.

There are several types of the sodium salts of polyphosphoric acid with an $\bar{n}$ = 4 or greater that are commonly utilized in food applications. All of the salts are mixtures of molecules of varying chain-length distributions. None of those commercially available are pure forms. This includes the so-called "reagent grade polyphosphates," such as sodium hexametaphosphate sold by some of the chemical supply houses. Paper chromatographic analysis of several sodium hexametaphosphate preparations labeled as "Reagent Grade" indicates that they contain sodium polyphosphates ranging from sodium orthophosphate to those containing very long chains of phosphate groups.[43] The average chain length does correspond to that of the commercial preparation commonly called sodium hexametaphosphate. However, because of the wide distribution of chain lengths of its molecules, it is incorrect to designate sodium hexametaphosphate as "chemically pure" or as a "reagent grade" chemical. It is also incorrect to assign a molecular weight to this compound, as one would to a pure compound. A number of publications in the food literature have erroneously based work and calculations on such molecular weights and specific chain lengths. The investigators would have been wiser if they had obtained their preparations directly from one of the manufacturers of these compounds and asked for a complete analysis of the specific preparation sent them.

Sodium tetrapolyphosphate, $Na_6P_4O_{13}$, sometimes called Quadrafos, cannot be prepared in the crystalline form by the usual precipitation procedures. The commercial preparation designated as sodium tetrapolyphosphate is a mixture containing both short and long-chain polyphosphates whose average chain lengths range between four and eight and correspond to the ratio of Na_2O/P_2O_5 between 1.5 and 1.25, as shown in Table 5.

Sodium hexametaphosphate (SHMP) is a common ingredient in food applications, and a considerable amount of research on its uses in foods has been reported in the literature. Other names for this salt are *glassy sodium metaphosphate, soluble sodium metaphosphate*, and *Graham's salt*. The latter name was given to all soluble long-chain sodium phosphate salts, because they were first discovered by Graham. The formula $Na_6P_6O_{18}$, or $(NaPO_3)_6$, was mistakenly given to the product by a later investigator on the basis of analyses that assumed the compounds were cyclic. Although erroneous, both the formula and name are most commonly used today.

Commercial sodium hexametaphosphate available today has an average chain length ranging between 10 and 15 phosphorus atoms. Carefully controlled processing can consistently produce a product with an average chain length between 11 and 13 phosphorus atoms in the chain. Chromatographic analysis has frequently demonstrated that freshly prepared sodium hexametaphosphate contains very little, if any, polyphosphates with chain lengths below four phosphorus atoms. However, like all polyphosphate glasses stored as the solid, it gradually "devitrifies," or hydrolyzes with time. The process of devitrification can be followed by the increasing amounts of ortho-, pyro-, and tripolyphosphate appearing in the product.[30]

Another type of "glassy" sodium metaphosphate, which occasionally is used in food applications but more usually as a dentifrice, is commonly known as *insoluble metaphosphate* (usually abbreviated IMP) or Maddrell's salt. This salt can be distinguished from the soluble sodium metaphosphates by the fact that it is insoluble in water and dissolves with difficulty in solutions of

TABLE 5

Molecular Composition of the Sodium Polyphosphates as Determined by Paper Chromatography[1,2]

Analysis			Percentage of total P_2O_5 as									
Weight percent P_2O_5	Na_2O/P_2O_5 R	$\bar{n}$	Ortho-	Pyro-	Tripoly	Tetrapoly	Penta-poly	Hexa-poly	Hepta-poly	Octa-poly	Nona-poly	Higher
53.383	2.00000	2.0	0.00	100.0								
54.803	1.88888	2.25	0.00	91.81	8.19							
55.994	1.80000	2.5	0.00	72.24	27:76							
57.007	1.72727	2.75	0.00	47.03	52.13	0.84						
57.881	1.66667	3.0	0.00	24.32	48.55	20.41	5.11	1.61				
59.308	1.57143	3.5	0.00	11.13	38.27	28.11	13.12	5.52	2.33	1.02	0.48	
60.426	1.50000	4.0	0.00	6.65	28.25	27.43	16.87	9.41	5.70	2.74	1.77	1.16
61.324	1.44444	4.5	0.00	4.74	22.17	23.99	17.30	11.52	8.24	4.60	3.15	4.29
62.063	1.40000	5.0	0.00	3.24	16.24	20.16	16.45	12.68	9.43	6.71	4.39	10.69
62.681	1.36364	5.5	0.00	2.65	13.94	19.14	17.92	14.12	8.92	6.40	4.76	12.15
63.205	1.33333	6.0	0.00	1.88	10.09	14.89	13.46	11.28	9.61	9.26	6.94	22.58
63.655	1.30769	6.5	0.00	1.81	8.71	13.07	12.32	11.70	11.12	9.11	6.32	25.84
64.389	1.26667	7.5	0.00	1.24	6.38	9.44	9.09	10.74	9.91	8.53	7.74	36.92
65.204	1.22222	9.0	0.00	0.95	4.35	7.63	6.52	9.74	8.00	7.49	6.13	49.19

Sources: Westman, A.E.R. and Gartaganis, P.A. *J. Amer. Ceram. Soc.*, 40, 293, 1957 and Van Wazer, J.P., *Phosphorus and Its Compounds*, Interscience Publishers, Inc., a division of John Wiley & Sons, Inc., New York, 1958. Reprinted by permission.
Potassium and other alkali metal polyphosphates have very similar molecular compositions.

ammonium and the alkali metal salts. The glassy characteristics of IMP distinguish it from a second type of insoluble sodium polyphosphate, usually designated *sodium Kurrol's salt.* The solid sodium Kurrol's salt is very similar to its potassium counterpart in that it forms fibrous crystals similar to asbestos. It also will dissolve in solutions of ammonium or alkali metal salts other than sodium. Sodium Kurrol's salt is occasionally mentioned in food applications.

The Cyclic Metaphosphates

The cyclic polyphosphates are correctly termed *metaphosphates.* They are prepared by the thermal dehydration of orthophosphates under very specific conditions; they are crystalline because they are composed of molecules of unvarying size. Only two of the possible compounds of metaphosphate are commercially available. Several potential food applications for the cyclic polyphosphates have been reported and some will be mentioned in the following sections. The only food application approved by the U.S. Food and Drug Administration (FDA), however, is the modification of starches by sodium trimetaphosphate. Possible toxicity was at first a concern, although recent work, reviewed in a following section, has demonstrated their safety. Applications to FDA for food uses shown to be advantageous would most likely be approved.

Metaphosphoric acid is not a commercial product and is therefore unimportant in food applications. The two compounds of importance are both sodium metaphosphates – sodium trimetaphosphate and sodium tetrametaphosphate. Their structures are shown in Table 2.

Sodium trimetaphosphate is a soluble, crystalline, cyclic, condensed phosphate. Its major food application is in the modification of starches to prepare starch phosphates (see the section on Phosphate Reactions with Starches). Although it is available as the anhydrous salt, the hexahydrate is also easily prepared.

Sodium tetrametaphosphate is also a soluble, crystalline, cyclic, condensed phosphate. It is not used in foods, although applications, such as the precipitation of proteins without forming a protein-phosphate complex, have been proposed.

Chapter 3

SOME GENERAL CHEMICAL CHARACTERISTICS OF PHOSPHATES

A study of the structural chemistry of the phosphate compounds aids in understanding of some of their reactions in food applications. An excellent review of this subject has recently been published by Corbridge.[44] In general, the phosphate ion is considered to be a tetrahedron in which the phosphorus is surrounded by four oxygen atoms. The linking together of phosphate ions in order to form the condensed phosphate chains, therefore, allows the tetrahedral structures to coil in the shape of a helix, and, in highly concentrated solutions of phosphates, the very long-chain polyphosphates are considered to form coiled chains.[30] Corbridge[44] provides a number of detailed diagrams demonstrating the geometric configurations of numerous ortho-, pyro-, straight-chain poly-, and cyclic polyphosphates. Structural diagrams are also given for a number of the alkali metal, alkaline-earth metal, and heavy metal compounds of each of the phosphates.

All of the phosphates from orthophosphate through the long-chain polyphosphates have the properties of highly charged anions.[30] Thus, they also can exhibit the properties of typical polyelectrolytes; the longer the chain length, the greater the polyelectrolyte properties become, as will be seen in discussions of their interactions with proteins. The pH titration curves of the various phosphates demonstrate some of their charged anion characteristics.[30] Trimeta- and tetrametaphosphoric acids are the strongest of all the phosphoric acids; their titration curves resemble those of hydrochloric acid. Titration curves of the straight-chain polyphosphoric acids have pH characteristics similar to those of orthophosphoric acid. The longer-chain phosphoric acids, however, become increasingly weaker acids after titration of the first hydrogen ion, and the weaker-acid characteristics increase with chain length. Thus, the shorter-chain length phosphoric acids are stronger acids than those having long-chain lengths. In addition, orthophosphoric and the short-chain pyro- and tripolyphosphoric acids exhibit three inflection points in their titration curves near pH 4.5, 7, and 10. Longer-chain polyphosphoric acids show only two inflection points – one near pH 4.5 and the other near pH 10.

Batra studied the ionization of DSP, TSPP, STP, sodium tetrapolyphosphate, and SHMP by use of sodium ion electrodes.[45] The data from this investigation demonstrate that the sodium orthophosphates ionize completely, even at high concentrations. The condensed phosphates exhibit decreasing ionization with increasing concentration, and the ionization of the individual phosphate decreases as the chain length increases. Batra also studied the exchange of sodium ions for calcium ions upon the addition of increasing concentrations of calcium ion. DSP continued to dissociate completely and formed a calcium phosphate precipitate at all concentrations of calcium (0.00125 to 0.00500 mol calcium ion/l.). No precipitate was formed when calcium ion was added to SHMP at any of the calcium concentrations used, but precipitates formed with the shorter-chain polyphosphates after a certain minimum level of calcium ion had been added. The level of calcium ion required for precipitates to form increased as chain length increased. Thus, the ability of a polyphosphate to complex metal ions, including the alkali metal ions and alkaline-earth metal ions, increases with the decrease in the degree of dissociation of the phosphate compound. Batra also found that the pH of the solution decreased as increasing quantities of calcium were added to the phosphate.

A number of investigators have added information concerning the complexing ability of the polyphosphates. From their investigations Van Wazer and Campanella[46] concluded that weak complexes were formed with alkali metal and similar cations and that strong complexes were formed with the alkaline-earth and heavy metal ions. An excellent review of the complexing of metal ions has been published by Van Wazer and Callis.[47] According to these authors, the long-chain polyphosphates are strong complexing agents; the ring phosphates are weaker complexing agents; and orthophosphate forms complexes only at very low concentrations and precipitates at higher concentrations.[47]

Sequestration, or soluble complex formation, is the result of a competitive reaction for the metal ion between a sequestrant anion and a precipitating anion in the same solution. Thus, the precipitation of a metal cation is prevented by the formation of a soluble cation-anion complex. For example, the addition of sodium hexametaphosphate will actually solubilize precipitated calcium oxylate. Another example is seen when SHMP is added to milk. The sequestration of calcium from the casein-calcium-phosphate complex results in the formation of a light-yellow, transparent solution due to the removal of calcium from the colloidal casein-calcium-phosphate and the formation of a soluble calcium-hexametaphosphate complex.

Formulas for the compounds of complexed heavy-metal-polyphosphate compounds can only be assumed in most cases.[47] It is believed that the bonds by which phosphate anions hold the complexed cations are not simple electrostatic attraction and that there are specific binding sites for the cation. Van Wazer and Callis[47] state that the phosphate tetrahedra in the chains of the polyphosphates are geometrically arranged in such a way that an oxygen atom from each of three neighboring phosphate groups can be in close proximity with the complexed metal ion. Van Wazer[30] believes that both ionic and covalent contributions are involved in the bonds between the complexed cation and sequestering anion. The amount of electrostatic attraction can range from 100% for the sodium anion to very low values for the transition group elements such as iron and cobalt. Thilo[48] has indicated that the complexing reactions of polyphosphates are very similar to ion-exchange reactions.

Figure 1 shows a comparison of the calcium ion remaining in solution in equilibrium with a 0.01 *M* solution of a number of sequestering agents and also compares the calcium ion in equilibrium with a number of precipitated calcium salts. Any of the complexing anions that appear below a precipitating anion on the scale will dissolve that precipitate completely. The chart also demonstrates the relative complexing ability of the shorter-chain polyphosphates with long-chain polyphosphates.

Sequestration at varying pH values and temperatures was studied by Irani and Callis.[49] The authors found that, as the temperature increased, the ability of sodium and potassium polyphosphates to sequester calcium and magnesium was also increased and that the amount of calcium or magnesium ions sequestered by any of the polyphosphates was affected by the pH. The amount of alkaline-earth metal cations sequestered increased as pH increased, reaching a maximum in the range of pH 8 to 11 and decreased as pH decreased below 8 or rose above 11. The sequestering ability of longer-chain polyphosphates demonstrated that at pH values below 8 they were more effective than shorter-chain polyphosphates, such as TSPP and STP. STP was,

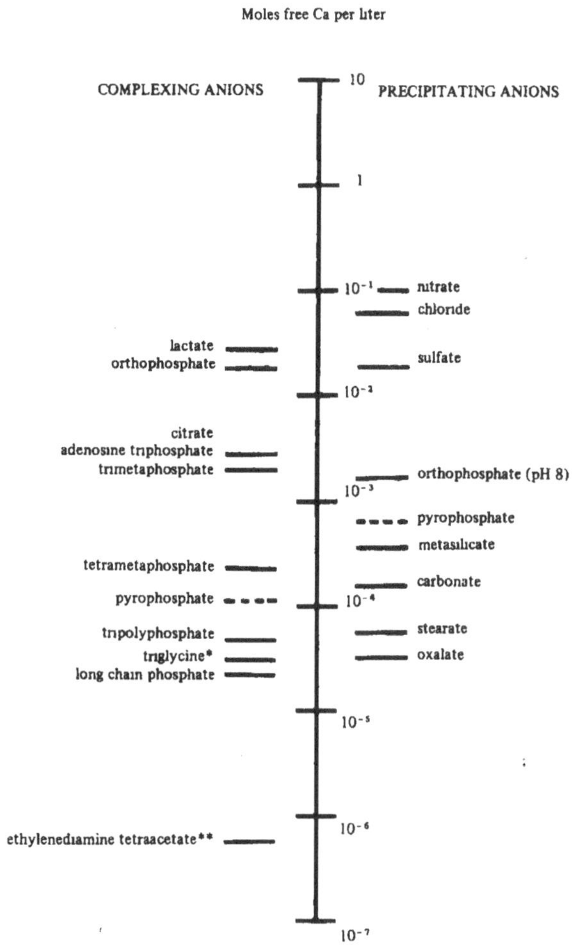

FIGURE 1. Comparison of complexing agents with precipitating agents for calcium. The free calcium for the complexing agent is computed for the dissociation of a 0.01 *M* solution of the 1:1 calcium complex. *Triglycine is also called ammonia triacetate or Trilon A.** A well known trade name for ethylenediamine tetraacetate (EDTA) is Versene®. (Source: Van Wazer, J. R. and Callis, C. F., *Chem. Rev.*, 58, 1011, 1958, with permission.)

however, more effective at pH values above 8 than the longer-chain polyphosphates, and this was especially true in dilute solutions. Irani and Morgenthaler[50] studied the sequestration of ferric and ferrous ions by polyphosphates at room temperature and various pH values. They reported that the ability of the polyphosphates at room temperature and various pH values to sequester iron decreases as the pH increases; it also decreases as the chain length of the polyphosphate increases. Tripolyphosphate was one of the most effective sequestrants for iron. In addition, the polyphosphate complexes with ferrous ion were weaker than those for the ferric ion.

Kutscher,[51] like Thilo,[48] reported on the similarity between calcium complexing of polyphosphates and ion-exchange reactions. The sodium polyphosphates are able to complex calcium, because the calcium is more firmly bound by the polyphosphate than sodium. The ability to form calcium complexes increases with dilution of the polyphosphate and decreases with decreasing pH or the addition of alkali metal salts.

Kutscher[51] also reports that the water-binding capacity of polyphosphate anions is due to their highly charged nature, leading to very strong hydration in water. This water-binding ability will be found highly important in later discussions of the interactions between polyphosphates and proteins, such as meat proteins. Van Wazer[30] has also discussed the ability of the polyphosphates to interact with other long-chain polyelectrolytes, such as proteins. The highly charged anionic nature of the polyphosphates causes them to be attracted to and to orient themselves along the charged sites of the protein.

In any food applications the stability of the phosphate is of prime consideration. A number of factors affect the rate of hydrolysis or degradation of both chain and ring phosphates to the orthophosphate anion. Hydrolysis is increased with increasing temperature; in the presence of enzymes, colloidal gels, and complexing cations, hydrolysis decreases with increasing pH. The presence of other anions or cellular material may increase or decrease the rate of hydrolysis, depending on the nature of the anion or cellular materials.[52]

Bell[53] demonstrated that the rate of hydrolysis of polyphosphate was greatly affected by temperature. In addition, under similar conditions shorter-chain polyphosphates hydrolyzed slower than longer-chain polyphosphates; thus, tripolyphosphate hydrolyzed more slowly than hexametaphosphate. The shorter-chain polyphosphates were more stable in the presence of excess alkali. Bell has reported that STP hydrolyzes to one mole each of ortho- and pyrophosphate. SHMP, however, hydrolyzes with two simultaneous reactions taking place. One portion of the molecule depolymerizes to form the cyclic trimetaphosphate, and the other portion hydrolyzes directly to orthophosphate. The cyclic trimetaphosphate first hydrolyzes to tripolyphosphate, the straight-chain compound, which then eventually hydrolyzes orthophosphate. At high pH values, however, trimetaphosphate is converted completely to tripolyphosphate. Pyrophosphates hydrolyze directly to orthophosphate. Thilo[48] studied the stability of water solutions of polyphosphates at 60°C; he reported the halflife of one of his solutions at various pH values to be 22 min at pH 1, 7.7 hr at pH 3, 4.4 days at pH 5, and 45.1 days at pH 8. Kutscher[51] reported that polyphosphates in neutral, weakly acid, or weakly alkaline solutions form trimetaphosphate and orthophosphates. In very acidic solutions the chains are randomly split to form a number of lower molecular weight polyphosphates that eventually hydrolyze to orthophosphate. Kutscher reports that the fact that trimetaphosphate is formed during hydrolysis of long-chain polyphosphates demonstrates that these chains exist as folded or spirally wound helixes, which allows formation of trimetaphosphate by close proximity of the first and third phosphate groups.

Further discussions of the influences of the chemical characteristics of the various phosphates, their interactions with food constituents, and their influences on food processing will be discussed in subsequent sections of this chapter.

Chapter 4

TOXICOLOGY OF THE FOOD PHOSPHATES

The toxicity of the phosphates commonly used in foods has been studied by numerous investigators. In common with all inorganic salts, it is reasonable to expect excess quantities to be toxic to an organism. Excessive amounts of inorganic salts not only will upset mineral balance in the body, but they also will affect the osmotic pressure of body fluids and, in the case of the phosphates, might be suspected of inactivating necessary nutrient compounds. An excellent review and discussion of the toxicology of the food phosphates has been published by the United Nations' World Health Organization.[37] This is the report of the seventh meeting of the FAO/WHO Expert Committee on Food Additives held in 1963; it lists the specifications for identity and purity of food additives and provides a review of the literature on their toxicology. The report has been used heavily in the preparation of this section of the Toxicology of the Food Phosphates; however, references not included in the WHO report have also been reviewed and are included.

Acute Toxicity

Table 6 summarizes data found in the literature reviewed that describe the studies of the acute toxicity of the phosphates commonly used in foods. In addition to providing a considerable extension of the literature review shown in Reference 37, the table also includes the acute toxicological data reported by Zipf[60] for sodium chloride, or common table salt, for comparison with the acute toxicity of the food phosphates.

The data in Table 6 demonstrate that orthophosphates and the shorter-chain polyphosphates are more toxic than sodium chloride when given orally, while the longer-chain and the cyclic polyphosphates are less toxic.[60] The table also demonstrates that all phosphates are significantly more toxic when introduced into the body in a manner that circumvents the digestive system. Sodium chloride also is more toxic when not administered through the digestive system. Intraperitoneal or intravenous injection of the phosphates, including the higher polyphosphates, produces only small differences in the LD_{50} levels, probably due to rapid enzymatic hydrolysis of the polyphosphate chains to orthophosphate in the blood. Several investigators who studied the orthophosphate levels that appeared in blood serum and urine after polyphosphate injection into the body have confirmed these observations.[37,51] There does appear, however, to be a decrease in acute toxicity of the medium chain-length polyphosphates (between 11 and 27), as demonstrated in an excellent study by Gosselin and Megirian.[61]

The introduction of the phosphate salts into the body by other than oral methods is meaningless if one is studying their toxicity as food ingredients., Injection directly into the bloodstream or into a body cavity does not allow for the changes which the salt undergoes prior to or during absorption through the intestinal wall. Therefore, the numerous studies of the phosphate feeding tests are reviewed in greater detail, as they represent the only accurate indication of the toxicity of phosphates as food ingredients.

Chronic Toxicity of Phosphates in Animal Diets

Table 7 summarizes the data from numerous reports of the effects of high phosphate levels in animal diets. These have also been reviewed in considerable detail in some of the references to be used for the following discussion. Although many of the investigations discussed in this section included studies of several of the phosphates, the data have been separated in Table 7 according to chain length of the phosphate.

Orthophosphates

Nazario[1] reviewed some early reports of feeding studies with phosphoric acid. These studies indicate that large quantities of dilute (36.4%) phosphoric acid could be given to animals and humans before any adverse effects would appear. Dogs tolerated as high as 13 g of the dilute acid before signs of enteritis appeared. Adult humans and their babies have been given dilute phosphoric acid in treatment of hypoacid diathesis. Babies showed no adverse effects when given 0.25 g of the dilute acid, and adults tolerated as high as 26 g

TABLE 6
Acute Toxicity Levels of Phosphates in Animals

Phosphate	Animal	Route[1]	LD_{50} mg/kg body weight	Approx. lethal Dose mg/kg body weight	Reference
H_3PO_4	Rabbit	i.v.		1 010	54
NaH_2PO_4	Mouse	oral		>100	55
NaH_2PO_4	Guinea pig	oral		>2000	56
NaH_2PO_4	Rat	i.p.		>36	57
Na_2HPO_4	Rabbit	i.v.		>985, ≤1075	56
NaH_2PO_4+Na_2HPO_4	Rat	i.v.	>500		58
$Na_2H_2P_2O_7$	Mouse	oral	2650		60
$Na_2H_2P_2O_7$	Mouse	s.c.	480		60
$Na_2H_2P_2O_7$	Mouse	i.v.	59		60
$Na_2H_2P_2O_7$	Rat	oral	>4000		59
$Na_4P_2O_7$	Rat	i.p.	233		59
$Na_4P_2O_7$	Rat	i.v.	100 to 500		58
$Na_4P_2O_7$	Mouse	i.p.		ca. 40	55
$Na_4P_2O_7$	Mouse	oral	2980		60
$Na_4P_2O_7$	Mouse	s.c.	400		60
$Na_4P_2O_7$	Mouse	i.v.	69		60
$Na_4P_2O_7$	Rabbit	i.v.		ca. 50	55
$Na_5P_3O_{10}$	Mouse	oral		>100	55
$Na_5P_3O_{10}$	Mouse	oral	3210		60
$Na_5P_3O_{10}$	Mouse	s.c.	900		60
$Na_5P_3O_{10}$	Mouse	i.v.	71		60
$Na_5P_3O_{10}$	Rat	i.p.	134		61
$Na_6P_4O_{13}$	Rat	oral	3920		60
$Na_6P_4O_{13}$	Rat	s.c.	875		60
$(KPO_3)_n$+ Pyro	Rat	oral	4000		58
$(KPO_3)_n$+ Pyro	Rat	i.v.	ca. 18		58
Hexametaphosphate[2]	Rabbit	i.v.		ca. 140	55
Hexametaphosphate	Mouse	oral		>100	55
Hexametaphosphate	Mouse	oral	7250		60
Hexametaphosphate	Mouse	s.c.	1300		60
Hexametaphosphate	Mouse	i.v.	62		60
$(NaPO_3)_{\bar{n}} = 6$[3]	Rat	i.p.	192		61
$(NaPO_3)_{\bar{n}} = 11$	Rat	i.p.	200		61
$(NaPO_3)_{\bar{n}} = 27$	Rat	i.p.	326		61
$(NaPO_3)_{\bar{n}} = 47$	Rat	i.p.	70		61
$(NaPO_3)_{\bar{n}} = 65$	Rat	i.p.	40		61
$(NaPO_3)_3$ cyclic	Mouse	oral	10 300		60
$(NaPO_3)_3$ cyclic	Mouse	s.c.	5 940		60
$(NaPO_3)_3$ cyclic	Mouse	i.v.	1 165		60
NaCl (table salt)	Mouse	oral	5 890		60
NaCl	Mouse	s.c.	3 000		60
NaCl	Mouse	i.v.	645		60

[1] Abbreviations designate the following routes of administration: i.p. = intraperitoneal; i.v. = intravenous; s.c. = subcutaneous; oral = by mouth (in diet, by stomach tube, etc.)

[2] Average chain lengths are not given.

[3] Polyphosphate preparations for which the $\bar{n}$ was determined.

TABLE 7
Dietary Levels of Phosphates Producing No Adverse Effects

Phosphate	Test animal	Length of test[1]	Maximum level tolerated	Effect of excess phosphate	Reference
Orthophosphate					
H_3PO_4 (36.4%)	Humans	Variable	17-26 g/d	No adverse effects[2]	1
H_3PO_4 (36.4%)	Rats	>12 mo	>0.75%	No adverse effects[2]	62
H_3PO_4 (36.4%)	Rats	44 d	<2.94%	Kidney damage	63
Na & K orthophosphates	Rats	44d	<3.93%	Kidney damage	63
MSP	Humans	Variable	5-7 g/d	No adverse effects	37
MSP	Rats	42 d	>3.4 g/kg/d	Kidney damage	57
MSP	Guinea pigs	200 d	>2.2%, <4.0%	Ca deposits, reduced growth	66
MSP	Guinea pigs	12-32 wk	4-8%[3]	Ca deposits, reduced growth	67
MSP + DSP	Rats	3 gen	>0.5%, <1%	Kidney damage	58
DSP	Rats	6 mo	>1.8%, <3.0%	Kidney damage	64, 68, 69
DSP	Rats	1 mo	<5%	Kidney damage	70
DKP	Rats	150 d	>5.1%	No adverse effects	65
Pyrophosphate					
TSPP	Rats	6 mo	>1.8%, <3.0%	Kidney damage	64, 68, 69
TSPP	Rats	16 wk	<1%	Kidney damage	59
SAPP + TSPP + $(KPO_3)_n$	Rats	3 gen	>0.5%, <1%	Kidney damage	58
Tripolyphosphate					
STP	Rats	6 mo	>1.8%, <3.0%	Kidney damage	64, 68, 69
STP	Rats	1 mo	>0.2%, <2%	Kidney damage	70
STP	Rats	2 yr	>0.5%, <5%	Kidney damage	71
STP	Dog	1 mo	>0.1 g/kg/d	No adverse effects	70
STP	Dog	5 mo	< 4.0 g/kg/d	Kidney, heart damage	70
Polyphosphate					
SHMP	Rats	150 d	>0.9%, <3.5%	Slight growth reduction	65
SHMP	Rats	1 mo	>0.2%, <2%	Kidney damage	70
SHMP	Rats	3 gen	>0.5%, <5%	Kidney damage	72
SHMP	Dog	1 mo	>0.1 g/kg/d	No adverse effects	70
SHMP	Dog	5 mo	< 2.5 g/kg/d	Kidney, heart damage	70
Graham's salt	Rats	6 mo	>1.8%, <3%	Kidney damage	64, 68, 69
$(KPO_3)_n$ + SAPP + TSPP	Rats	3 gen	>0.5%, <1%	Kidney damage	58
Cyclic phosphate					
$(NaPO_3)_3$	Rats	1 mo	>2%, <10%	Kidney damage	70
$(NaPO_3)_3$	Rats	2 yr	>0.1%, <1%	Retarded growth of males	73
$(NaPO_3)_3$	Rats	3 gen	>0.05%	No adverse effects	73
$(NaPO_3)_3$	Dog	1 mo	>0.1 g/kg/d	No adverse effects	70
$(NaPO_3)_3$	Dog	5 mo	<4 g/kg/d	Kidney, heart damage	70
$(NaPO_3)_4$	Rats	1 mo	>2%, <10%	Kidney, heart damage	70
$(NaPO_3)_4$	Dog	1 mo	>0.1 g/kg/d	No adverse effects	70
$(NaPO_3)_4$	Dog	5 mo	<4 g/kg/d	Kidney, heart damage	70

[1] Some reports did not give definite time periods.
[2] "No adverse effects" = no physiological damage noticeable at any level of phosphate tested.
[3] Guinea pigs tolerated higher levels of phosphate only when increased levels of Mg^{++} and K^{+} were presentin their diets.

Abbreviations:

d = day	kg = kilogram	yr = year
g = gram	mo = month	% = percent phosphate in diet (often calculated from % P)
gen = generation	wk = week	

without difficulty. Based on results with dogs and lethal doses for rabbits, Nazario calculated that a 70-kg man could tolerate 17 g of 100% phosphoric acid, while approximately 22 g would be a lethal dose.

Bonting[62] reported that more than 700 rats were used in an investigation of the tolerance of phosphoric acid up to levels of 0.75% in their diet. Studies of the effects on reproduction were carried through three generations of rats; no evidence of any adverse effects of the acid in the diets was found in young or adult rats. The study included investigations of the blood, tissues, mineral balance, nitrogen retention, and the acidic conditions in the digestive tract.

MacKay and Oliver[63] reported the results of a study in which 2.94% phosphoric acid and from 3.39 to 7.86% sodium or potassium orthophosphates, and combinations of these, were fed to white rats. Addition of these levels caused extensive pathological damage and significant increases in size and weight of the kidneys of the rats.

Hahn and co-workers[64] fed diets to rats that contained 1.8, 3.0, and 5.0% DSP, TSPP, STP, and SHMP (Graham's salt). Calcification was found in tissues of the kidneys of all animals fed 3 and 5% of the four phosphates. Graham's salt (SHMP) produced lower levels of calcification than the other phosphates. Slight increases in kidney weight and renal calcification in a few rats were found for the phosphates at the 1.8% level. Studies of the blood and of the weight and ash analyses of the organs showed no significant changes in the levels of phosphorus, copper, iron, and calcium for any level of phosphate.

Dymsza and co-workers[65] fed diets to rats that contained 0.87% and 5.1% DKP and 0.93% and 3.5% SHMP for 150 days. Growth and protein efficiency were somewhat reduced at the high level of SHMP, although the results were not conclusive. Autopsy, histological, and chemical studies showed no adverse physiological effects for either the type or level of phosphate. The animals on the metaphosphate diet stored more calcium and phosphorus in their bodies, although this was not evident from analysis of the bones. This study also indicated that there was no change in absorption or utilization of calcium, phosphorus, or iron with either type or level of phosphate.

McFarlane[57] studied the site of renal damage in rat kidneys when they were given 0.1 to 0.36 g/100 g body weight of MSP by intraperitoneal injection and 0.34 to 1.2 g/100 g body weight per day by diet. The rats were sacrificed at various periods ranging from 7 hr to 42 days after administration. Both methods of administration gave similar results. Only the lowest level of oral administration produced no adverse effects; evidence of tissue damage occurred at all higher levels.

A diet containing a mixture of MSP and DSP and another containing a mixture of equal parts potassium metaphosphate (Kurrol's salt), SAPP, and TSPP were studied in rats by van Esch and co-workers.[58] Up to three generations of rats were studied to determine the effects of the two mixtures of phosphates on reproduction, as well as their short-term effects. The levels used in the study were 0, 0.5, 1.0, 2.5, and 5.0% of each of the two mixtures in the respective diets. The mixtures of orthophosphates and polyphosphates showed similar physiological effects, principally, tissue calcification and renal damage in the kidneys. No physiological effects were noticed at the 0.5% level. Damage was found in rats receiving the diets that contained 1% or more of either type of phosphate. In addition, the animals receiving 5% polyphosphate in their diets had shortened life spans, impaired growth, and reduced fertility.

Hogan and co-workers[66] reported that calcium phosphate deposits appeared in the soft tissues of 90% of the foot pads of guinea pigs consuming a diet containing 0.8% calcium and 0.9% phosphorus (equivalent to about 4% MSP). By reducing the phosphorus content of the diet to 0.5% (equivalent to about 2.2% MSP), the number of guinea pigs showing the deposits was reduced to less than 10%. Therefore, the investigators suspected that the ratio of calcium to phosphorus was involved. Further work reported and summarized by House and Hogan[67] demonstrated that, if the levels of magnesium and potassium were low, the amount of phosphorus in a diet was the major factor determining whether calcium phosphate would deposit in the soft tissues. If the diet contained adequate levels of magnesium and potassium, however, few or no deposits were observed. A diet containing 0.9% calcium, 1.7% phosphorus (equivalent to about 8% MSP), 0.04% magnesium, and 0.41% potassium produced severe reductions in weight gain, increased stiffness of leg joints, and numerous visible deposits of calcium phosphate in the foot pads of the guinea pigs. These symptoms were significantly reduced when the diet was altered to contain 0.35% magnesium and 1.5%

potassium with the same levels of calcium and phosphorus.

These latter two papers probably explain the wide variations in the levels of phosphate reported to cause adverse physiological or biological effects. Optimum ratios of calcium to phosphorus, or of magnesium and potassium to phosphorus most likely explain why the diets of some studies required higher levels of phosphates in order to demonstrate adverse physiological effects. It is impossible to determine these ratios from the data given in most of the papers reviewed for this chapter.

Pyrophosphates

Hahn and Seifen[68] and Hahn[69] reported the results of six-month studies on rats fed 1.8%, 3%, and 5% "tetrasodium diphosphate" ($Na_4P_2O_7$). As with the DSP diets, 1.8% of the TSPP produced a slight increase in weight of the kidneys and renal calcification in some, but not all, of the rats to which it was fed. Later, a diet containing 1.1% of added TSPP was compared with a similar level of DSP. The growth again was slightly retarded, but it required 39 weeks on the diet in order to produce a slight degree of calcification in the tissues.

Datta and co-workers[59] studied the effects of 1%, 2.5%, and 5% levels of TSPP in the diets of rats for 16 weeks. Even the lowest level, 1%, of the TSPP caused a reduction in kidney function and an increase in its weight. They reported that the rat kidney is capable of concentrating salts to a greater degree than the kidney of other species. Therefore, they questioned the validity of using the information obtained from animals and calculating the effects on humans.

Van Esch and co-workers[58] reported the results of feeding studies with rats through three generations. Levels of 0.5%, 1%, 2.5%,and 5% TSPP were added to their diets. Kidney damage was observed at all levels of phosphate from 1% and above, and a significant reduction in growth, fertility, and life span was found at the 5% level.

Polyphosphates

Hodge[70] compared diets of rats containing 0.2%, 2%, and 10% STP with diets containing 10% sodium chloride or 5% disodium phosphate. Diets with 10% sodium chloride and 10% polyphosphate decreased the growth, increased the kidney weights, and damaged the kidneys, although no rats died. The rats fed the 2% polyphosphate diet had normal outward appearances, but their kidneys were found to be damaged upon histological examination. The 0.2% polyphosphate diet caused no changes at all.

Hahn and co-workers[64,68,69] reported that rats fed 1.8% STP or SHMP showed no evidence of physiological change; however, 3% and 5% of either phosphate caused severe kidney damage. The 5% phosphate diet also caused a reduction in growth. The amount of kidney damage in the animals was practically identical to that of animals fed similar levels of DSP, which indicated to the authors that the polyphosphates must be hydrolyzed to orthophosphate before they can be absorbed through the intestinal walls.

Hodge[70] fed diets that contained 0.1 g/kg/day of STP and SHMP to four dogs. Another group of four dogs was given the same phosphates in increasing dosages, beginning at 1.0 g/kg/day and increasing to 4.0 g/kg/day for the last month of the five-month feeding period. The dog given 0.1 g/kg/day showed no adverse effects. The dog given increasing dosages of SHMP began to lose weight when the daily dose reached 2.5 g/kg/day or more, while the dog fed STP did not begin to lose weight until his diet reached 4.0 g/kg/day.

Gassner and co-workers[74] fed 750 mg of potassium Kurrol's salt to rats for six days. Urine was collected for 10 days and analyzed by paper chromatography for evidence for the various types of phosphates. The results indicated that approximately 25% of the daily intake of the potassium metaphosphate was found in the urine as orthophosphate. After the first two days, when no condensed phosphate was found in the urine, an average of about 4% of the daily dose was found in the form of condensed phosphates. Approximately 80% had an average chain length of 4 to 5, and the balance was tripolyphosphate or pyrophosphate. A few samples contained trimetaphosphate in quantities below 0.2%.

Cyclic Polyphosphates

Hodge[70,73] studied the toxicity of the cyclic polyphosphates. Rats fed 0.2%, 2%, and 10% sodium trimetaphosphate and others fed similar quantities of sodium tetrametaphosphate showed fewer physiological effects than with any of the other phosphates in the study.[70] The 10% tetrametaphosphate caused slight increases in kidney weights and less evidence of kidney injury than with the straight-chain polyphosphates. Only transient kidney injury resulted from the 10% trimetaphosphate. No adverse effects were found in diets containing lower levels of these two phosphates. Dogs were also fed the cyclic phosphates; no adverse effects were noticed until the level of the phosphates had reached 4 g/kg/day.

Hodge[73] reported the results of a two-year study in which 100 rats were fed 0.1%, 1.0%, and 10% sodium trimetaphosphate in their diets. Male rats failed to grow as rapidly on 1% trimetaphosphate diets, although female rats grew normally. At 10% trimetaphosphate growth of all rats was retarded, although excessive mortalities were not observed. Histological studies and studies of reproduction through two generations showed no further abnormalities.

Biochemical Aspects of Phosphate Toxicity

As stated in the introduction to this chapter, the phosphate anion is essential to the metabolism of all living organisms. However, since no living organism is known to synthesize the phosphate anion, it must be supplied in the diet.[1,75] Special mechanisms for maintaining the required level of phosphate in the blood serum of higher organisms are therefore required; these stabilize the level of inorganic phosphate in the blood through mechanisms controlled by the parathyroid gland. Hormones from this gland, regulated by the blood calcium level, are able to trigger the release or deposition of phosphate from and into the skeleton. It should not be necessary to point out that phosphate is essential in the metabolic processes; this has been demonstrated many times by the various schemes showing the metabolism of protein, fat, and carbohydrate. All living tissue must contain a certain level of phosphate, which is regulated by mechanisms that either prevent absorption or cause elimination of excess phosphate.[37,75] A further regulatory mechanism in the metabolism of phosphates is that of the phosphatase enzymes. Divalent cations, such as magnesium and calcium, are required for activation of these enzymes.

It has been estimated that adult humans require a minimun of 0.88 g of phosphorus in their daily diets. The blood serum of human adults normally carries approximately 2.5 to 4.5 mg phosphorus per 100 ml, while that of children carries a higher level.[75]

Phosphates, and particularly polyphosphates, are known to precipitate or sequester calcium, iron, magnesium, copper, and similar ions essential to human metabolism. However, numerous studies have demonstrated that absorption and metabolism of these ions are not inhibited in any way, even by excessive levels of the various phosphates in the diet.[37,62,64,65,76] Several investigators have shown that the addition of polyphosphates to the diet increases retention of essential mineral ions.[37,65,77], Borenstein and Schwartz[77] found that the addition of 7% STP or SHMP to the diets actually increased the absorption and retention of calcium in the bodies of rats, and SHMP caused higher calcium retention than STP. They also reported evidence indicating that SHMP complexed with calcium was absorbed intact through the intestinal walls. After absorption, SHMP was hydrolyzed to orthophosphate. In contrast, STP and calcium are absorbed independently, and the STP is then hydrolyzed to orthophosphate after absorption.

Orthophosphates are readily absorbed through the intestinal wall. However, the chain length of the condensed phosphates determines the amount that will be absorbed. The pyrophosphates are readily absorbed intact but then are completely hydrolyzed to orthophosphate.[78] Mitchell[79] demonstrated that synthetic digestive fluids were capable of hydrolyzing polyphosphates to orthophosphate in a step-wise fashion. Several investigators have shown that feeding condensed phosphates to animals resulted in absorption of only the short-chain polyphosphates or the orthophosphates resulting from their hydrolysis.[21,75,79-84] Furthermore, only 10 to 40% of hexametaphosphate and the longer-chain, Graham's, Tammann's, and Kurrol's salts are absorbed from the digestive tract.[61,76,80-82] Studies in which radioactive polyphosphates, from tripolyphosphate through Kurrol's salts, were fed to animals showed that these compounds are absorbed only after hydrolysis to ortho- or pyrophosphate in the gut and that only radioactive orthophosphates could be recovered from the urine. Approximately half of the administered radioactivity was found to be absorbed; the other half was recovered in the feces as polyphosphate.[78]

Acceptable Daily Intakes for Humans

The FAO/WHO Expert Committee on Food Additives has evaluated much of the experimental data on the toxicity of phosphates and has issued estimates of acceptable levels in human diets.[37] The fact that levels of calcium and other minerals in the diet affect the levels that will produce the earliest signs of adverse effects was also considered. The experimental data indicated that the total amount of phosphorous in the diet, rather than only that added as a food ingredient, must be considered. Furthermore, the toxic effects of the condensed phosphates are directly proportional to the extent of their hydrolysis to ortho- or pyrophosphate. The extent of hydrolysis in the intestine depends on many unpredictable variables. Therefore, for the purpose of setting acceptable dietary limits, the Committee assumed that all of the condensed phosphate would hydrolyze to orthophosphate.[37]

With these considerations in mind, the Committee recommended an unconditional acceptance level for total daily dietary phosphorus and a conditional acceptance level. The *unconditional acceptance level* is the level considered safe under all conditions of diet. The *conditional acceptance level* is acceptable only when the dietary calcium level is high. These levels are as follows[37]

	mg/kg body wt
Acceptable levels for all diets	<30
Acceptable levels for high calcium diets	30 to 70

Summary

The results of animal feeding studies reported in the scientific literature indicate that levels of 0.5% of the phosphates could be tolerated in the diet without adverse physiological effects. Higher levels could possibly be tolerated if a proper dietary balance of other ions, particularly calcium, magnesium, and potassium, is maintained. Few, if any, applications for the phosphates require over 0.5% to obtain the desired effect, and higher levels often produce adverse physical and chemical effects and off-flavors. In addition, the phosphates produce adverse effects in numerous food systems so that it is not possible that they would be added to all foods. As a result, the level of phosphates likely to be found in the diet, even though they are fully utilized in all their potential applications, is self-limiting. It is highly unlikely that a level of 0.5% phosphate could ever appear in the total human diet.

Chapter 5

THE ATTITUDES OF REGULATORY AGENCIES

The U.S. Food and Drug Administration

The U.S. Food and Drug Administration (FDA) considers the orthophosphoric and polyphosphoric acids and their calcium, potassium, sodium, and ammonium salts as GRAS (Generally Recognized As Safe). Lehman[83] referred to the work of Gosselin and co-workers[80,84] and Mitchell[79] as evidence that the straight-chain polymeric phosphates are hydrolyzed to orthophosphate. This work is used as the basis for the GRAS status of these food additives. Based on the same references, Lehman[83] stated that the cyclic metaphosphates may not be safe food additives. Gosselin and co-workers[80,84] found that the injection of cyclic phosphates resulted in recovery of most of the intact cyclic phosphate in the urine. Lehman failed to note that the same authors[84] and Hodge[70,73] have shown that little, if any, cyclic metaphosphates were absorbed through the intestinal wall when they were administered orally. The metaphosphates must first be hydrolyzed to tripolyphosphate and next to orthophosphate, which then can be absorbed. It would thus appear that the use of the cyclic phosphates in foods would be no more, or even less, toxic than any other condensed phosphate.

All of the phosphate food additives recognized as GRAS by FDA are listed in Table 8; each food additive is listed under functional categories for its applications.[85] No tolerance levels or limitations are listed, other than that the quantity of the substance added to the food must not exceed the amount required to accomplish its intended purpose.

There has been some confusion in the past regarding sodium metaphosphate being listed under the category of sequestrants. This seems to have been clarified recently to mean that any compound that meets the specifications for sodium metaphosphate in the *Food Chemicals Codex* is GRAS.[38] It is also understood that the ammonium, calcium, potassium, and sodium salts are equally acceptable.

The Meat Inspection Division, U.S.D.A.

The Meat Inspection Division, Consumer and Marketing Service, U.S. Department of Agriculture also provides regulations for the use of phosphates as additives to meat products. All of the substances acceptable for addition to meat are listed in the *Code of Federal Regulations*;[86] these phosphate additives are summarized in Table 9. Certain applications for the phosphates in meat products have limitations for the quantity that can be used; these are clearly listed in the table. The longest chain polyphosphate allowed by MID is SHMP. Therefore, none of the longer chain sodium metaphosphates or the highly polymerized potassium polymetaphosphates are allowed to be used in meats under federal inspection. Many state regulations do not prevent the use of the longer chain polyphosphates, and some of the intrastate meat packers have been using potassium polymetaphosphate (Kurrol's salt).

TABLE 8
Phosphate Food Additives Recognized as GRAS by FDA

Emulsifying Agents	
Monosodium phosphate derivatives of mono- and diglycerides	
Nutrients and/or dietary supplements	
Calcium glycerophosphate	Magnesium phosphate (di-, tribasic)
Calcium phosphate (mono-, di-, tribasic)	Manganese glycerophosphate
Calcium pyrophosphate	Manganese hypophosphite
Ferric phosphate	Potassium glycerophosphate
Ferric pyrophosphate	Sodium phosphate (mono-, di-, tribasic)
Ferric sodium pyrophosphate	

TABLE 8 (continued)

Sequestrants	
Calcium hexametaphosphate	Sodium metaphosphate
Calcium phosphate (monobasic)	Sodium phosphate
Dipotassium phosphate	(mono-, di-, tribasic)
Disodium phosphate	Sodium pyrophosphate
Sodium acid phosphate	Sodium pyrophosphate, tetra
Sodium hexametaphosphate	Sodium tripolyphosphate
Miscellaneous and/or general-purpose food additives	
Ammonium phosphate	Sodium aluminum phosphate
(mono- and dibasic)	Sodium phosphate
Calcium phosphate	(mono-, di-, tribasic)
(mono-, di-, tribasic)	Sodium tripolyphosphate
Phosphoric acid	
Sodium acid pyrophosphate	

Source: Paragraph 121.101 (D), Part 121, *Code of Federal Regulations, Title 21*, (revised as of January 1, 1971), U.S. Government Printing Office, Washington, D.C., 1971.

TABLE 9

Phosphate Food Additives Acceptable for Use in Meat Processing

Class of substance	Phosphate	Purpose	Products	Amount
Cooling and retort water treatment agents	TKPP	To prevent staining on exterior of canned goods	Various	Sufficient for purpose
	SHMP	To prevent staining on exterior of canned goods	Various	Sufficient for purpose
	TSPP	To prevent staining on exterior of canned goods	Various	Sufficient for purpose
	STP	To prevent staining on exterior of canned goods	Various	Sufficient for purpose
Hog scald agents	SHMP	To remove hair	Hog carcasses	Sufficient for purpose
	STP	To remove hair	Hog carcasses	Sufficient for purpose
	TSP	To remove hair	Hog carcasses	Sufficient for purpose
Phosphates	DSP	To decrease amount of cooked-out juices	Cured hams, pork shoulder picnics, and loins; canned hams and pork shoulder picnics; chopped ham; and bacon	5.0% of phosphate in pickle at 10% pump level; 0.5% of phosphate in product (only clear solution may be injected into product).

TABLE 9 (continued)

Class of substance	Phosphate	Purpose	Products	Amount
	MSP	To decrease amount of cooked-out juices	cured hams, pork shoulder picnics, and loins; canned hams and pork shoulder picnics; chopped ham; and bacon	5.0% of phosphate in pickle at 10% pump level; 0.5% of phosphate in product (only clear solution may be injected into product).
	SHMP	To decrease amount of cooked-out juices	cured hams, pork shoulder picnics, and loins; canned hams and pork shoulder picnics; chopped ham; and bacon	5.0% of phosphate in pickle at 10% pump level; 0.5% of phosphate in product (only clear solution may be injected into product).
	STP	To decrease amount of cooked-out juices	Cured hams, pork shoulder picnics, and loins; canned hams and pork shoulder picnics; chopped ham; and bacon	5.0% of phosphate in pickle at 10% pump level; 0.5% of phosphate in product (only clear solution may be injected into product).
	TSPP	To decrease amount of cooked-out juices	Cured hams, pork shoulder picnics, and loins; canned hams and pork shoulder picnics; chopped ham; and bacon	5.0% of phosphate in pickle at 10% pump level; 0.5% of phosphate in product (only clear solution may be injected into product).
	SAPP	To decrease amount of cooked-out juices	Cured hams, pork shoulder picnics, and loins; canned hams and pork shoulder picnics; chopped ham; and bacon	5.0% of phosphate in pickle at 10% pump level; 0.5% of phosphate in product (only clear solution may be injected into product).
Rendering agents	TCP	To aid rendering	Animal fats	Sufficient for purpose
	TSP	To aid rendering	Animal fats	Sufficient for purpose
Synergists (used in combination with antioxidants)	H_3PO_4	To increase effectiveness of antioxidants	Lard and shortening	0.01%

Source: Paragraph 318.7(4), *Code of Federal Regulations, Title 9* (Revised as of January 1, 1971), U.S. Government Printing Office, Washington, D. C., 1971.

Chapter 6

FUNCTIONS AND APPLICATIONS OF PHOSPHATES IN FOOD SYSTEMS

Some General Functions of Phosphates in Foods

Phosphates, through their interactions with some of the constituents of food systems, have very useful and important functions in food processing.

Complexing Metal Ions

Phosphates are capable of acting as ion exchangers[25] and, thus, of forming soluble complexes with polyvalent ions that would interfere with food-processing operations. A significant advantage to the use of phosphates to complex such nutritionally important ions as calcium, magnesium, and iron is that the ions can still be absorbed through the intestinal walls and can be utilized by the body.[25]

The orthophosphates are capable of precipitating iron and similar transition metal ions. They are also reported to form complexes with calcium, although the usual reaction observed is that of precipitation of the calcium.

The polyphosphates, including those ranging from two to many phosphorus units per molecule, form complexes with nearly all of the metallic ions. Very stable soluble complexes are formed with the ions of the transition metals, such as copper, nickel, and iron; less stable complexes are formed with the alkaline-earth metals, such as calcium and magnesium. These latter complexes are dissociated to some extent, a characteristic that leads to some very useful applications in the formation of food gels. The polyphosphates form very weak soluble complexes with the alkali metals and ammonium.

The ability to form soluble complexes with metal ions apparently does not depend on the chain length of the polyphosphate, since complexes are reported for the pyrophosphates as well as for the long-chain polyphosphates. The strength of the complexes increases with the phosphate chain length. A precipitate is often formed initially, as one begins to add a polyphosphate to a solution containing the metal ion. This precipitate dissolves upon the addition of more of the polyphosphate. The formation of soluble complexes that prevent the formation of precipitates with the metal ions is often called *sequestration.*[32]

The cyclic condensed phosphates, sodium trimetaphosphate and sodium tetrametaphosphate, form very weak, complexes, if any, with metal ions of all types. They, therefore, are not used for this purpose.

The treatment of water used in food processing is an important application of the metal-complexing ability of the phosphates. The mineral constituents often present in natural waters can have devastating effects on the foods processed with those natural waters. It is, therefore, common practice to treat the waters in such a way as to "soften" them. The treatment usually involves the addition of a phosphate, such as SHMP, SHMP plus alkali, shorter-chain polyphosphates, and, in some cases, orthophosphate or mixtures of these, in order to complex the heavy metal ions such as iron and copper, and the alkaline-earth metal ions, calcium and magnesium. Some of the specific phosphate applications, such as the use of polyphosphates to tenderize the skins of vegetables and fruits, have the objective of complexing the natural calcium and magnesium ions present in the water so that they will not precipitate or form gels with natural constituents of the skins, such as pectins.

The treatment of waters used in food processing, and also in drinking water, has been thoroughly reviewed in many publications. Examples are those of Navet[87] and Hofer.[88] The phosphates can also, however, be used for removing fluorides[89] and for such applications as preventing the contamination of water by the materials from the containers.[90] Further information on water treatment can be obtained from the literature on that topic.

Other functions of phosphates in water used in food processing are the prevention of "milk stone" deposits in dairy equipment, the reduction of corrosion and scale in processing and heating equipment, and the stabilization of dissolved iron to prevent formation of "red water."[91]

Complexing Organic Polyelectrolyte Food Constituents

Some functions of phosphates in food processing depend on their ability to form complexes and reaction products with constituents of foods other than mineral elements. This is demonstrated by the ability of phosphates to interact with many organic polyelectrolytes, such as protein, pectin, starch, etc. Numerous complicated protein-phosphate or protein-phosphate-salt complexes are known and have been studied and reported.[92-94] The most frequently studied of these complexes is probably the casein-calcium phosphate complex in milk. One has only to review the numerous publications of Leviton, Morr, Rose, Waugh, Zittle, their co-workers, and numerous others to appreciate the enormous effort that has been expended in research on this complex. Even so, the exact structure, its chemistry, and how it can be modified to obtain desired effects are still unknown. Further discussions of these important complexes will appear under specific sections of this chapter.

Direct Chemical Reactions with Food Constituents

In addition to the formation of complexes between organic polyelectrolytes and phosphates, phosphates may undergo esterification reactions and cross-linking between molecules of proteins and starches. (This will be discussed further in subsequent sections.)

Buffering or pH Stabilization

Another important function of the phosphates is their buffering ability, which can be used to stabilize the optimum pH required for processing or stabilizing a food.

Dispersion of Food Constituents

Phosphates have the ability to promote dispersion and peptization of relatively insoluble food constituents, such as the proteins in concentrated milk and pasteurized, process cheese. Entire food particles are dispersed in some cases, much the same as phosphates act as dispersants for clay particles.[30]

Emulsion Stabilization

Phosphates have the ability to stabilize emulsions. Their ability to stabilize sausage emulsions, for example, will be discussed in the section on meat applications.

Increasing Hydration and Water Binding

The phosphates are also able to interact with animal and vegetable proteins in such a way that hydration is promoted. This important function is utilized in preventing freeze-thaw drip in meats, poultry, and fish.

Mineral Supplementation

Calcium, iron, sodium, and potassium phosphates are also used to improve the nutritional properties of numerous cereal products e. g., flour, prepared mixes, baked goods, and breakfast cereals.

Acidification or Lowering pH

Phosphoric acid and its acidic salts are used in food acidification. The acidic phosphate salts are used as acidifying agents to react with baking soda in baking powders, prepared mixes, self-rising flour, and corn meal and to produce acidic pH values in many other foods. Orthophosphoric acid is the commonest acidic constituent used in cola and root beer soft drinks to produce tartness.

Alkalization or Raising pH

The alkaline phosphate salts function to produce alkaline pH values where these are necessary. DSP and TSP, for example, are used in the preparation of pasteurized, process cheese to produce the slightly alkaline pH values necessary for optimum protein dispersion as well as for interacting with the proteins themselves to improve their emulsification and water-binding abilities.

Prevention of Caking

Insoluble phosphates are commonly used as anticaking agents, as they can be prepared in a very fine powdered form that can be used to separate crystals of substances that normally adhere to each other. They are used also as water-absorbent anticaking agents. Warning and Schille[95] used a mixture of aluminum phosphate and TCP as an anticaking agent for salt. Lubeck[96] proposed the use of DCP and dimagnesium phosphate to produce a free-flowing table salt. TCP has also been proposed as an anticaking agent for dietary seasoning compositions.[97] A somewhat unrelated example of the ability of TCP to maintain free flow of particles is that patented by Sorgenti, Nack, and Sachsel,[98] who advocated the use of TCP as the solid particles in a fluidized bed cooking process for food products. Other applications of insoluble phospates as anticaking agents will be discussed in subsequent sections.

Food Preservation

The phosphates, and especially the polyphosphates, have been found to prevent or retard the oxidation of unsaturated fats in aqueous food systems and to inhibit the growth of some of the microorganisms involved in the spoilage of foods. Although some investigators relate these functions to the ability of the phosphates to complex metal ions or organic polyelectrolytes essential to the fat oxidation or spoilage organisms, others report evidence of additional factors involved in this important application. The antioxidant and microbiological preservation applications of the phosphates will, therefore, be treated separately in subsequent sections.

The following sections consist of further discussions of the various functions of phosphates and their applications in the processing of foods and food ingredients. The sections are arranged according to general food categories, with a discussion of the functions of the phosphates in the processing of specific foods and their reactions with food ingredients under each category.

Phosphate Applications in Beverages

Phosphates have been found to function as buffers, metal-complexing agents, microbiological inhibitors, acidifiers, mineral supplements, and flow agents in alcoholic, carbonated, powdered, and nutritional beverages.

Applications in Alcoholic Beverages

Buffering

A combination of DSP·$2H_2O$ and of MKP was proposed by Heyer[99] as a buffering system for use in obtaining maximum conversion of humulon to *iso*-humulon in preparation of a dried hops extract. Heyer discovered that the boiling hops suspension had to be buffered between pH 8.0 and 8.3 for maximum conversion and yields.

Bacteriological Control

Bah and McKeen[100] reported that the bacteria that contaminate yeasts used in brewing beer and ale could be controlled by washing the yeast with a solution of H_3PO_4 and sodium persulfate at 5°C. The treated yeast performed satisfactorily during fermentation of the beer and ale. Malkov and Kormushkina[101] found that treating yeast with a solution of TKP prior to its addition to the beer wort increased its viability and resistance toward contaminating bacteria.

Kohl and Ellinger[102,103] have described the inhibition of bacteria and fungi, including yeasts, by polyphosphates in beer and other foods. The most effective polyphosphates for the inhibition of bacteria were found to be those with average chain lengths in the range of 18 to 34 phosphate units. Shorter- and longer-chain phosphates, although effective, failed to inhibit the bacteria to the same extent. Fungi and yeasts, including *Saccharomyces cerevisiae*, were inhibited by polyphosphates of the same chain length as well as the longer-chain-length and highly polymerized potassium polyphosphates (Kurrol's salts). The polyphosphates can be added to draft beers to inhibit the last traces of microorganisms. The short-chain polyphosphates were weak inhibitors of microbiological growth unless they were combined with high pH.

Complexing Metal Ions

The polyphosphates are especially useful in alcoholic beverage production because of their ability to form stable, soluble complexes with heavy metal and alkaline earth metals. Iron, copper, and calcium ions are especially troublesome and fortunately form such complexes readily. The formation of the complexes with phosphates inactivates the metal ions so effectively that they need not be removed from the beverage, which reduces processing costs. The polyphosphates are also used to soften the water used in brewing beer. Plain SHMP and SHMP made alkaline with sodium hydroxide are most frequently used for this purpose.

Libbey[104] reported that the addition of 0.1 to 1 g of SHMP per gallon directly to beer or wine also prevented clouding or haze formation. Clouding or haze formation occurs when heavy metal and alkaline earth metal ions react with organic compounds normally present in the filtered beverage. The polyphosphate complexes with the interfering metal ions and prevents the haze-forming interactions.

Peynaud[105] reported data showing the reduction of iron content in wines upon the addition of 150 to 450 mg of SHMP. Significant reduction occurred from precipitate and complex formation. The same author showed that iron content was also reduced by addition of 100 to 300 mg/l. of STP and that this phosphate aids in formation of cream of tartar, thus stabilizing young wines.

The unwanted metals, i. e., copper and iron, can also be removed by filtration of the beverage through insoluble calcium phosphates.[106] A method for the direct precipitation of dicalcium phosphate as a means of coprecipitating impurities in a beverage was proposed by Skrivanek.[107] This is accomplished by the addition of phosphoric acid to the beverage and subsequent neutralization with calcium carbonate to precipitate dicalcium orthophosphate.

Ohara et al.[108] proposed the use of ammonium phosphate to reduce the tannin and acid content of wines, thus reducing the chances for formation of cloud or haze.

Applications in Carbonated Beverages

Complexing Metal Ions

Heavy metal ions in the water used in the preparation of these beverages can serve as nucleii for the formation of carbon dioxide bubbles. This results in more rapid dissipation of the carbonation of the beverage. A patent was obtained by Karlson[209] for the use of soluble sodium polyphosphate glasses having a Na_2O/P_2O_5 ratio of 1.2:1 to sequester these ions and thus prevent the loss of carbonation. Since iron was the most troublesome metal ion involved, the amount of polyphosphate required was approximately 3ppm polyphosphate for each 1ppm of iron in the water. The author recommends adding an extra 2ppm over this amount to assure complexing of all the iron. The degree of carbonation required to be added to a beverage thus treated was found to be reduced from between 25-90% of that required if the beverage were made with untreated water.

Acidification

H_3PO_4 is the acidulant used in many of the bottled soft-drink beverages. This is especially true of those flavored with leaf, root, or nut extracts, such as the carbonated cola beverages.[29,110,111] Phosphoric acid, being a natural constituent of many fruits, blends well with their flavors in carbonated drinks, as it adds a natural tartness to the beverage. The quantity used is often determined by the type of fruit flavor and varies according to the degree of tartness and acidity desired. The acidic, crystalline phosphate salts are also often included as a portion or all of the acidulant in powdered beverages. The salts used include MCP · H_2O, MSP, and hemisodium phosphate.[29,30] Diller and Brout[39,112,113] have received patents for a soft, carbonated drink dry-mix that, upon addition to tap water, produces a prolonged effervescence. The basis for the prolonged effervescence is a combination of hemisodium phosphate, $NaH_5(PO_4)_2$, and mono-ammonium acid carbonate. The authors reported that hemisodium phosphate had, in the past, been considered to be a double salt of H_3PO_4 and MSP, but that they had found it to actually be a complex having the formula $(HPO_3)_n \cdot NaH_2PO_4 \cdot H_2O$. The effervescence was reported to last up to one half hour or at least as long as that of a pressure-packed, bottled, carbonated soft drink. An undesirable precipitate that formed during this reaction adhered to the side of the glass from which the beverage was consumed. The addition of minute but critical amounts of glycine and/or SHMP to the dry mix prevented its formation.

Improving Flow of Powders

In order to obtain proper flow properties of a powdered, instant beverage mix, Common[114] incorporated TCP. The phosphate was found to influence the particle-size distribution of the dry mix, since more of the powder containing TCP flowed through a standard sieve than when TCP was absent.

Applications in Other Beverages

Acidification

Two patents have been issued that utilize phosphates as acidifying agents in the preparation of dry instant coffee and imitation coffee products. DiNardo[115] has described the use of phosphoric or sulfuric acid as a hydrolyzing agent for a portion of the extracted coffee grounds. He reported that he obtained increased yields of the coffee extract.

Stayton[116] used acidic phosphates and pyrophosphates as well as other acidic inorganic and organic salts to neutralize histidine compounds in chickory for use in a coffee-chickory blend. Neutralizing the histidine compounds removed characteristic aromas and flavors of chickory so that the blend tasted like high-grade coffee.

Mineral Supplementation

Beverages designed to provide improved nutrition or for certain dietary purposes have become popular in recent years. These beverages must be fortified with various mineral and vitamin elements necessary in the diet. Sodium iron pyrophosphate, ferric pyrophosphate, and ferric orthophosphate are used as sources of iron in some of these beverages.[29] Calcium phosphates also are often added to provide the proper levels of calcium and phosphorous.

Complexing Metal Ions

The presence of heavy metal ions can cause the oxidation of naturally present or added ascorbic acid and, therefore, its inactivation. Polyphosphoric acids have been found to inactivate the metal ions and to stabilize ascorbic acid, as well as added food colorings in vitamin-C fortified beverages.[117-119] At least 90% of the ascorbic acid remained after 10 days according to these reports. The optimum conditions for stabilization of the ascorbic acid were found to be a pH between 2.5 and 3.0 and between 0.1 and 0.2% polyphosphoric acid.

Phosphate Applications in Cereal Products

Phosphates were probably first used as direct food additives in cereal products, beginning with their applications as leavening acids. Other applications include pH adjustment, buffering, dough conditioning, and mineral enrichment. The applications of phosphates in cereal products have been reviewed by Stahl and Ellinger.[120]

Applications as Leavening Acids

The prepared mix and baking industries probably are the largest consumers of phosphates through their applications in chemical leavening systems. Since orthophosphoric and polyphosphoric acids contain three or more hydrogens, they can exist as partially neutralized, acidic salts. These acidic phosphate salts have the ability to neutralize sodium bicarbonate, or common baking soda. The gas, CO_2 (carbon dioxide), is released during the neutralization of the soda and forms the basis for most chemical leavening reactions.

Characteristics of Leavening Systems

Leavening, in general, refers to the introduction and expansion of some type of gas in the batter or dough system. Mixing or kneading the batter or dough also incorporates air in proportion to the energy expended in the mixing action. These gas bubbles then expand during baking and provide part of the leavening in the baked product. The formation of steam from the water used in the preparation of the batter or the dough also contributes to the leavening of the system. Yeast is added to a dough in order to produce CO_2 as the organisms metabolize the carbohydrates. This is a biological leavening system.

The type of leavening to be discussed in this section, however, is that provided by the chemical reaction between soda and an acidic phosphate salt. Several functions of phosphates in chemical leavening systems in a baked product follow:

1. *acidification* to release CO_2 from soda to create new gas bubbles and to expand existing gas bubbles,
2. *buffering* to provide the optimum pH in the baked product, and
3. *interactions with the protein* constituents of flour to modify the elastic and viscosity properties of doughs and batters.[30,121]

As a dough or batter system is mixed, the leavening ingredients are thoroughly dispersed throughout and are dissolved in the water added to the dough or batter system. With proper mixing the phosphate and soda are distributed as individual molecules throughout the system. Carbon dioxide is released upon reaction of an acid molecule with a soda molecule. After sufficient CO_2 has dissolved to saturate the dough or batter, subsequent gas molecules diffuse into the system. Some enter the nearest gas bubble; others form a new bubble nucleus when sufficient gas molecules have accumulated in an area so that they form a gas pressure greater than the cohesive forces at the weakest point in that area of the dough or batter. Further CO_2 molecules then diffuse into this tiny bubble to further expand it.[120-132]

It is desirable, in most baking applications, to have numerous tiny leavening gas bubbles uniformly distributed throughout the system. Properly designed chemical leavening systems can make significant contributions toward accomplishing this goal. Numerous other factors also influence the size and distribution of the bubbles to a significant degree. For example, the quality of the gluten in the flour used, the viscosity of the system, and the degree of emulsification play major roles. Although a discussion of these factors is outside the scope of this chapter, excellent reviews have been published, and the interested reader should consult them.[120-131]

Two terms must be defined in order to prepare for a general discussion of the leavening acids now of commercial importance – *dough rate of reaction* and *neutralizing value*. Since soda (sodium bicarbonate) is highly soluble, it dissolves as soon as it is immersed in water. Therefore, the rate at which CO_2 is released in a chemically leavened system depends on the rate of dissolution of the leavening acid system used.[121] The *dough rate of reaction*, or *DRR*, is a test designed to measure the reactivity of the leavening acid with soda during the mixing stage and during the bench action (the holding period after the mixing is complete). The bench action is important to commercial bakers who normally prepare large batches of a batter or dough that is then portioned into pans of proper sizes for baking in individual loaves. The DRR test involves the preparation of a simple dry biscuit mix containing the proper amount of the leavening acid and soda. The mix is placed in a sealed bomb that contains mixing blades, and the bomb is attached to an apparatus that is a closed, gas tight system. Water is added through the closed system, and the dough is then mixed for a uniform amount of time. The amount of gas released is measured automatically and calculated as the percent of the total gas that could be expected from the amount of sodium bicarbonate present in the mix. The percent of the gas evolved is plotted against time.[121,133-136] Figure 2 represents the DRR's for typical commercial leavening acids. The amounts of gas evolved after 2 min, which represents the mixing stage, and after 10 min, which represents 8 min of bench action, are usually used as important DRR values for each leavening acid.[121,133-138]

The *neutralizing value*, sometimes called neutralizing strength of a leavening acid, represents the number of grams of sodium bicarbonate that will be neutralized by 100 g of the leavening acid. It is determined by titration of the acid and can be expressed as an equation:

$$NV = \frac{a}{b} \times 100$$

where a = pounds of sodium bicarbonate neutralized, and b = pounds of leavening agent required.[121,134,137]

Characteristics of Commercial Leavening Acids

Some of the properties of the common commercially available phosphate leavening acids are

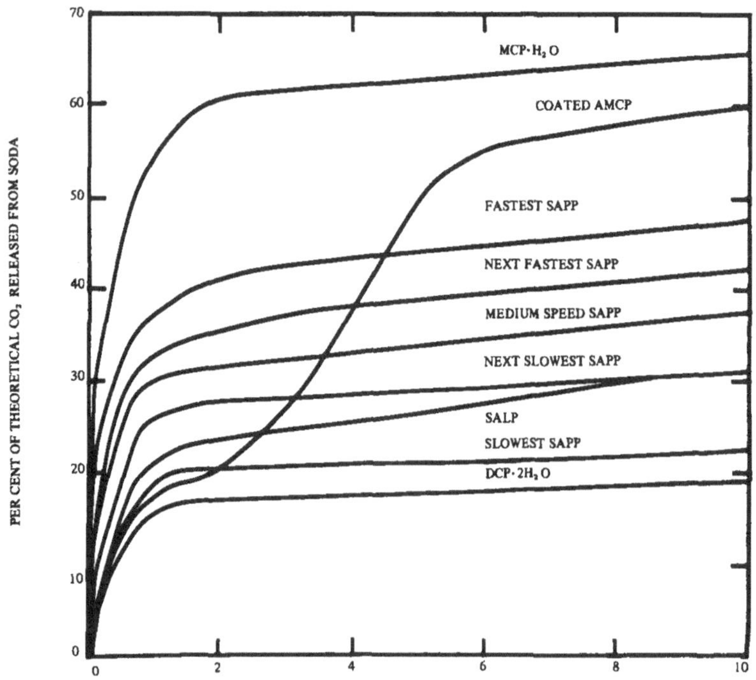

FIGURE 2. Dough reaction rates for commercially available phosphate leavening acids.[120,121,130-136]

shown in Table 10. Other acidic phosphates have been proposed, and patents for their manufacture and use have been issued. Some of these will be mentioned for interest, but only those in common use will be discussed in detail. Some of the chemical characteristics of the three acidic calcium phosphates in commercial use are shown in Table 2.

Monocalcium phosphate monohydrate – $MCP{\cdot}H_2O$ was the first acidic phosphate used as a leavening acid. A patent for its use in baking powders was issued in 1956. The early $MCP{\cdot}H_2O$ was prepared by ashing bones to form a crude product. This phosphate is classed as a fast-acting leavening acid; approximately 60% of the theoretical CO_2 is released by reaction of the $MCP{\cdot}H_2O$ with soda during the mixing stage of batter or dough preparation. No further gas is then released until the product is placed in the oven and the temperature has reached approximately 140°F. The reason for this is that $MCP{\cdot}H_2O$ disproportionates upon contact with water to form some dicalcium phosphate according to the following equation:

$$Ca(H_2PO_4)_2 \cdot H_2O + H_2O = CaHPO_4 \cdot 2H_2O + H_3PO_4$$

Because of the speed of its reaction, $MCP{\cdot}H_2O$ has limited applications by itself. Although it is commonly used in the preparation of phosphated flour at levels of 0.25 to 0.75% of the weight of the finished product and in the preparation of cookies, most of its applications are in combination with other slower-acting phosphate leavening acids. It is expected to release gas during the mixing stages to assist in forming bubble nucleii, while the other slower-acting acids release gas during bench action or baking.[126,132,134,135,137-141]

Anhydrous monocalcium phosphate – AMCP is used to a greater extent than the monohydrate. However, since the pure crystalline AMCP would react just as rapidly as the monohydrate, the crystals of the commercial product are coated with slightly soluble, alkali and polyvalent metal phos-

TABLE 10

Properties of Commercially Available Phosphate Leavening Acids[1]

Phosphate leavening acid[2]	Chemical formula	Neutralizing value	% Leavening gas released: 2 Min mix stage	8 Min bench action	During baking	Uses[3]
$MCP{\cdot}H_2O$	$Ca(HPO_4)_2 \cdot H_2O$	80	60%	None	40%	AC, CBP, Ck, HBP, PM, CM
AMCP, coated	$Ca(HPO_4)_2$	83	15%	35%	50%	CM, SRF, SRC, PM
$DCP \cdot 2H_2O$	$CaHPO_4 \cdot 2H_2O$	33	0%	0%	100%	CM, FD
SALP-1:3:8	$NaH_{14}Al_3(PO_4)_8 \cdot 4H_2O$	100	22%	9%	69%	FD, PM, CM, BM
SALP-3:2:8	$Na_3H_{15}Al_2(PO_4)_8$	100	21%	4%	75%	FD, PM, CM, BM
SALP + AMCP	Blend	80	27%	20%	53%	SRF
SAPP – fastest	$Na_2H_2P_2O_7$	72	40%	8%	52%	Adjust rates of other SAPP
SAPP – second fastest	$Na_2H_2P_2O_7$	72	36%	8%	56%	DM
SAPP – medium	$Na_2H_2P_2O_7$	72	30%	8%	60%	CM, BM, PM
SAPP – second slowest	$Na_2H_2P_2O_7$	72	28%	8%	64%	RD, CBP
SAPP – slowest	$Na_2H_2P_2O_7$	72	22%	11%	67%	RD

[1] Data obtained from References 126, 132, 134, and 136 to 138.

[2] $MCP{\cdot}H_2O$ = monocalcium phosphate monohydrate; AMCP = coated, anhydrous monocalcium phosphate; $DCP{\cdot}2H_2O$ = dicalcium phosphate dihydrate; SAS = sodium aluminum sulfate; SALP = sodium aluminum phosphate; SAPP = sodium acid pyrophosphate.

[3] AC = angel food cakes and mixes; BM = biscuit mixes; CBP = commercial baking powder; HBP = household baking powder; Ck = cookies and cookie mixes; CM = cake mixes; DM = doughnut mixes; FD = frozen doughs; PM = pancake and waffle mixes; RD = refrigerated doughs; SRF = self-rising flour; and SRC = self-rising corn meal.

phate compounds. In order for the leavening acid to react with the soda, the liquid in the system must first dissolve the coating. This results in a delay in the release of its acid, as can be seen from the data in Table 10 and the curve for coated AMCP in Figure 2. Only a small quantity of the gas, approximately 15%, is released during the mixing stage, with a further release of approximately 35% during the bench-action stage. The balance of the leavening is retained for reaction when the temperature of the system reaches approximately 140°F, again due to formation of some dicalcium phosphate.

Coated AMCP was developed in 1939.[137,138] It completely replaced the monohydrate in self-rising flour, since it provided a major improvement in the leavening stability of this product. Some self-rising flours still contain AMCP as the sole leavening acid. Most, however, now contain a blend of AMCP and SALP (sodium aluminum phosphate). AMCP is used as the sole leavening acid in most of the self-rising corn meal preparations on the market. It is also used as the sole leavening acid in one commercial household baking powder. AMCP is used in combination with sodium acid pyrophosphate and sodium aluminum phosphate in the manufacture of the numerous prepared dry mixes, i.e., cake, pancake, muffin, biscuit, and others.

Dicalcium phosphate dihydrate – $DCP \cdot 2H_2O$ is only slightly soluble at ordinary temperatures of mixing and holding doughs and batters. As a result, it does not release acidity for reaction with soda until late in the baking stage, when the temperature reaches 135 to 140°F. Since $DCP \cdot 2H_2O$ does not begin to react below 135°F, and the interior structure of a baked product begins to firm at about 160°F, a product that bakes rapidly may not provide sufficient time for complete release of all the CO_2.[30,126] $DCP \cdot 2H_2O$, therefore, cannot be used in biscuits, pancakes or any baked product that is completely baked in less than 20 min.

$DCP \cdot 2H_2O$ is seldom used by itself in leavening systems but is usually combined with faster-reacting acidic phosphates. Its major applications are in cake mixes, frozen bread doughs, and other products requiring a half hour or more to complete baking. It has a low neutralizing value, and therefore more $DCP \cdot 2H_2O$ is required to neutralize a given amount of soda than for other phosphate-leavening acids.[126,138-142]

Sodium aluminum phosphates – Two chemically different acidic *sodium aluminum phosphates* (SALP) are commercially available, although several others have been patented as potential leavening acids.[126,132,135,137-140,143] Some of their chemical characteristics are shown in Table 2. These phosphates can be distinguished from each other by using numerals corresponding to the number of sodium, aluminum, and phosphate units in each molecule. The two acidic sodium aluminum phosphates, SALP-1:3:8 and SALP-3:2:8, have been introduced within the last ten years. As shown in Table 10, the sodium aluminum phosphates provide a low initial reaction during the mixing stage of preparation and release low levels of CO_2 during bench action, especially during prolonged holding of a dough or batter. The latter characteristic is especially useful in restaurants that prepare quantities of pancake batter during slack work periods or prior to rush periods and hold them in the refrigerator until needed. An adequate amount of baking response, 70 to 75%, is still retained in the batter for leavening the pancake while it fries. Refrigerated pancake and cake batters have retained adequate leavening for up to two weeks.

Further advantages to the sodium aluminum phosphates are their bland flavors and tolerance to variations in flour and other ingredients. They are especially useful in batter systems that are prepared with highly emulsified shortenings; these systems require little or no leavening action prior to baking, because the emulsified shortening incorporates and retains large amounts of air in the batters. Introduction of the SALP's made possible the development of the "one-step" cake mix, which allows the housewife to add all of the liquid and eggs to the batter prior to beating it 2 to 4 min. The previous procedure required addition of the liquids in three separate steps with beating between successive steps. Self-rising flours, prepared mixes of all types, and frozen bread doughs are especially improved by leavening systems containing the SALP's. As can be seen from Table 10, the two SALP's differ very little in their reaction characteristics and therefore are interchangeable in all but a few applications.[126,132,135,137-140,143,144] A specially modified sodium aluminum phosphate has been developed for use with highly emulsified shortenings in prepared cake mixes. The modified SALP improves volume, grain, moisture retention, and texture of the cakes. The cakes are also more tolerant to variations in formulas, ingredients, and mixing times and baking temperatures in the home kitchen, which cannot be controlled by the manufacturer.[132,137-139]

Blends of SALP and AMCP are marketed as separate leavening acid systems for use in self-rising flour, self-rising corn meal, and biscuit and pancake mixes.[137-140,143,144] The blends usually contain between 50 and 70% SALP, a small quantity of tricalcium phosphate as a conditioning or flow agent, and the balance is ADCP.

Sodium acid pyrophosphate – At least five grades of *sodium acid pyrophosphate*, or SAPP, are commercially available. Each has a different rate of reaction and can be prepared through variations in manufacturing processes and by coating the surfaces of the crystals with slowly soluble phosphates.[126,134,135,137-141] All react very quickly with sodium bicarbonate in water solution. However, in dough or batter systems, they apparently interact with other ingredients of the system to exhibit very specific controlled rates of reaction typical of each grade. All grades of SAPP react rapidly with soda in doughs if no source of calcium is present. Calcium ion, either added as a soluble calcium salt or from milk, is often used to further delay the rate of reaction of the SAPP's. It is believed that the calcium ion reacts with the pyrophosphate anion at the surface of the SAPP crystal to form a porous coating of minute insoluble calcium pyrophosphate crystals. The initial reaction of the SAPP with soda during the mixing stage therefore continues until the coating is sufficient to retard further reaction during the holding or bench-action stage. Further release of substantial quantities of leavening gas requires heating, as in the baking stage, or holding for an hour or longer. The retarding action of calcium ion can be increased by reducing the particle size of the SAPP. This provides further evidence in favor of the "calcium coating theory" and is one of the factors used in preparing the "slow SAPP's."[30,126,137-141]

The *fastest reacting SAPP* can probably be termed a pure SAPP with no additives or manufacturing modifications. It is highly soluble and releases about 40% of its acidity during the mixing stage. It then reacts slowly during the holding stage and retains more than 50% of its acidity for reaction during baking. Because of its rapid reaction time, it is often blended with the slower SAPP's in order to increase their rates slightly. The *next fastest SAPP* releases about 36% of the theoretical CO_2 during the mixing stage and again retains well over 50% for reaction in the oven. This grade is frequently used in doughnut mixes intended for use in automatic doughnut machines.

A *medium speed SAPP* releases approximately 30% of the leavening gas in the mixing stage and retains about 60% for release in the oven. This is the most frequently used SAPP in prepared mixes. It is usually combined with one of the monocalcium phosphate leavening acids and at times also with dicalcium phosphate for further oven spring. The *second slowest SAPP* releases about 28% of the CO_2 during the mixing stage. A so-called "summer grade" having a 26% 2 min mixing rate is prepared at times. This grade of SAPP retains over 60% of the leavening gas for reaction in the oven and is most often used in commercial baking powders and in refrigerated doughs that are packed in round, cardboard containers and sold from dairy cases. This, and the *slowest SAPP*, which releases 22% of the CO_2 in the first 2 min and retains almost 70% for oven spring, are the only leavening acids that have been successfully used in commercial refrigerated doughs. Their very slow release of leavening acid during the preparation of the doughs allows the manufacturer to pack the doughs in the cans and seal them before further reaction occurs. A faster-reacting leavening would cause the dough to rise sufficiently so that the proper number of doughs could not be packed in the cans prior to capping them. The selection of the optimum one of these two slow SAPP's depends on the dough formula and the manufacturing conditions. Some formulas require the very slowest, and others would find this a bit too slow.

Some consumers are highly sensitive to a disagreeable flavor found in baked products containing SAPP leavening acids. This flavor is described as "astringent", a "coating of the teeth", or "coating the roof of the mouth." The flavor is often called the "pyro taste" and is typical of that of the orthophosphates produced by enzymatic hydrolysis of the pyrophosphate in the dough system.[30] Studies show that 60 to 70% of the pyrophosphate in a refrigerated biscuit dough is hydrolyzed to orthophosphate in 2 hr at room temperature, and only 2% pyrophosphate remains after 24 hr.[30,126]

Other Potential Phosphate Leavening Acids

In addition to those acidic phosphate salts now in commercial use as leavening acids, other acidic phosphates have been proposed for this application. These have probably not been developed as commercial products, because they offer no advantages over the existing leavening acids. Patents have claimed that *calcium acid pyrophosphate* can be manufactured to have a DRR such that it releases between 40 and 65% of the carbon dioxide in 10 min at 26°C.[145] A later patent claimed that a combination of calcium acid pyrophosphate with a minor amount of an alkali metal hexametaphosphate made an improved leavening acid that eliminated the taste of the calcium acid pyrophosphate and allowed the use of milk in products containing this acid.[146]

A number of *alkali metal-polyvalent metal complexes of orthophosphoric acid* similar to the SALP's have also been proposed as leavening acids. These include ammonium and potassium aluminum phosphates, sodium, ammonium and potassium iron phosphates, and SALP's with ratios of sodium aluminum and phosphate differing from the two in commercial use.[147-150]

The *acidic salts of sodium metaphosphate*[151] and the *acidic ammonium, alkali metal and alkaline earth metal metaphosphates*[152] have also been proposed as leavening acids. These are reported to provide some rapid initial release of CO_2 as well as a substantial reserve for release of CO_2 in the oven. Neither product described has been offered commercially.

Leavening Acids in Baking Powders

The manufacturer of prepared mixes can economically purchase and maintain inventories of sodium bicarbonate and the individual leavening acids. He then adds these ingredients separately to his prepared mixes. However, the housewife and the small baker cannot economically store the individual ingredients. They, therefore, purchase and use baking powder, a complete leavening system consisting of a mixture of the proper quantities of soda, leavening acid, and a diluent, usually starch, that separates these ingredients so that they will not interact with each other during storage. The baking powders in commercial use contain sufficient soda to yield not less than 12% of their weight as CO_2 in use.

Although the total volume of baking powder sold each year is gradually decreasing as fewer housewives bake in their homes and the number of small bakeries decline, baking powders are nevertheless important to those who still depend on them. Typical formulas for baking powder compositions that have been or still are marketed are shown in Table 11. Some additional formulas for commercial baking powders were published by Conn.[153]

There are two types of household baking powders marketed today. The type sold in largest volume is called *double-acting baking powder*. It contains $MCP \cdot H_2O$ and sodium aluminum sulfate. The $MCP \cdot H_2O$ provides rapid reaction during the mixing stage. The sodium aluminum sulfate does not react until the dough or batter is placed in the oven.[121,126] The second type might be called single-acting but is usually called *phosphate or tartrate baking powder*. As shown in Table 11, these contain one of the MCP's or cream of tartar and tartaric acid as the single leavening acid. Very little baking powder containing $MCP \cdot H_2O$ or tartrate is now sold in this country. The largest proportion of single-acting phosphate baking powder sold contains the AMCP leavening acid. The phosphate or tartrate baking powders are usually sold on a regional basis rather than nationally, whereas double-acting baking powders are nationally distributed.

Commercial baking powders are used by bakers, restaurants, and other large-volume consumers. The most popular formulas contain $MCP \cdot H_2O$ and SAPP as the leavening acids. All of these baking powders are double-acting, since there is an initial rapid reaction during the mixing stage with very slow release of gas during a bench-action stage and a retention of most of the acid for release during the baking stage.[30,126,153] Formulas for those in commercial use are shown in Table 11.

The SALP's could be used to replace either the sodium aluminum sulfate in household baking powders or the SAPP in commercial baking powders to provide additional stability during storage and improve performance during use. Formulas for baking powders containing the SALP's have been proposed, but they have not yet been accepted by the industry.[137,138,153]

Leavening Acids in Prepared Mixes

The growth of convenience foods since the end of World War II has caused a spectacular growth in the annual sales of prepared baking mixes. A few contain all of the dry ingredients needed to

TABLE 11

Typical Baking Powder Formulas[1]

Constituents	Household baking powders						Commercial baking powders		
	Pure phosphates		Phosphate-SAS, double-acting types			Cream of tartar			
	1	2	1	2	3		1	2	3
Soda, granular	28.0	28.0	30.0	30.0	30.0	27.0	30.0	30.0	30.0
Monocalcium phosphate monohydrate (MCP·H_2O)	35.0		8.7	12.0	5.0		5.0		5.0
Monocalcium phosphate, anhydrous (AMCP)		34.0							
Corn starch, redried	37.0	38.0	26.6	37.0	19.0	20.0	24.5	26.0	27.0
Sodium aluminum sulfate (SAS)			21.0	21.0	26.0				
Sodium acid pyrophosphate (SAPP)							38.0	44.0	38.0
Calcium sulfate			13.7						
Calcium carbonate					20.0				
Cream of tartar						47.0			
Tartaric acid						6.0			
Calcium lactate							2.5		

[1] From References 120, 126, 138, and 153.

prepare the final food product and require only the addition of liquid. Some require the addition of liquid and eggs, as is the case with most prepared cake mixes. Others require the addition of liquid and shortening, as in self-rising flours for use in preparation of biscuits.

Prepared baking mixes offer the consumer the advantages of convenience, economy, and uniformity of quality that he cannot equal by himself. The manufacturer of prepared mixes is able to purchase and stock special ingredients, often tailored to contribute highly desirable qualities in the final product, that no housewife would be able to purchase or afford. The manufacturer of the mixes also is able to maintain a higher degree of uniformity than is possible in the home. He buys large quantities of his ingredients under rigid specifications and tests them to assure conformity to the specifications. Furthermore, the manufacturer is able to use special processing methods not available to the consumer in the manufacture of the mix in order to introduce special convenience and quality factors. An example is the use of high shear mixers to incorporate shortening into the dry ingredients of the modern one-step cake mixes so that all the liquid can be added at once and not in three stages.

Cake mixes – Cake mixes represent a large portion of the prepared mix market. Most require only the addition of liquid and eggs by the consumer, although at least one contains dried eggs so that only liquid needs to be added. The modern cake mixes are formulated to take the greatest advantage of all characteristics of the ingredients so that the finished cake is light, tender, and moist, with uniform grain and maximum volume. Flour, milk, and eggs furnish structure to the finished cake, and sugar, shortening, and chemical leavening furnish tenderness. The formulas are adjusted so that mixing for 2 to 4 min on a household mixer will incorporate a large volume of air that is dispersed throughout the batter as minute bubbles. Special, highly emulsified shortenings cause the batter to incorporate and retain numerous minute air bubble nuclei during the mixing stage that can later be expanded by the thermal expansion of the entrapped air, vapor pressure of the batter water, and the chemical leavening.[131]

Proper leavening is a critical part of any cake-mix formula. The leavening must react at the right point in the preparation and baking of the cake. If it reacts too soon, the gas may be lost from the batter. If it reacts too late, the cake structure will be set by heat, after which no further expansion can occur without rupturing the air cells and weakening the structure of the cake.[120,121,126,131,132,135,137-141,143,153,154] Modern cake mixes are prepared with highly emulsified shortenings that reduce the requirements for early leavening action during the mixing stage. It is, therefore, common practice to use combinations of leavening agents that retain much of the leavening reaction for the baking stage. If the release of gas is too slow, the result is likely to be large gas cells, coarse grain, and thick cell walls that lead to crumbliness and toughness of the cake.[120,121]

The leavening system also controls the pH of the cake. This should be around 7.0 or slightly above. Optimum pH values range between 6.9 and 7.2 for white cakes, 7.2 and 7.5 for yellow cakes, and between 7.1 and 8 for chocolate or devil's food cakes. The pH of the cake will affect crumb and crust color, intensity of flavor, and other properties.[121] These characteristics are usually optimum when the optimum pH for the formula is maintained.

The shape of the cake also can be affected by the leavening system. The use of too fast a leavening system can result in cakes that fall easily, while a leavening system that reacts too late in the baking stage may result in peaked centers.[121]

Handleman and co-workers[154] demonstrated that the amount of leavening gas retained and the number and uniformity of air cells in the final cake could be increased as the number of bubble nuclei incorporated during the mixing stage was increased. These experiments demonstrated the importance of shortening emulsification. As the efficiency of the emulsifier in the shortening improved, the amount of air incorporated increased and the fat was dispersed to an increasing degree so that it did not surround bubbles and prevent them from being nucleating sites for diffusion of leavening gas.

Several investigators demonstrated the interrelationship of the leavening system with shortening emulsification.[122-132,154] As the number of bubble nuclei into which the leavening gas can diffuse increases, the air cells in the final cake will be smaller, and fewer air bubbles will coalesce to reach a sufficient size so that they can rise out of the batter. In addition, as the efficiency of the emulsifying system increases, the need for an early reacting leavening system during the mixing stage decreases and the need for a leavening system reacting largely in the baking stage is increased. This explains the improved performance of SALP's in modern highly emulsified cake-mix systems.[121,132]

Most modern cake mixes contain combinations of leavenings in order to obtain maximum volume, maximum uniformity of grain and texture, and optimum shape of the baked cake. The most important factor to be considered in designing a leavening system for a cake mix or any other baked product is the quantity of leavening gas required to provide the optimum characteristics. When this is known, the amount of soda that must be added to the product is calculated. Once this is established, the necessary quantity of the leavening acids can be calculated from their neutralizing values and the requirements of the pH of the final system.[121] Calculations based on neutralizing value provide approximate levels for each acid. One or two trials will establish whether the optimum pH has been reached. If not, minor adjustments can be made in the quantity of leavening acids.

The most common leavening acid blends in use are combinations of AMCP and SALP in white and yellow cakes and AMCP and SAPP or $DCP \cdot 2H_2O$ in chocolate cakes, since the SALP's tend to dull the chocolate color. The AMCP can be replaced by $MCP \cdot H_2O$, and the SALP can be replaced by SAPP or DCP. The usual quantitites of the acids used range from 10 to 20% of the fast-acting $MCP \cdot H_2O$ or AMCP and 80 to 90% of the delayed-reaction leavening acids.

Pancake and waffle mixes – Pancake and waffle mixes are usually somewhat simpler ingredient systems than prepared cake mixes. They normally contain wheat flour, shortening, a small amount of sugar, nonfat dried milk, salt, soda, and the proper amount of leavening acid. Some mixes also contain one or more of the following: corn flour, rice flour, buckwheat flour, corn sugar, whey solids, or dried buttermilk. The most common blend of leavening acids is 20 to 30% $MCP \cdot H_2O$ or AMCP and 70 to 80% SALP, a combination of leavening acids that allows the cook in a pancake house or a housewife to prepare pancake batter the night before it is to be used. If kept in the refrigerator, the batter will retain adequate leavening action to prepare excellent pancakes or waffles for up to a week. The batter normally deteriorates through bacterial souring before there is a serious loss of leavening capacity. A recent trend is toward the sale of frozen pancake batters. Again, combinations of the MCP's and SALP's or combinations of the MCP's, SALP's, and SAPP's provide proper stability under conditions of occasional thawing and refreezing, as may occur in transit between the manufacturer and the final consumer.

Prepared biscuit mixes – Prepared biscuit mixes normally contain flour, shortening, salt, and leavening. Milk is added to such a mix, although if the manufacturer has included dried milk, the consumer merely adds water. Very small quantities of sugar are sometimes added for flavor and for optimum crust color. These mixes normally contain blends of 30 to 50% AMCP and 50 to 70% SALP or SAPP.

Self-rising flour and self-rising corn meal mixes – These have been very popular since about 1910 in the southeastern part of the U.S.;[126] they are beginning to appear more frequently in national markets. Typical formulas recommended for these products are shown in Table 12. As shown in the table, self-rising flour and corn meal mixes contain the flour or meal, salt, soda, and leavening acid. The consumer must add shortening and milk or water in order to prepare biscuits or muffins. The self-rising flour mixes can also be used, if recipes are properly modified, to prepare cakes, pancakes, and waffles and similar baked products by the addition of milk, eggs, and shortening.

The earliest self-rising products were prepared with $MCP \cdot H_2O$ and had a very short shelf-life. The coated AMCP, developed in 1939, replaced $MCP \cdot H_2O$ in these products because of its superior stability, and this leavening acid is still in use in self-rising corn meal products.[120,126] One of the slow SAPP's began to replace AMCP in self-rising flour in the middle 1950's, but this was in turn replaced by SALP and AMCP blends about 1960 because of improved flavor and greater storage stability under high humidity.[155]

Cake doughnut mixes – Cake doughnuts differ from yeast doughnuts in that the cake doughnuts contain chemical leavening systems, while the yeast doughnuts contain yeast leavening systems. Most cake doughnut mixes are formulated to contain SAPP leavening acids. Most doughnuts now marketed are prepared by machine. Even small hand operations use a type of mechanical cutter to place the round doughs into the hot fat. After the dough leaves the cutter, it sinks beneath

TABLE 12

Typical Commercial Formulas for Self-Rising Flour and Self-Rising Corn Meal[1]

	Self-rising flour		Self-rising corn meal		Self-rising corn meal mix	
Ingredients	Percent	lbs	Percent	lbs	Percent	lbs
Wheat flour	94.899	100.000			9.434	10.00
Corn meal			94.340	100.00	84.906	90.00
Soda	1.305	1.375	1.651	1.75	1.651	1.75
SALP/AMCP blend[2]	1.661	1.750				
AMCP			1.651	1.75	1.651	1.75
Salt	2.135	2.250	2.358	2.50	2.358	2.50
Total	100.000%	105.375	100.000%	106.00	100.000%	106.00

[1] Information obtained from References 138, 139, 144, 155, and 156.

[2] Commercial blends of SALP and AMCP are available for use in self-rising flour. See references 137, 138, 139, and 144.

the surface of the hot fat. A slight crust forms completely around it, and the leavening reaction causes the dough to expand. The dough rises to the surface of the hot fat with approximately half still submerged in the fat. This half continues to bake and the crust turns brown. The machines are programmed so that the doughnut is "flipped over" at the proper time so that the unbaked portion also has an opportunity for its crust to bake and to brown on the surface. The proper formulation of the leavening system is critical in order that the dough will expand at the correct times for the machine used. The optimum DRR for a cake doughnut leavening system is 36% release of the theoretical CO_2 in 2 min.

In addition to wheat flour, doughnut formulas may contain soy flour, potato flour, cottonseed flour, various sugars, egg yolk, some shortening, milk, salt, soda, and the leavening acid. Complex interactions among the ingredients require very careful leavening formulation, as the amount and speed of leavening are highly influenced by the type and quality of ingredients in the formula. It is common practice, therefore, to make adjustments in the leavening acid system by the addition of proper proportions of a faster or slower SAPP when tests indicate this is necessary.[30] Although the most common leavening system contains SAPP, SALP and at times small quantities of MCP can also be used. Combinations of all three, or of the MCP and SALP, are also possible.

Leavening Acids in Refrigerated Doughs

Refrigerated biscuits, dinner rolls, sweet rolls (cinnamon rolls, caramel nut rolls, etc.), and similar products are used in large quantities in the U.S. These products are packed in cylindrical, cardboard, pressurized containers and are sold from refrigerated dairy cases held at 35 to 40°F. The consumer needs merely to open the can, place the biscuits on a baking sheet, and bake them for 10 to 20 min before serving them hot. Between 60 to 75 million dozen containers of refrigerated doughs are sold annually.

The basic biscuit and roll doughs contain flour, water, shortening, nonfat milk solids or dried whey solids, sugar or corn sugar, salt, soda, and SAPP. Examples of biscuit dough formulas are shown in Table 13. The proportions of ingredients vary according to whether a biscuit or a dinner roll is being prepared. The sweet doughs normally contain more sugar and slight differences in proportions of some of the other ingredients. Roll doughs often also contain fillings or toppings typical of the product, i.e., a cinnamon filling rolled into the dough and a can of white icing for cinnamon rolls. All products, including the biscuit doughs, are variations of chemically leavened roll doughs.

The normal baking powder biscuit, as made in the home, is carefully prepared in such a way that the ingredients are hydrated without developing the gluten in the flour. The canned refrigerated biscuits are prepared by first mixing the ingredients in large mixers similar to those used in the manufacture of conventional bread. The period of mixing is determined by the type of flour used. Normally a strong, high-quality bread flour is required for all of these doughs. After optimum mixing, the dough is dumped from the mixer and cut into pieces weighing between 30 and 50 lb. These are then carefully sheeted, folded, and resheeted two or three times. The doughs are overlapped during the final sheeting so that a continuous sheet of dough is formed. This travels along a belt as a continuous ribbon to the biscuit cutters, which automatically cut and stack ten biscuits into the cardboard can. The container is capped, and one to two dozen cans are packed in cartons. The cartons are then "proofed" in a high-temperature room or tunnel, or at ambient temperatures, until sufficient leavening reaction has occurred to cause the doughs to expand, completely fill, and seal the cans with the dough under about 15 lb/in.^{2} pressure. The cases of biscuit doughs are then refrigerated and marketed.

Elaborate equipment and processes have been developed for manufacture of these products. Most manufacturers maintain strict secrecy in regard to their formula, equipment, and process. Dinner rolls and sweet rolls are manufactured by slight variations of the biscuit manufacturing process. Between 1.0 and 1.55% soda and 1.4 and 2.0% SAPP are used in these formulas. Only the two slowest SAPP's are suitable for refrigerated dough systems. The slowest SAPP can be utilized where the dough temperatures during manufacture are 70°F or above. The next slowest SAPP can be used in those processes in which the dough temperatures are controlled at about 60°F. A SALP leavening system was proposed several years ago but has never been accepted in commercial practice.

The end product of the reaction between SAPP

TABLE 13

Some Published Formulas for Refrigerated Biscuit Doughs

Ingredients	Formula number 1[1] %	2[2] %	3[3] %	4[4] %
Flour	54.9	52.5	55.7	54.8
Water	33.0	30.8	32.4	32.8
Shortening	5.5	9.0	5.5	4.5
Nonfat milk solids	1.6	–	–	2.2
Buttermilk solids	–	3.2	–	–
Whey solids	–	–	1.3	–
Sucrose	–	0.5	–	–
Dextrose	1.1	–	1.0	2.0
Salt	1.1	1.4	•1.2	1.4
Soda	1.2	1.2	1.3	1.1
SAPP	1.6	1.4	1.6	1.0
Fumaric acid	–	–	–	0.2
Total	100.0%	100.0%	100.0%	100.0%

[1] Data from Van Wazer, J. P., *Phosphorus and Its Compounds*. Interscience Publishers, Inc., a division of John Wiley & Sons, Inc., 1958. Reprinted by permission.
[2] Data from The Pillsbury Co., Minneapolis, Minn. (Reference 157), with permission.
[3] Percentages calculated from Baker, J. S., and Lindeman, C. G., U.S. Patent 3,297,449, 1967, with permission.
[4] From Erekson, A. B. and Duncan, R. E., U.S. Patent 2,942,988, 1960 and Matz, S. A., U.S. Patent 3,397,064, 1968. Used with permission of Borden Food Products Co.

and soda is a trisodium monohydrogen pyrophosphate as shown in the following formula:

$$Na_2H_2P_2O_7 + NaHCO_3 = Na_3HP_2O_7 + H_2O + CO_2$$

This end product, as well as unreacted SAPP, can also be hydrolyzed by the action of enzymes naturally present in the dough to form orthophosphates, as shown in the following two reactions:

$$Na_3HP_2O_7 + H_2O = NaH_2PO_4 + Na_2HPO_4$$

$$Na_2H_2P_2O_7 + H_2O = 2NaH_2PO_4$$

The tendency for orthophosphate to accumulate in both refrigerated doughs and refrigerated batters is inherent to the SAPP leavening acid in any system that will be held for several hours before baking. Tests have shown that from 60 to 70% of the pyrophosphate in refrigerated biscuit doughs held at room temperature is hydrolyzed to orthophosphate in 2 hr and that only 2% pyrophosphate remains in the dough after 24 hr of refrigerated storage.[30]

Two problems are caused by the conversion of the pyrophosphate to orthophosphate. First, a disagreeable, astringent flavor is noticeable to some consumers who are particularly sensitive to it. This flavor can be reduced to some extent by proper formulation, but it can never be eliminated. A second problem is that under certain conditions of formulation and storage, the orthophosphates can form large visible crystals that have the appearance of pieces of glass. Although they are harmless and dissolve readily during baking, the consumer often mistakes them for glass and discards the biscuits prior to baking.[30,126,134] Erekson and Duncan[159] and later Matz[159] reported, that the addition of fumaric acid to the refrigerated biscuit dough prevents dough discoloration and deposition of phosphate crystals.

Baker and Lindeman[158] have proposed the use of potassium bicarbonate alone or in combination with soda in order to eliminate these crystals.

Some refrigerated cake, pancake, and muffin batters have been marketed for short periods of time. These, however, have never been commercially successful, since the quality of the resulting baked products has not equaled that obtained from prepared mixes. Combinations of $MCP \cdot H_2O$, AMCP, DCP, SAPP, or SALP have been used in these batters. The SALP's are especially effective as the leavening acids, particularly in refrigerated pancake batters delivered on some local dairy routes.

Non-leavening Applications in Cereal Products

Dough Conditioning

Because of the variability in quality of gluten, the flour protein, many manufacturers of baked products take advantage of the interaction between the phosphates and proteins to improve gluten characteristics in doughs. The type of interaction and its effect on the final dough or batter system depend on the chemical characteristics of the phosphate and of the protein. At times the effect is nothing more than control of the pH within an optimum range for the desired physical or chemical effect. At other times advantages are taken of the polyelectrolyte characteristics of polyphosphates and proteins.[30]

Numerous investigators report that ***bread dough*** can be improved by the addition of calcium acid phosphates. The effect of the alkaline earth and polyvalent-metal ions on the stiffening or toughening of bread doughs has been utilized by the baking industry for a number of years.[30,134,135] Calcium and aluminum ions are particularly effective, with the aluminum ion having the greater effect.[30,135,149] Products marketed to the baking industry as "dough conditioners" normally contain one or more of the following phosphates: $MCP \cdot H_2O$ or $DCP \cdot 2H_2O$, monoammonium phosphate, and the potassium phosphates.[30] The value of the calcium phosphates in production of yeast doughs was reviewed by Barackman.[160] The calcium phosphates were reported to provide beneficial effects upon the colloidal properties of the dough and to stimulate yeast activity by maintaining an optimum pH for its growth.[30,160] A further advantage is that the low pH prevents growth of unwanted bacteria, such as those that cause "ropiness."

A detailed study of the advantages in maintaining proper brew and dough pH in continuous bread formulas was reported in a series of papers by Bayfield and Lannuier.[161] SALP's, as well as calcium acid phosphates, were found to be beneficial in buffering brew and dough to optimum pH levels. Of several phosphate salts tested, the addition of SALP to flour brews produced the greatest improvement in bread volume, score, and crumb color.

The stabilization of refrigerated doughs against microbial deterioration is usually accomplished by means of the pressurized container in which they are marketed. The doughs, however, undergo a nonmicrobial deterioration that limits their shelf-life. Matz[159] reported that this nonmicrobial deterioration could be delayed for a significant period of time by the addition of 0.1 to 3.0% potassium polymetaphosphate with a molecular weight of 200,000 or greater. The pieces of refrigerated biscuit doughs thus treated were reported to hold their shape, were easier to separate, and produced higher baked volumes than controls. Matz speculates that the polyphosphate linked with the gluten of the dough to stiffen it and to maintain its structure.

Inhibition of Enzyme Activity

Phosphates have been reported to inhibit undesirable enzyme activity. Takebayashi[162] reported that the pyrophosphate anion inhibited lipoxidase activity in sponge-cake batter. The addition of between 0.01 and 0.1% of TSPP, $TSPP \cdot H_2O$, or calcium acid pyrophosphate inhibited the enzyme without affecting the quality of the cake. Carter and Hutchinson[163] reported that inorganic phosphates, especially DSP, inhibited α-amylase activity. This was demonstrated by Brabender amylographs of mixtures of starch and malt to which the phosphate was added.

Antioxidant Activity

The phosphates have been reported to inhibit oxidative rancidity in cereal products. Muhler[164] reported that the addition of a mixture of MSP and DSP to cereals increased their oxidative stability and inhibited development of stale, rancid flavors.

Inhibition of Microbiological Growth

The ability of polyphosphates to stabilize cereal products, such as doughs, batters, and baked products, against microbiological growth was reported by Kohl and Ellinger.[102,103] The most effective polyphosphates were found to be those having an average length of 18 to 34 phosphate units. The shorter-chain and longer-chain polyphosphates were also found to inhibit the growth of yeasts and molds.

Mineral Enrichment

Although the inorganic salts, commonly called minerals, make up a small percentage of total body weights, they are as important to the well-being as the proteins, carbohydrates, and fats that provide energy. In fact, it is impossible to obtain energy from these foodstuffs without the mineral elements. The elements calcium and iron are two of the minerals required for the total well-being of all organisms. Calcium and iron salts and often their phosphates are commonly added to foodstuffs in order to provide these essential mineral elements. Since phosphorus, as the phosphate anion, is also an essential mineral, the phosphate salts of these mineral elements are doubly useful.

Calcium and phosphorus – Calcium and phosphorus are present in living organisms in the largest quantities of all the minerals. Children require calcium and phosphorus for normal growth. Most of the calcium and phosphorus is incorporated in the formation of the skeleton. A newborn child contains approximately 24 g of calcium and 14 g of phosphorus. Adults require calcium and phosphorus for maintenance and replacement. A normal adult contains approximately 1,150 g of calcium and 650 g of phosphorus, which, based upon total ash, represents 39% calcium and 22% phosphorus. Pregnant and lactating women require larger quantities of calcium and phosphorus than other adults, because a portion of their intake is utilized by the growing fetus or child.[165,166]

In addition to being a necessity for formation and maintenance of skeletal tissue, calcium and phosphorus are both needed for growth and replacement of blood and soft tissues. Most of the calcium in blood and tissues is associated with protein. Most of the phosphorus is incorporated in the phosphate anion, which functions as a portion of the overall buffering system in the body as well as in the mechanism of energy transfer during metabolism of foodstuffs.[166]

Present knowledge indicates that the optimum ratio of calcium to phosphorus for children is between 1:1 and 2:1; that for adults is slightly less than 1:1. Calcium deficiency in children results in stunting of growth, bone malformation, and poor quality bones and teeth. When the human diet lacks sufficient calcium and phosphorus, or has improper ratios of calcium to phosphorus, these elements are removed from the skeletal tissue as required.[166]

Since most modern methods of food processing tend to remove portions of the calcium in the food before processing, it is common practice to supplement the food product with this element through the addition of calcium phosphates. However, foodstuffs such as meat, fruits, and vegetables are difficult to supplement with calcium, since the calcium ion often introduces undesirable physical characteristics to the processed foods. As an example, calcium toughens the skins of fruits and vegetables and precipitates pectins.

Cereals, however, are easily fortified with calcium phosphates, since they are seldom adversely affected by them. In fact, it is often possible to improve some of the physical characteristics of doughs and batters by the addition of these compounds. As a result, federal definitions for flour and many baked products allow the optional addition of certain minimum levels of calcium compounds, including the phosphates. These levels are therefore provided by the addition of calcium phosphates to provide the minimum levels of these elements required in the definitions. The type of calcium phosphate and the quantity required to provide the minimum level depend on the type of cereal product, the amount of naturally occurring calcium, and the amount of the element contributed by the selected compound. Table 14 shows the levels of calcium and phosphorus in each of the three calcium

TABLE 14

Levels and Ratios of Calcium and Phosphorus in Commercial Calcium Phosphates

Compound	% Ca	% P	Ca/P ratio
Monocalcium phosphate	16.4	24.2	0.68/1.0
Dicalcium phosphate	23.6	18.0	1.31/1.0
Tricalcium phosphate	36.7	17.3	2.12/1.0

Source: *The Importance of Calcium and Phosphorus Supplements in Foods*, Technical Service Bulletin, Stauffer Chemical Co., New York, with permission.

orthophosphates commercially available for food use as well as the calcium to phosphorus ratio in each. By using this table, it is simple to calculate the required amount of the proper calcium phosphate to be added to any cereal product to provide the necessary minimum level of each element. Excellent reviews of enrichment of flour and baked goods have been provided in publications of the National Research Council (e.g., 167).

It is common practice in the cereal and baked-goods industry to enrich flour. Enrichment can also be added during the manufacture of baked goods in the form of complete mineral enrichment compositions. Tablets, or powdered mixtures, in soluble, edible packages containing the proper weight of enrichment for standard weights of doughs or batters are available for use in enriching baked products. The FDA considers such direct enrichment of bakery products as equivalent to the use of enriched flour.[168]

Supplementation of foods with calcium orthophosphate salts has often been proposed as a means of *controlling dental caries*. Numerous investigations have been reported in the literature, and a number of the phosphate salts have been proposed. Among these are metaphosphoric acid,[169] DCP·$2H_2O$,[170-172] MCP·H_2O,[173,174] TCP,[174] DSP,[171,173,175-178] a "fluffy sodium phosphate,"[179] and calcium pyrophosphate.[170] The investigators speculate that the increase of soluble phosphate, especially calcium phosphate in the saliva, prevents the dissolution of calcium phosphate from the tooth enamel and therefore aids in inhibiting development of dental caries. Support for this theory is found in the fact that feeding calcium phosphates by stomach tube had little or no effect on reduction of dental caries.

Nizel and Harris[180] have published the results of more than 100 studies conducted on rodents to determine the cariostatic activity of phosphates in their diets. They found that the phosphates differ widely in their cariostatic activity in the following decreasing order of effectiveness of the anions: trimeta- (cyclic), tripoly-, hexameta-, ortho-, and pyrophosphate. The type of cation affects the cariostatic activity in the following order of decreasing activity: H, Na, K, Ca, and Mg. These investigators found that sodium chloride increases the cariostatic activity of soluble phosphates, especially when these were baked in bread used in the diets. All methods of feeding the phosphates orally were effective, but the investigators found that dissolving the phosphates in the water supply and delivering them into the stomach by means of stomach tubes were less effective than feeding them mixed in the diet, embedded in fat used in the diet, or baked in bread used in the diet. No relationship was found between the solubility of the phosphate in water, its buffer activity, or its ability to chelate alkaline-earth or heavy metals. In general, the investigators found that it was necessary to double the phosphorus content of the diet in order to obtain cariostatic activity; this amounted to the addition of approximately 0.4 g of phosphorus for each 100 g of diet. Their studies also indicated that there seemed to be some type of morphological change in the surface of the tooth when some phosphates were fed, which indicated that enamel was first demineralized and later remineralized by deposition of calcium phosphate to change the surface character of the tooth. The cariostatic effect of the phosphates was also reported to exert an additive effect to the administration of fluorine in preventing dental caries.[180]

Iron – Also an essential mineral, iron is found in numerous iron-containing enzymes and in a number of iron-protein complexes, such as blood hemoglobin and myoglobin. Approximately 80% of the iron in the human body is found in the hemoglobin of red blood cells.[181] The functional iron level in the human body is very stable. It is lost only through bleeding and through reproduction processes, since the fetus and newborn infant require iron for the same purposes as its parent.[181]

Some functional iron can be stored. Studies of humans have demonstrated that approximately 1,000 mg is stored in bodies of men as ferritin and hemosiderin and from 200 to 400 mg is stored by women. Women lose iron through menstruation and reproductive processes, rather than there being differences in iron absorption or storage between the male and female bodies.[181]

The human body maintains a sophisticated regulatory system for absorption and utilization of iron.[182] Loss of iron through bleeding stimulates the production of red blood cells, which then require iron. This iron is obtained from the iron stores in the marrow, liver, or plasma. Since the body attempts to maintain these stores at a constant level, the depleted iron stores are replenished by removal of iron from the mucosa of the small intestine. As the concentration of iron in the mucosa falls, the epithelial cells in the small intestine begin to absorb iron from the available food supply. When the iron level in the intestinal mucosa is restored, the epithelial cells again refuse to absorb iron from the food. No more iron is absorbed until the next loss of iron that requires replacement.[182]

The total dietary iron is not necessarily available from absorption by the intestinal mucosa. Little of the iron in vegetable and meat products is released without digestion by acid-pepsin. Cooking increases the availability of the natural iron in foods. In general, iron in vegetables is less available than in foods derived from animal products. Approximately 30 to 50% of the iron in natural food products is absorbed. Other items in the diet, such as egg, can reduce the amount of iron available for absorption.[183] Studies indicate that approximately 5 to 10% of the iron available in the diet is assimilated by the body of a normal adult. The amount varies among individuals and is probably increased in certain iron deficiency diseases. In addition, there are differences in the availability of the iron in the different foods. Studies of the "available" iron in diets and the amount of this iron absorbed indicate that the human diet should contain between 20 and 40 mg of available iron. This is at least twice the level of available iron in present human diets.[184]

Iron from iron salts is more readily absorbed from the diet than the iron naturally present in animal and vegetable tissues. The present trend toward iron supplementation in bakery products, flour, rice, and cereals has aided in combating iron deficiency, especially among low-income families who use these products in large quantities. However, nutritionists believe that the supplementation of these products should be at least doubled and perhaps tripled.[184,185] Supplementation of iron in baked products, flour, and cereals can be accomplished through the addition of iron phosphate compounds to these foods.

Studies of the availability of iron from various iron phosphate salts provide conflicting data. Nakamura and Mitchell[186] showed that SIPP (sodium iron pyrophosphate), ferrum reductum, and ferric phytate regenerated hemoglobin as well as ferric chloride. Street[187] found that SIPP was less than 50% as available to rats as ferrous or ferric sulfate. Street, however, was unable to explain a discrepancy between his results and those of Nakamura and Mitchell. Freeman and Burrill[188] reported similar qualitative and quantitative absorption of iron by anemic rats, whether fed the iron salts directly in the diet or baked in bread used in the diet. The effectiveness of the various iron salts tested in their study in respect to iron retention was as follows, in order of decreasing retention: ferric chloride > sodium ferric orthophosphate = ferric orthophosphate > reduced iron > SIPP.

Steinkamp and Dubach[189] reported that iron is effectively absorbed from iron salts used in bread in the human diet. Radioactive iron salts were used to enrich bread fed to 32 healthy human subjects. In these tests 28 of the individuals absorbed between 1 and 12% of the iron regardless of the form of salt used. The other four, thought to have suboptimal iron stores, absorbed between 26 and 38% of the radioactive iron. Additional supplementation of the iron enriched bread with asorbic acid increased iron absorption two to three times. Radioactive ferrous sulfate, reduced iron, ferric orthophosphate, and SIPP were used in the study. Three iron-deficient patients absorbed 45 to 64% of the iron in bread enriched with radioactive ferric sulfate, ferric orthophosphate, and SIPP.

Fritz[190,191] studied the availability of iron from various iron supplementation salts added to low-iron diets of young rats and chicks. Ferrous sulfate was used as the standard and was given an arbitrary relative biological value of 100. Other iron salts were rated according to the percent availability when compared to the relative biological value of the ferrous sulfate. Those having good availability, i.e., a biological value of 70 or above (ferrous sulfate = 100), were ferric ammonium citrate, ferrous sulfate, ferrous gluconate, ferrous fumarate, ferrous tartrate, and ferrous ammonium sulfate. Iron salts with relative biological values of 20 to 70 were reduced iron, ferric chloride, ferric pyrophosphate, a flour enrichment mixture, and a corn meal enrichment mixture. Those salts having poor availability, i.e., a relative biological value compared to ferrous sulfate of 20 or less, were ferric oxide, ferric orthophosphate, ferrous carbonate, and SIPP.

Numerous other studies have also been reported in the literature. All seem to bear out the discrepancies noted in the few references quoted above. An excellent review of the literature can be found in the publication of the British Ministry of Health entitled *Iron in Flour*.[192] In Part I of this report, Dr. P.C. Elwood described the results of his studies with four sources of radioactive iron baked in bread. These four compounds were reduced iron, ferric ammonium citrate (either green or brown), powdered iron, and ferrous sulfate. The results showed that the common form of iron enrichment in flour, powdered iron, was less available to the human volunteers than the other three forms. These results differed from those reported by Hinton and co-workers[193] and Hinton and Moran[194], who found that powdered iron, ferrous sulfate, SIPP, and green ferric ammonium citrate baked into bread were all absorbed equally well and that the differences were unimportant. Dr. Elwood[192] summarized the work by questioning the benefits of adding iron to flour. He also questioned some of the tests used to determine whether the iron was absorbed from the diet. He called for long-term, large-scale, community-based trials for effectiveness of iron supplementation.

Such trials are under serious consideration by the Committee on Iron Deficiency, formed by the Council on Foods and Nutrition of the American Medical Association.[195,196] The University of Washington in Seattle has recently received a research grant from the Agricultural Research Service of the U.S. Department of Agriculture to study the availability of iron in foods for human beings.[197] The study will include iron supplements used in food processing.

In 1970 the National Millers Federation and the American Bakers Association jointly petitioned the Food and Drug Administration for approval to triple the iron supplementation in flour and bread to overcome iron deficiency in the U.S. diet.[198] At the time of this writing, approval is expected for substantial increases in the levels of thiamine, niacin, riboflavin, and iron that can be added to flour and/or baked products.

One very recent development that may aid in providing iron supplementation in baked products is the development of a whey-protein-iron complex by Jones of the U.S. Department of Agriculture.[199] The whey protein is treated with a "ferri-polyphosphate" to form the complex and probably involves the ability of polyphosphates to function as polyelectrolytes. Studies must yet be made of the oxidative stability of the foods thus supplemented and of the ability of the human body to utilize the iron from this complex before the complex should be used for iron supplementation.

Decreasing Cereal Cooking Time

The addition of phosphates has been shown to decrease the cooking time of several of the grains used as breakfast cereals. Billings[30,200] reported that he decreased the cooking time of ground cereals when he treated them with 0.2 to 2.0% of a phosphate, such as DSP, sufficient to raise the pH of the cooked cereal between 0.3 and 1.0 units. This was particularly true of corn and wheat cereals. The phosphate cooking solution decreased the hydrogen-ion concentration and buffered the pH in a range optimum to setting up and maintaining a Donnan equilibrium with respect to the semi-permeable membranes of the cereal particles. This resulted in more rapid hydrolysis and solution of the starch and protein of the grain. Rice, barley, rye, and oats were also reported to respond to this treatment.

Rao and co-workers[201] reported that the cooking time of pulses could be decreased by the addition of 0.5% TSP or 1.0% DSP to the cooking water. Tanaka and Yukami[202] prepared a precooked, dry rice by soaking the milled rice in a phosphate solution prior to cooking in the presence of a sugar and a surface active agent, which caused partial gelatinization of the starch. Steaming completed gelatinization, and the rice was then dried.

The cooking rate of macaroni and related products was reported to be increased by the addition of 0.5 to 1.0% DSP, and the FDA regulations now allow the addition of this phosphate to these products.[30]

Miscellaneous Applications

Phosphates have been used to improve the color of cereal products. Malkov and co-workers[203] reported that they were able to reduce the rate of respiration of barley malt and the accompanying loss of carbohydrates by treating the soaked barley seed with a solution of phosphate. Others reported that the treatment of germinating barley with 0.01 to 0.75% of aqueous pyrophosphoric acid produced a brighter, more acceptable color and decreased rootlet development during its germination.[204]

The color of precooked "cocoa noodles" was reported by Payumo et al[205] to be improved by soaking them in a solution of 0.35% SAPP for 20 min prior to drying. The storage qualities of soybeans were reported to be improved by treating them with steam under vacuum,[206] and the color could be preserved by adding 4 lb of MCP · H_2O or 10 lb of MKP per 1,000 lb of soybeans.

Other miscellaneous applications include the addition of soda and an acidic phosphate salt, such as SAPP, MCP, or DCP, to a gravy mix so that the evolution of gas during its preparation would prevent lumping of the mix and increase the dispersion of the particles.[207]

The phosphates also are useful in the nixtamalization process for cereal grains. The process, as normally practiced in Central and South America, involves steeping the grain (particularly corn used for tortillas and similar products) in a large quantity of a saturated solution of calcium hydroxide or limewater. Madrazo and Cortina[208] found that they could accomplish the nixtamalization of cereal grains with significantly less aqueous alkali by treating them under pressure with 15 to 35% (based on the weight of the grain) of a solution of MSP, DSP, or TSP or other alkaline compounds. Since less water was present, the grain was much easier to dry. Other advantages were that the softened hull was partially hydrolyzed and was ground with the grain. The improved process retained more of the nutrients of the whole grain associated with the hull and layers of the endosperm immediately beneath the hull.

Monoammonium and diammonium phosphates are used as nitrogen sources in the so-called "yeast foods" that commercial bakers add to the doughs of yeast-raised baked goods. These mixtures are composed of the nutrients necessary for rapid yeast leavening that are not found in the doughs. The ammonium phosphates provide both nitrogen and phosphorus. Potassium phosphates are also often added to provide potassium.[29] Although the orthophosphates promote microbiological growth, all of the polyphosphates from the pyrophosphates through long-chain polyphosphates will tend to inhibit growth of yeast to degrees depending on chain length.[103] They must, therefore, be used with extreme caution in any food in which growth of yeast is important.

Phosphate Applications in Dairy Products

The functions and number of applications of the phosphates in milk processing are nearly equal to those in meat processing. The literature reporting the investigations of phosphate functions and applications in milk products is so voluminous that only a portion of it will be reviewed here. The greatest amount of work reported is that of the interactions between the phosphates and inorganic cations, especially calcium, and between phosphates and the milk proteins.

Interactions of Phosphates with Milk Constituents

The chemistry of the casein-calcium phosphate complex has been studied by numerous investigators. Calcium orthophosphate is intimately involved in the formation and stability of the casein micelles that provide the white, opaque appearance of milk. A review of the current knowledge and four proposed structures for the casein micelles of milk has been published by Rose.[94]

Interaction with Milk Calcium

Phosphates have certain effects on milk that are very useful in its processing. Although many investigators have proposed that all the effects of added phosphate involve precipitation or complexing of the calcium ions in the milk, others have provided evidence that phosphates, especially polyphosphates, also undergo important interactions with milk proteins.

Sommer and Hart[209] first proposed that the addition of phosphates and/or citrates changed the "salt balance" of the milk through formation of complexes with the calcium and magnesium ions. Although the complexing of calcium and magnesium ions is an extremely important function of the phosphates in food processing,[210] especially in obtaining improved quality in dairy products,[209-216] recent work has shown other effects of equal or greater importance. Casein, in milk, acts as a multivalent ion that can undergo ion-exchange reactions.[210,214] This results in a reduction in the levels of free calcium ions and an increase in bound calcium or "colloidal calcium" as phosphates are added.[210,211,213,215,216]

The addition of MSP has no effect on the levels of soluble and ionic calcium and insoluble phosphate. The level of soluble phosphate, of course, is increased, and the pH is decreased slightly. The addition of DSP reduces the soluble and ionic calcium, probably by precipitation of insoluble calcium phosphates, because the amount of insoluble phosphate also increases.[211,213] The addition of pyrophosphate and polyphosphate to milk reduces the levels of soluble and ionic calcium and increases the level of insoluble phosphate. The calcium and phosphate that disappear from solution appear in the colloidal casein-calcium-phosphate complex.[211,217,218] Orthophosphates are, therefore, believed to precipitate calcium,[213] while the polyphosphates form casein-calcium phosphate complexes from which the calcium cannot be easily removed.[210,211,217] Pyne and McGann[215] proposed that colloidal, or bound calcium phosphate, occurs in a casein-calcium-phosphate-citrate complex and suggested a casein-apatite-citrate formula.

The effects of SHMP and other calcium complexing agents on milk have also been studied. SHMP reduces the turbidity of milk at much lower levels than other calcium complexing agents.[219] Studies by the author and co-workers[220] show that as little as one part of SHMP per 20 parts of the casein in fluid skim milk causes almost complete loss of opacity, and the milk becomes a yellow-green, translucent serum.

Odagiri and Nickerson[219] studied the difference in the reaction between SHMP and other calcium complexing agents, including citrate, oxalate, and EDTA. They speculated that the casein combined with a portion of the SHMP added to form a stable, large complex. If, however, an excess was added above that required for formation of this casein-SHMP complex, the excess caused the caseinate particles to release their attached colloidal inorganic salts, which resulted in the complete dispersion of the casein micelles with the resultant loss in the normal opacity of the milk.

Effect of Phosphates on Heat Stability of Milk

The addition of orthophosphates increases the heat, rennet, and alcohol coagulation times of milk.[212,221-229] This effect is very useful in the manufacture of cheese and stabilization of evaporated milk; it could also be useful in stabilizing the effects of seasonal variations in calcium levels in milk.[229,230] Studies by Zittle and Pepper[222] showed that the aggregation of calcium caseinate is reduced within any given time and that the effect of heat is also lowered by the addition of phosphate. Rose[223,224] found that the amount of calcium and phosphate ions absorbed on the surface of casein micelles has a direct bearing on the heat stability of milk. Other studies have shown that the degree of heat stability of milk is inversely proportional to the ratio of soluble calcium/soluble orthophosphate.[225-229]

Interactions with Milk Proteins

Orthophosphates, pyrophosphates and longer-chain polyphosphates (tri- and higher) differ in their reactions with casein. The addition of small quantities (1 to 15 m*M*) of orthophosphates to milk causes little change in the viscosity of the milk. Fox et al.[231] found that the addition of 0.14 to 0.225 *M* orthophosphate would produce gelation, the rate of gelation being directly proportional to the level of phosphate. The rate of gelation increased in proportion to the calcium ion concentration and the level of milk solids-not-fat.

Vujicic and deMan[232] studied the interactions between ortho-, pyro-, tripoly-, tetrapoly-, and hexametaphosphates with casein. A 1% suspension of casein in solutions of each of the ortho- and polyphosphates was prepared, then acidified to pH 4.6 with acetic acid-sodium acetate to precipitate the casein or casein-phosphate complex. The precipitate was removed by filtration or ultrafiltration. Analysis of the filtrates demonstrated that only the orthophosphates, MSP and DSP, passed into the filtrates, while the polyphosphates remained in the precipitated casein fractions. Paper chromatography of the filtrates confirmed the absence of the pyro- through polyphosphates and the presence of orthophosphates.

Casein is precipitated by pyrophosphate to form thick gels.[217,218,233] Zittle[217] believed the negatively charged pyrophosphate anion became bound to the positively charged casein at the normal pH values existing in milk. The resulting reduction in its net charge caused the casein to precipitate in the same manner as it does at its isoelectric point. Heating the pyrophosphate-treated milk eliminated the gelling effects.[218] Pyrophosphate did not precipitate the soluble β-lactoglobulin under these conditions.

Longer-chain polyphosphates, however, precipitate both casein and β-lactoglobulin.[217] The polyphosphates are well-known general protein-precipitating agents.[217,234-240] It is believed that the acidic polyphosphate anion forms ionic linkages with the basic amino groups of the protein molecules, thus lowering the net charge and causing precipitation. Because they are polyvalent anions, the polyphosphate molecules are capable of reacting with more than one basic group to form cross-links between protein molecules. Thus large, compact particles, or micelles, are formed.[241,242]

Differences between the action of added pyrophosphate and that of added polyphosphate were reported by Leviton.[240] It is well documented that gels are formed when pyrophosphate is added to dilute or concentrated milk.[217,218,233] Leviton found that polyphosphates are rapidly hydrolyzed to pyrophosphate in the milk. The rate of hydrolysis was significantly increased by heat, since Leviton found two thirds of the polyphosphate converted to pyrophosphate after heating for 15 sec at 280°F. Yet, although a substantial quantity of pyrophosphate was now present, the milk was stabilized against gelling. Leviton believed this to be evidence that the polyphosphates became stabilizers against milk gelling by formation of cross-linkages, most likely involving the pyrophosphate formed by hydrolysis of the polyphosphate. Furthermore, studies involving the addition of TSPP to milk with subsequent heating showed that the colloidal proteins were highly dissociated rather than compacted, as is the case when polyphosphate had been added,[218,241,242] even though the polyphosphate was rapidly converted to pyrophosphate.

Further study of the reactions of the polyphosphates with the natural protein systems in milk and other foods is needed in order to enable them to be fully utilized. Better knowledge of these complex interactions should result in many new, practical applications. It should be noted that similar studies of the reactions of polyphosphate on meat proteins are under way. Preliminary results reported by Hamm and others show the similarity between their reactions with meat and with milk proteins. It appears that polyphosphates may be most useful in manipulating food-protein systems in order to assist in improved processed foods having greater storage stability. Some evidence has also been reported that addition of STP or SHMP can improve the digestibility of milk and milk foods for babies and convalescents.[243]

Applications in Milk Beverages

The phosphates have been found to have several functions in the processing of milk-based beverages. Kawanishi[244] prevented the formation of a blue-purple color in a milk drink containing strawberry juice by chelating the traces of iron that cause the discoloration with < 0.3% polyphosphate. Wouters[245] used a mixture of DSP and trisodium citrate to adjust and buffer the pH of pasteurized milk to 6.3 in preparation of solid milk-drink products. Jackson[246] sensitized the spoilage bacteria to easier sterilization in cocoa and cocoa malt mixtures intended for use in chocolate-flavored milk products. He accomplished the sterilization by (1) acidifying a syrup or suspension of these ingredients to pH 4 or 5 with H_3PO_4, (2) heating to sterilizing temperatures (190°F or higher) for the time (5 or more min) necessary to destroy the bacteria, and (3) restoring the original pH (between 6 and 7) by addition of anhydrous DSP. The sterilized syrup, or slurry, was then added to milk, given a high-temperature short-time heat treatment, and canned. Sterilization of the flavoring syrup prior to addition to the milk was essential, since the spoilage organisms were not destroyed during the high-temperature short-time heat treatment of the flavored milk. Caramelization and cooked or burned flavors were produced if the flavored milk was heated sufficiently to destroy the bacteria in the untreated flavoring ingredients.

Roland[247] used long-chain, insoluble potassium and sodium salts of polyphosphates to thicken aqueous dispersions of milk, such as chocolate milk. The thickened milk dispersion could then be added to untreated, flavored, aqueous milk dispersions at levels that would promote and maintain proper viscosities in the final product.

Van Wazer[30] reported that the addition of a small quantity of powdered TSPP to malted milk powders improved the wetting of the particles and the color and flavor. The TSPP may also add to stabilization and viscosity of the drink by its interaction with casein to produce a thin gel.

Applications in High Butterfat Dairy Products

Butter

The ability of phosphates to precipitate iron and copper ions from solution was utilized by Malkov et al.[248] to stabilize butter against oxidative spoilage during storage. The butter was washed for 1 hr with an 8.8% (2% P) solution of MKP, then carefully washed with water until all phosphate was removed. The level of free fatty acids and the peroxide value were significantly lowered, and the iodine values were stabilized in the phosphate-treated butter as compared to a water-washed control.

North[249] found that acids, including phosphoric acid, would aid in the clarification of butter oil by coagulating the suspended materials that caused cloudiness.

Meyer and Weck[250] found that buffering the lactic acid in slightly or completely sour, whole milk allowed the buffered milk to be used for the preparation of stable, sweet-tasting butter from the cream and for the production of cheese from the milk. Alkali salts, such as DSP, TSPP, and sodium polyphosphates, were among those mentioned as being effective. A 1:1 mixture of DSP and basic sodium citrates was especially effective, since the combination dissolved coagulated milk casein faster than either salt alone.

Buttermilk

In order to overcome the short shelf-life of bacterially fermented buttermilk, Corbin and Long[251] developed a method for producing buttermilk by direct acidification in a two-stage process. The resulting buttermilk had about four times greater refrigerated shelf-life than cultured buttermilk. The addition of 0.01 to 1.0% water-soluble ammonium or alkali metal salt of an edible phosphate prior to acidification produced enhanced flavor, viscosity, body, stability to whey-off, and appearance of the buttermilk. The examples used to describe the process demonstrated that the addition of ammonium or alkali metal salts of orthophosphates and pyrophosphates was essential to production of a stable buttermilk. The authors also reported that the ammonium, sodium, and potassium salts of the various longer-chain and cyclic polyphosphates were equally effective. In addition, phosphoric acid can be used alone or in combination with other edible acids in the acidification steps. A great advantage to the process is that it reduces the time necessary for the culture process from 15 to 48 hr to 1 to 5 hr. Such a reduction in time in buttermilk production could enable the processor to design a continuous process.

Van Wazer[30] reports that, although TSPP produces thickening or gelling in sweet milk, it has been used to prevent excessive thickening of buttermilk. The TSPP apparently acts as a dispersing or deflocculating agent for the curd that would normally be formed by the casein in the acidic medium of the buttermilk. In contrast, TSPP acts as a gelling agent for the milk at its normal pH.[217,218,233]

Cream Products

Rothwell[252] reported that whipping cream gave improved overrun and stability against leakage when 0.1 to 0.3% sodium alginate or TSPP was added just prior to whipping. The TSPP was not as effective as the sodium alginate, except at the 0.3% level, where it produced a soapy taste.

Wilson and Herreid[253] reported that flavor of sterilized, 50% fat cream containing 9% solids-not-fat could be stabilized by the addition of 0.15% sodium polyphosphate (N = 5 to 7) and ascorbic acid and/or α-tocopherol. Flavor scores for the canned, sterile cream were very high, and peroxide values of the cream containing ascorbic acid were almost zero for three to five weeks.

Sodium phosphates are used in some commercial half-and-half products, especially those sold for use in restaurants. One container of "Sterilized Half-and-Half with Stabilizers Added" served the author in a restaurant just prior to the writing of this section declared the following ingredients: milk, cream, 2% nonfat dry milk, disodium, sodium tetraphosphates. Another declared its ingredients to be half-and-half with 2% nonfat dry milk and sodium citrate, sugar, and disodium phosphate added.

Applications in Condensed and Evaporated Milks

Condensed and evaporated milks are prepared by concentrating the milk, then canning and sterilizing it by heating the cans to approximately 240°F. Unless the casein of the milk is stabilized against heat coagulation by the addition of DSP or sodium citrate, it tends to gel during storage.[29,30,110,111] The Federal Standard of Identity for evaporated milk allows the processor to add up to 0.1% of anhydrous DSP.[254] The processor, however, must determine the quantity that is actually necessary in each batch of milk, since milk differs in its calcium to phosphorus ratio according to the time of the year and the diet of the cow from which it is obtained. It is, therefore, common practice to add between 0.02 and 0.06% of DSP to the evaporated milk.

Development of aseptic canning techniques has created a great deal of interest in use of this process for production of evaporated milk. Edmondson[255] studied the effect of additives and sterilization procedures on the viscosity and shelf-life of evaporated milk. He found it was necessary to control the initial viscosity of the sterilized milk through addition of calcium chloride and DSP at a point just below the level of additives that would cause immediate coagulation. The highest levels of added phosphates studied produced milks with the lowest initial viscosity and shortest shelf-life. Martynova[256] reported that the addition of 0.009 to 0.011% DSP or sodium citrate to milk prior to pasteurization increased the shelf-life and stabilized the evaporated milk against age-thickening for at least six months of storage at room temperature.

The presence of sugar in sweetened, condensed milk increases the susceptibility of the milk to formation of Maillard-type pigments.[257] Ascorbic acid, vitamin A, glucose oxidase, and 0.15% SHMP were found to be effective inhibitors of these discoloration compounds during an eight-month shelf-life test, while DSP and vitamin E were found to be ineffective. Muller[258] found that DSP prevented the coagulation of evaporated milk prepared from skim-milk powder recombined with butter oil.

Applications in Sterile Concentrated Milk

Condensed and evaporated milks discussed previously are characterized by being sterilized in the can. The resulting product contains a high level of cooked flavor and often develops caramel flavors. It is also subject to discoloration, ranging from deep cream through yellow to brownish shades.

Recent developments in food-processing equipment have resulted in the development of an aseptically canned, sterile milk concentrate. The milk, after concentrating, is given a high-temperature, short-time sterilization treatment and is then canned under aseptic conditions in sterilized cans. Most work has been done with 2:1 or 3:1 concentrates, which means that the concentrate is reconstituted to the solids content of whole milk with either two parts water or three parts water to one part concentrate. Naturally, the most desirable concentrate is the 3:1 concentrate, since less water is being transported.

Seehafer[259] defined three major problems associated with the production of sterile milk concentrate in an excellent review of the topic. Heat stability, due to the coagulation of the milk proteins during heat processing of the concentrate, is a problem that varies with the season of the year, the geographical location, the breed and period of lactation of the cows producing the milk, the acidity and the degree of concentration of the milk, and other factors. Storage stability is a second problem, because concentrated milk has a tendency to form gels ranging from slight increases in viscosity through soft to very firm gels. A second type of storage stability problem can be defined as sedimentation, or the separation of the milk constituents into a solids and a serum phase, with the solids often forming granular particles that precipitate to the bottom of the container. Sedimentation problems give rise to the description of "chalkiness" or "sandiness" in the milk when it is reconstituted. The third type of problem is that of flavor stability; this usually results in the development of cooked or stale flavors in the reconstituted milk.

Polyphosphates have been reported by numerous investigators to improve the heat and storage stability of sterile concentrated milks.[259-267] Leviton and Pallansch[260] reported that they observed a sixfold increase in storage life of the 3:1 and a threefold increase in storage life of the 2:1 concentrates when they added 0.05% of either sodium polyphosphate or manganese sulfate to the milk after concentrating it. The sodium polyphosphate used had an average chain length of 4.8 phosphorus atoms per chain. Since sodium polyphosphates were a more acceptable food additive than mangannous salts, the authors recommended use of the polyphosphates for future work. They also reported that the polyphosphates stabilized the concentrated milk against heat coagulation during sterilization. Furthermore, they had found that polyphosphates increased in antigelation activity as their chain length increased and that cyclic sodium tetrametaphosphate and adenosine triphosphate were more effective than the straight-chain sodium tetrapolyphosphate and STP, respectively.

Leviton and co-workers[261] reported that they were able to increase the shelf-life of sterile concentrated milk from 45 to 441 days when they added approximately 0.1% of a commercial polyphosphate with an average chain length of 4.8 phosphorus atoms to sterile concentrated milk. They investigated various sequences of processing steps and found that the sequence – forewarming, homogenization, addition of additive, sterilization, and concentration of the milk – increased shelf-life from 7 days for the control to 240 days for the concentrate containing polyphosphate; the sequence – forewarming, concentration, addition of additive, sterilization, and homogenization – increased shelf-life from 45 days for the control to 441 days for the concentrate containing phosphate when stored at 70°F. Therefore, they recommended the latter sequence of processing steps for commercial utilization. These authors also reported that it was necessary to add the polyphosphate just prior to sterilization and to use high pressure (6,000 to 8,000 psi) homogenization in order to obtain sufficient heat stability to survive the high-temperature sterilization process. The maximum stability to development of off-color, off-flavor, thickening, gelation, and sediment formation was found to occur when the concentrates were stored at 70°F or lower.

The work of Leviton and his co-workers was verified by Wilson and co-workers[262] who studied various phosphates as stabilizers for sterile concentrated milks at levels between 0.01 and 0.1%. They found that polyphosphates with chain lengths above five gave best stability, with stability increasing with chain length. Storage temperatures appeared optimum between 40 and 70°F. These authors reported that titratable sulfhydryl groups decreased during storage and with onset of gelation. Further studies with sulfhydryl reagents indicated that disulfide bonds were definitely involved in the formation of gel structure in the concentrates.

As a result of their work, Leviton and Pallansch[263] were granted a public patent for the use of polyphosphates in processing sterile concentrated milks. They reported the effective levels to be from 0.1 to 1.0% but that 0.6% (6 g of polyphosphate per kg dry weight of milk solids-not-fat) was an upper limit for practical purposes.

Herreid and Wilson[264] reported that the polyphosphates penetrate the casein micelle to combine with the proteins and calcium. They bind the calcium and form a covalent linkage to the stable calcium caseinate-calcium orthophosphate complex and thus retard gelation.

Leviton[240] reported that the tri- and higher polyphosphates were first hydrolyzed to pyrophosphate during their interaction with the casein micelle. He reported that at least two polyphosphate molecules were involved and that pyrophosphate and orthophosphate were produced from this interaction. The hydrolysis of the higher linear, condensed phosphates was rapid upon addition to the milk; Leviton found that approximately 60% of the phosphorus in added hexametaphosphate was converted to pyrophosphate phosphorus in 15 sec at 137.8°C. He further speculated that the pyrophosphate formed by this hydrolysis formed cross-linkages within the casein micelle, thereby stabilizing the micelle. Leviton further proposed that when pyrophosphate was added directly to the milk, it interacted with the magnesium and calcium ions to form insoluble colloidal complexes. Therefore, it was not available to stabilize the casein micelles in the same manner as the pyrophosphate derived from hydrolysis of polyphosphate. Leviton found that cyclic sodium tetrametaphosphate was also hydrolyzed to pyrophosphate but at a much slower rate than the linear condensed phosphates. He speculated that this was due to the planar configuration of the cyclic compound in contrast to the "snake" configuration of the linear polyphosphates.

Swanson and Calbert[265] reported that milk produced at different seasons of the year reacted in different ways to heat sterilization. They reported that a computer was being used to correlate the characteristics of the milk with the necessary processing conditions to produce stable, sterile milk concentrate. The results of the computer work have since been reported by Magnino[266] and by Bixby and Swanson.[267]

Stewart[269] also reported that milk varied in its reaction to polyphosphates and other processing variables. He reported that although sodium orthophosphate, added to the concentrated milk prior to sterilization, would stabilize the milk against heat coagulation, it increased the rate of gelation of the product after the sterilization and aseptic canning step. Similarly, addition of polyphosphates prior to sterilization would, in some cases, cause similar results because the polyphosphate was hydrolyzed to produce considerable quantities of orthophosphates at the temperatures used in sterilization and aseptic canning. Stewart reported that attempts to use polyphosphate solutions that were heat sterilized prior to addition to sterile concentrated milk also were ineffective in preventing gelation; these phosphate solutions may actually have increased gelation slightly because the polyphosphate was hydrolyzed to produce considerable amounts of orthophosphate during the high-temperature, short-time sterilization step. He, therefore, developed a method of sterilizing polyphosphate solutions by ultrafiltration and addition of the solution to previously sterilized, concentrated milk after cooling so that maximum use could be obtained from the polyphosphate. He reported uniform stabilization against gelation by this process.

The work of Bolt and Kastelic[270] demonstrated that young, growing rats were able to utilize sterile concentrated milks as effectively in their diets as did rats fed spray-dried whole milk or concentrated milk without added polyphosphates. The rats appeared to ultilize phosphorus, calcium, magnesium, and protein from the polyphosphate-treated milk as well as they did from untreated milk.

The reactions of consumers to sterile concentrated milk were reported by Swanson and Calbert,[265] Calbert and Jackson,[271] and two industrial reports.[272,273] Approximately 70% of those who used the product liked it and reported they would continue to buy it if it were available, especially since it appeared that it could be sold at a lower price than milk delivered to the home. Alwes and Swanson[274] reported that they attempted to whip sterile milk concentrates, since consumers found them an acceptable substitute for cream in coffee. The presence of polyphosphate was found to be detrimental to the whipping properties of the concentrate, and the concentrate, therefore, could probably not be used as a replacement for whipping cream.

Nakanishi and Itoh[275] reported that the addition of 0.025 *M* MSP or 0.02 *M* calcium chloride reduced the solubility of the caseins in frozen concentrated milk and allowed then to flocculate during storage. The addition of 5% lactose, 2% "hexametaphosphoric" acid, or 0.01 *M* sodium citrate improved the stability of the caseins to flocculation during frozen storage. Other Japanese researchers also reported that unspecified sodium or potassium phosphates did not improve the storage life of frozen milk, but that addition of 0.1 or 0.2% sodium or potassium citrate or the removal of more than 9.8% of the calcium ion from the milk did improve its storage life.[276]

The effects of irradiation on concentrated milk were studied by Hoff and co-workers[277] in order to determine whether the concentrated milk could be sterilized by this means instead of by high-temperature, short-time sterilization processes studied by others. It was found that irradiation ruptured pyrophosphate linkages that are essential to stabilize the casein micelles against gelation. Only concentrated milk that had been stabilized by added manganese or magnesium ions was stable for the test period of 60 days.

Since flavor is a problem in the high-temperature sterilization of milk, Leviton and co-workers[278] and Johnson and Winder[279] studied the stabilization of frozen-milk concentrates. These authors reported that the polyphosphates effectively stabilized the frozen-milk concentrates to resist heat and gelation. They further found that the polyphosphates had some effect on inhibiting lactose crystallization, a major problem with frozen-milk concentrates. Johnson and Winder[279] reported that the addition of a pancreatic enzyme prevented the crystallization of lactose, while the addition of Tenox®* antioxidants stabilized the milk against flavor deterioration for more than 40 weeks, which is considered to be an acceptable shelf-life.

In spite of the work reported above, which indicates that successful processes have been developed for production of concentrated milks, no large successful commercial venture has been established. Although the technical difficulties seem to be capable of being solved, it does not appear to be the right time for marketing such a product. The demand does not yet exist. If increases in labor and delivery costs continue, these products may soon be in demand.

Applications in Milk Gels and Puddings

Prior to the development of instant milk puddings, the preparation of a pudding by cooking starch was a tedious process. Attempts to prepare puddings from pre-gelatinized starch were unsatisfactory, because the quantity of starch necessary produced starchy flavors, heavy, pasty textures, and sticky consistencies. The first instant-pudding mix was patented by Kennedy and Castagna in 1952.[30,280] This instant-pudding mix contained TSPP and calcium acetate as a source of soluble calcium ion with which to cause the milk to gelatinize. The mix, therefore, required only a small quantity of pre-gelatinized starch to prepare the firm pudding gel desired. Preparation consisted

*Tenox is a trademark of Eastman Chemical Products, Inc., subsidiary of Eastman Kodak Co., 260 Madison Ave., New York, N. Y., 10016.

of adding cold milk to the mix, blending it thoroughly, and allowing it to stand for a short time at refrigerator temperatures. After about 10 min, the calcium acted as a coagulation or gelation accelerator and caused the mixture to set up to pudding consistency.[280]

A later patent by Breivik et al.[281] reported improvements on Kennedy and Castagna's mix by addition of an acetylated monoglyceride to the formulation. Clausi[282] patented the use of TSPP to gel the milk with the addition of orthophosphates as gelation accelerators, in place of the calcium salts. Typical formulations from these three pudding mix patents are summarized in Table 15.

Charie and Savage[283] reported that the production of acceptable puddings required a pH ranging between 7.5 and 8.0. Since milk has a normal pH of about 6.6, it required the addition of an alkalizing agent. However, the alkalizing agent could not be an alkaline-reacting orthophosphate or a calcium-sequestering agent, such as SHMP or sodium citrate. They recommended the use of anhydrous sodium carbonate.

These authors also reported that, since the gelling ability of TSPP varied, not all batches manufactured would produce firm, nonlumping puddings in a reasonable setting time of 10 to 15 min after the addition of the milk.[283] This fact has required TSPP manufacturers to test all batches of TSPP in order to select those that will provide successful puddings. At least one manufacturer now labels the acceptable batches as "Pudding Grade."

A recent patent makes use of the ability of carrageenan and milk to form complexes in the presence of phosphates in order to prepare flan-type milk puddings.[284] Phosphates selected from the group SHMP, STP, TSPP, and TSP are added to decrease the influence of the milk calcium on gelation of the carrageenan. The resulting pudding is reported to have no syneresis when the heated milk-carrageenan mixture is treated with the phosphate.

A German patent[285] reports the use of combinations of TSPP and sodium hydroxide having the composition $Na_4P_2O_7 \cdot 0.2NaOH$ to $Na_4P_2O_7 \cdot 2NaOH$ with a calcium phosphate having a calcium to phosphorus ratio between 0.9:1 to 1.2:1.

Partridge[286] found that he could prepare reaction products between milk protein and water-

TABLE 15

Instant Pudding Mix Formulas for Addition to One Pint of Cold Milk

	Pat. 2,607,692[1]		Pat. 2,801,924[2]		Pat. 3,231,391[3]
Ingredients	Vanilla (g)	Chocolate (g)	Vanilla (g)	Chocolate (g)	Unflavored[4] (g)
Sucrose	82.4	81.2	66.5	65.5	78.3
Pre-gelatinized starch	20.0	22.0	20.0	26.0	16.7
Cocoa	–	16.0	–	16.0	–
Vanilla flavor	2.0	2.0	2.5	1.4	–
Salt	2.4	2.4	–	–	0.6
Color	2.6	–	0.3	0.4	–
TSPP	2.8	2.8	1.5	1.5	2.2
Calcium acetate	1.2	1.2	–	–	–
$MCP \cdot H_2O$	–	–	–	–	1.6
DSP, anhydrous	–	–	1.5	3.0	–
Acetylated monoglyceride	–	–	–	–	0.6
Total mix weight	113.4	127.6	92.3	113.8	100.0[4]

[1] From Kennedy, M. H. and Castagna, M. P., U.S. Patent 2,607,692, 1952,[280] with permission.
[2] From Clausi, A. S., U.S. Patent No. 2,801,924, 1957,[280] with permission.
[3] From Breivik, O. N., Slupatchuk, W., Carbonell, R. J., and Weiss, G., U.S. Patent 3,231,391, 1966,[281] with permission.
[4] The addition of flavors and colors to this mix is usually required.

insoluble, crystalline potassium metaphosphates that increased the viscosity and water-binding capacity of the foods in which the treated milk proteins were used. In most cases it was necessary to solubilize the insoluble potassium metaphosphate by the addition of a compatible, edible sodium salt, such as sodium citrate, sodium chloride, or sodium phosphate. Upon the addition of 0.1 to 2% of the potassium metaphosphate with its sodium solubilizing salt, Partridge increased the stability of chocolate milk preparations, improved the viscosity of market cream, improved ice cream and sherbet mixes, improved custards and puddings, and improved milk-containing confectionery products.

Canned, refrigerated, and frozen ready-to-eat puddings are a recent marketing innovation.[287] These puddings are prepared by use of ultra-high-temperature, short-time (UHTST) heat treatment and, if they are to be stored at room temperature, aseptic canning. It is possible to prepare such ready-to-eat puddings only by sterilizing them; the UHTST heat process holds them for so short a time at the high sterilizing temperatures (285 to 305°F for 8 to 12 sec) that they do not discolor or develop overheated flavors, as would be the case with regular retort canning processes. The ready-to-eat puddings utilize TSPP and sometimes DSP to provide proper gelling of the milk proteins so that the desired firmness is obtained. McGowan[287] has published an excellent review of the processing of these puddings. Typical ready-to-eat pudding formulas are shown in Table 16.

TSPP is also useful in the preparation of an "instant" cheese cake filling that requires only that the dry ingredients be blended thoroughly with milk and then poured into the prepared, baked shell. Allowing the cheesecake to set for about 15 min at refrigerator temperatures results in an instant cheesecake. A typical cheesecake filling formula is shown in Table 16.

Other applications have also been developed for the gelling ability of phosphates on milk proteins. Roland[288] received a patent for the preparation and applications of milk protein stabilized with long-chain sodium or potassium metaphosphates (Kurrol's salts). Such a stabilizer prepared from metaphosphate-stabilized skim milk can be used for the suspension of chocolate particles in chocolate-milk drink. It is also possible to make an

TABLE 16

Formulas for Ready-To-Eat Puddings and a Cheese Filling

Ingredients	Firm-bodied[1] pudding, %	Creamy[1] pudding, %	Cheesecake[2] filling, %
Cheese, dried			9.61
Sugar (sucrose)	6.44	10.00	7.69
Corn sugar (dextrose)	6.87	4.35	3.08
Corn starch, modified	4.94	2.61	3.84
Tapioca starch, modified	1.07	2.61	
Nonfat dried milk	7.78		
Buttermilk, dried			7.69
Whole milk, liquid		66.95	61.51
Shortening, dry	2.68		5.77
Salt	0.11	0.22	0.38
TSPP (anhydrous)	0.05	0.22	0.31
$MCP \cdot H_2O$			0.12
DSP (anhydrous)	0.04		
Water	70.02	13.04	
Flavor	q.s.[3]	q.s.[3]	q.s.[3]
Color	q.s.	q.s.	q.s.
Total	100.00	100.00	100.00

[1] Adapted from McGowan, J., *Food Prod. Devel.*, 4(5), 16, 1970, with permission.
[2] From Stauffer Chemical Co., 1968, with permission.
[3] A sufficient quantity (*quantum sufficiat*).

aqueous solution or dispersion containing milk and metaphosphate that has any desired viscosity higher than that of the desired viscosity of the finished product; the thickened, metaphosphate-treated milk is then diluted until the desired viscosity is reached.

Freund and Danes[289] received a patent for the preparation of proteinaceous gels prepared from casein or sodium caseinate, orthophosphate, sodium pyrophosphate, and a soluble calcium salt. These gels were reported to aid in improving food compositions, such as milk, foams, fat-containing products, and similar systems by binding water and reducing migration of fat.

Studies of the gelling phenomenon produced by addition of phosphates to milk gave certain indications of the action involved. Fox and co-workers[231] found that polyphosphates retard milk gelation when they are present below certain critical concentrations, but that they have the opposite effect at higher concentrations. They found, too, that although orthophosphates have little effect on milk viscosity at low concentrations, high concentrations of orthophosphates will gel the milk. The rate of the gelation process was found to be dependent on the concentration of phosphate, calcium, and the casein complex as well as on the pH, temperature, amount of agitation, and type of phosphate counter ion present. Their studies also indicated that the gelation process occurs in two phases. The first is an initial "null phase" during which the viscosity changes very little; the second is a sudden, very rapid change in viscosity. Since calcium ion is also required for the gelation process, Fox et al. speculated that calcium phosphate may be bound in the interior of the casein molecules.

As previously discussed, Zittle[217] hypothesized an interaction between the positively charged casein and negatively charged pyrophosphate, resulting in reduction of the casein net charge and its precipitation as it occurred at its isoelectric point. If, however, the pyrophosphate-treated milk was heated, the gelling effect was reversed.

Lauck and Tucker[233] reported the results of their investigations of the gelling phenomenon between skim milk and TSPP. They prepared starch-free cold milk gels with various combinations of TSPP, $MCP \cdot H_2O$, alginate, and milk and then measured the viscosities of the systems. As shown in Table 17, they produced the highest viscosities with the presence of all four ingredients; they were also able to produce a gel with substantial viscosity by eliminating the milk and maintaining monocalcium phosphate as a calcium source in the system. Gels were also obtained when only TSPP and milk were present but were strengthened when additional calcium was added in the form of $MCP \cdot H_2O$. Alginate alone in milk produced no gel, and the alginate by itself produced no gel. The authors interpreted these results to mean that the gels formed in such TSPP-treated milk systems were those formed between the calcium and the TSPP rather than the milk protein and the pyrophosphate. However, the data in Table 17 and that of Fox et al.[231] and of Zittle[217] do not support this conclusion in all respects.

TABLE 17

Effect of Various Ingredients on a Starch-Free Cold Milk Gel[1]

TSPP[2]	MCP[3]	Alginate	Milk	Viscosity cps
X	X	X	X	33,700
X	X	X		19,200
X	X		X	6,400
X			X	3,200
		X	X	0
		X		0

[1] From Lauck, R. M. and Tucker, J. W., *Cereal Sci. Today*, 7(9), 314, 1962, with permission.
[2] Tetrasodium pyrophosphate.
[3] Monocalcium phosphate monohydrate.

These phenomena should be highly useful in food-processing systems in which gelation would be an advantage in handling the product with subsequent heating to eliminate the gelation. Further applications will certainly result from the continuing investigations of the gelling effect of phosphates on milk and other proteins.

Applications in Nonfat Milk

The phosphates have been found to improve the properties of dried and liquid skim-milk products through their interactions with milk proteins and their ability to complex calcium. Salzberg[290] reported that he was able to increase and improve the solubility of dried milk solids in water by treating the skim milk, either before or after drying, with acids or their soluble salts, which would yield calcium salts less soluble in water than the tricalcium phosphate normally present in the milk. Only oxalates, fluorides, and metaphosphates were capable of producing such insoluble calcium salts. The soluble salt had to be added to precipitate all of the calcium. For example, dried skim milk containing 1.88% calcium as calcium oxide and dried buttermilk containing 2.89% calcium as calcium oxide were found to be 71% and 53% soluble in water, respectively, before treatment with sodium oxalate. Their solubilities were increased to 96% and 84%, respectively, after treatment. In addition to improved solubility of the milk solids, Salzberg reported that the treated milk solids could be used as emulsifiers for oil in water systems. Dilute emulsions of oils in water with low stability could be obtained with untreated roller-dried milk powder. Salzberg was able to prepare very stable emulsions of equal parts of oil and water merely by stirring the oil with a solution of the treated milk solids. Since oxalates and fluorides would be unacceptable food additives, only the metaphosphate-treated dry-milk solids could be used in foods.

Wouters[291] discovered that he could improve the taste and the stability toward flavor deterioration of dried milk products by the addition of combinations of sucrose and reducing sugars such as glucose. The reducing sugar was believed to lower the redox potential of the system by 30 to 50 mV, a quantity sufficient to prevent the formation of oxy-fatty acids from the unsaturated fatty acids. However, the solubility of such treated milks was often reduced. Wouters found that he could improve the solubility of the treated milk product by making certain that the milk proteins were capable of swelling upon hydration. He accomplished this by the addition of 0.01 to 0.02% of DSP, sodium citrate, or sodium-ammonium phosphate compounds similar to those present in the milk. By such treatments he obtained milk powders that had excellent stability, even though stored for long periods of time.

Curry[292] was able to prepare a skimmed milk colloid, which could be used as an emulsifying agent of fats and oils for various food purposes, by forming a stiff gel with TSPP; he treated the gel with an acid to hydrolyze lactose to its constituent simple sugars while heating at 60 to 65°C. The mixture was later neutralized to a pH of 7 and again held at 75 to 80°C for approximately 2 hr. The solution was then refrigerated until used. It was reported to have excellent fat-emulsifying properties for use in preparing table spreads, ice cream, and similar dairy foods.

Lindewald and Kimstad[293] reported they were able to prepare a replacement for egg white in baking and other food applications by treating milk protein with alkali phosphates. A suspension of casein was prepared either from the hydrated casein or by treating skim milk with rennet to coagulate the protein. The casein suspension was then treated with one of the phosphates that would either precipitate or complex the casein in the suspension. Such phosphates are orthophosphate and pyrophosphate. The pH is maintained between 6.3 and 6.7. After proper treatment, the product can either be whipped as a suspension or dried for later hydration and whipping. The foams prepared from these products were reported to have greater heat resistance than those of egg albumin.

The ability of sequestering agents to improve the whipping properties of nonfat milk solids was investigated by Lewis and co-workers.[294] SHMP and STP were found to substantially improve the whipping properties of nonfat milk solids when added to a level of 1% based on the milk solids. Shorter-chain and longer-chain polyphosphates had much less effect on whipping properties and formed very unstable foams. Sodium citrate and EDTA were also found to improve foaming properties but not to the extent of the hexametaphosphate and tripolyphosphate, and the foams were found to be very unstable. Table 18 demonstrates the effects of various calcium-sequestering agents on the whipping ability of nonfat milk solids. Table 19 demonstrates the effects of hexametaphosphate and tripolyphosphate on the whipping properties and stability of a high quality preparation of nonfat milk solids.

Kempf and Blanchard[295] were able to prepare a dried milk product for use as a coffee whitener by treating the milk with DSP, heat-treating the milk, then drying it. They found that levels of DSP between 2 and 3.7% for fluid milk and 1.4 and 3.1% for concentrated milk (containing between 28 and 30% total solids) were the most effective levels. They were able to find specific heat/time of heating relationships for each level of DSP and provided curves demonstrating these relationships. The ratios of milk fat to solids-not-fat could be varied between 1:3 and 3:1 in the product to

TABLE 18

Effect of Various Sequestering Agents on the Whipping Ability of Nonfat Milk Solids[a]

Sequestering agent added (1% in nonfat milk solids)	Increase in volume, %	Drainage in 20 min, %
Control	330	100
Sodium orthophosphate	370	100
Sodium tripolyphosphate	750	64
Sodium hexametaphosphate	750	56
Maddrell's salt	560	100
Sodium citrate	660	100
Versene	610	100

[a]Source: Lewis, M. A., Marcelli, V., and Watts, B. M., *Food Technol.*, 7, 261, 1953, Table 1, with permission.

provide any degree of fat desired in the coffee whitener.

Kumetat[296] prepared a replacement for whole eggs for use in baked products such as sponge cake by treating concentrated skim milk or a similar suspension of casein with calcium-sequestering polyphosphates or citrates at pH values below 7. The product was held at an elevated temperature of not more than 60°C for a period of time, then the pH was raised to substantially above 7, for example to pH 9. The product was again held at an elevated temperature for a brief period of time and then dried. The resultant dried milk product was reported to have all of the properties of whole egg and could replace part or all of the egg in sponge cake formulas.

Hayes and co-workers[297] demonstrated that STP could be used to disperse calcium caseinate. The caseinate containing 0.25% calcium became soluble at pH 7.0 and maintained its solubility with the addition of up to 1.5% more calcium chloride. A caseinate containing 1.9% calcium could not be dispersed except at pH 9 to 10 or unless 2.5% or more STP had been added.

Prodanski[298] studied the effects of adding varying levels of MSP, DSP, diammonium citrate, and sodium citrate to milk both prior to roller drying and during reconstitution. Roller-dried milk was found to have a solubility between 58 and 62% without additives. Sodium citrate at approximately 0.1% increased the solubility the most – 87.8% for that in which the citrate was added prior to drying and 85% when added during reconstitution. The other additives were less effective.

Applications in Frozen Dairy Desserts

The beneficial effects of phosphates on properties of ice cream have been known for some time.[299,300] Processors were prevented from using the phosphates in production of frozen desserts, because they were not included in the standards of identity. It was felt by the regulatory

TABLE 19

Effect of Polyphosphates on the Whipping Ability of High Quality, Nonfat Milk Solids[1]

Added polyphosphate (4% of milk solids)	Milk Solids in 50 ml water	Volume increase	Drainage in 2 hr
	g	%	%
Control	10	560	60
Hexametaphosphate	10	800	50
Tripolyphosphate	10	750	52
Control	31	390	10
Hexametaphosphate	31	420	0
Tripolyphosphate	31	420	0

[1]Source: Lewis, M. A., Marcelli, V., and Watts, B. M., *Food Technol.*, 7, 261, 1953, Table 2, with permission.

authorities that the phosphates could be used to neutralize acidic or spoiled cream and milk products that then could be used in the preparation of frozen desserts. Until 1967, phosphates and citrates were allowed only in chocolate ice cream and similar frozen desserts, because they were necessary in maintaining stable chocolate suspensions. It was, however, known that phosphates could stabilize frozen desserts against churning. *Churning* is the phenomenon of agglomeration of fat particles or globules during excessive whipping to form sufficiently large agglomerates so that they are detectable as "sandy" particles in the ice cream. This is an especially serious problem in the case of the soft frozen desserts that are kept recirculating in a freezer until they are drawn for sale to customers. Weinstein[301] received a patent for the incorporation of SHMP and sodium citrate in various stabilizer formulas for use in soft-serve ice creams. Frazeur[300] reported that the addition of DSP, TSPP, SHMP, and sodium citrate to frozen desserts substantially prolonged their stability to churning. Keeney[302] also investigated the effects of polyphosphates on churning and found they had a significant beneficial effect. Keeney also reported that they had an insignificant effect on the pH of the mixes at the effective levels and stated that the effects were so small that they could not possibly be used as neutralizing agents for acidic or spoiled cream. As a result of these and other investigations, applications were made to FDA for inclusion of sodium citrate, DSP, TSPP, and SHMP at levels up to 0.2% in frozen desserts, in order to stabilize them against churning. This application was approved in 1967,[303] and these salts are now commonly used to stabilize frozen desserts against churning. Either TSPP or SHMP is usually added along with other thickening substances, such as the alginates, at a level of approximately 0.1% of the final frozen dessert. These compounds cause the protein to swell and remain highly dispersed (probably through polyelectrolyte interactions), and they keep the fat particles apart from each other so that they cannot coagulate and they will remain as small fat globules. In this way, a very smooth frozen dessert is produced.[30,300-302]

Wallander[304] reported that lipase activity remained in sterilization heat treatments in whole milk, skim milk, and a 4% fat, 12.5% solids-not-fat model system used in preparation of sterilized iced milk mixes and eggnog. The addition of sodium tetraphosphate, increased fat level, storage at 7.2°C, or post-sterilization holding treatment would retard the lipase activity in these products.

Applications in Whey, Lactose, and Lactalbumin

Whey is the residue that remains when cheese curd is removed from milk during the processing of natural cheese products. A recent survey[305] indicates that approximately 14 billion pounds of whey is produced annually. About 4.2 billion pounds of whey is processed, while the balance is disposed of as liquid whey. Approximately 67% of the whey that is processed is dried for use in human or animal food; 26% is used in isolation of lactose and other minor constituents of the whey; and 11% is used for the manufacture of the whey proteins.

Whey was recognized at an early date as a good source of the milk sugar, lactose; its solids contain from 70 to 80% of this carbohydrate. As a result, lactose is isolated from whey. A common practice is to heat the liquid whey until the proteins are coagulated. The heat-coagulated protein is then filtered off, the residual liquid is concentrated to approximately 60% solids, and seeded to precipitate a first crop of lactose. Additional lactose can be obtained if the residue is again heated, the coagulated protein is filtered off, and the filtrate is concentrated and seeded with lactose to precipitate a second crop.

One application of polyphosphates in the production of lactose from liquid whey was patented by Zboralski.[306] Approximately 0.1% of a polyphosphate was added to the filtrate resulting from the removal of heat-denatured whey protein. The polyphosphate maintained the residual protein in the filtrate in a soluble state, allowing the processor to concentrate the filtrate at higher temperatures without precipitating further proteins. The polyphosphate was also found to cause the whey protein, which was heat coagulated in the second stage of the process, to form coarse, flocculent, readily filtered precipitates.

Other processes have also utilized the ability of the polyphosphate to solubilize the whey proteins. In one process the calcium is precipitated from the whey by means of a sodium pyrophosphate, removing the precipitate by centrifugation and adding SHMP to allow concentration and crystallization of the lactose without precipitating the protein. A second process calls for the addition of polyphosphates, such as STP, sodium trimetaphosphate, or SHMP to the liquid whey. It is then concentrated and seeded, and the precipitated lactose is removed by filtration and washed with sodium metaphosphate solution.[307]

Dried whey has been prepared for use in human and other animal foods for some time. Although whey has been proposed for use in baked products, especially as a replacement for skim milk, it has caused serious problems through weakening the dough, lowering the pH to levels detrimental to the properties of the dough, and producing too little buffering action to replace that of milk. Ward, Johnson, and Robertson[308] reported that dried whey could be improved in its ability to replace milk solids in bakery products if, prior to drying, a soluble calcium compound, a soluble phosphate compound, and a neutralizing agent were added to liquid whey which had been adjusted to a pH ranging between 6.8 and 7.5. They reported that the resulting dried product permitted the addition of greater amounts of moisture to the dough, while maintaining good structural properties. The product also increased the initial pH of the dough, provided pH control for optimum fermentation, and improved the crust color of the baked product to which it had been added.

The 14 billion lb of liquid cheese whey produced in the U.S. annually represents the potential source for approximately 70 million lb of whey protein of a nutritional quality at least equivalent to that of casein.[305] In fact, Arena[309] reports that human milk, in contrast to cow's milk, has much more lactalbumin than casein and that lactalbumin is nutritionally more valuable than casein. He also reports that lactalbumin is more easily digested and is "the best protein for growth." The characteristics Arena describes, of course, refer only to undenatured lactalbumin. The heat-coagulated lactalbumin commonly available as a by-product of lactose production is probably less digestible and nutritious. It is, therefore, wasteful practice to dump liquid whey when the world faces a protein shortage.

Considerable interest has been shown in salvaging this valuable source of nutritious protein. It has been common practice to dry and use the heat-denatured lactalbumin obtained from lactose production. Most of the lactalbumin products available in the past have been this by-product. The denatured, heat-coagulated lactalbumin is very difficultly soluble, and there have been some reports that is is not as digestible as it is in its undenatured state in milk. This has led to investigations to determine how undenatured lactalbumin can be separated from the milk whey.

The ability of long-chain polyphosphates to react as polyanions and of large protein molecules to react as polycations has been utilized for this purpose.[30,234,235,310] Gordon[310] described a process for the recovery of undenatured lactalbumin from liquid whey that involves precipitation of the protein as a complex with long-chain polyphosphates. Gordon provided an example of isolation of undenatured whey protein. Approximately 0.2 g SHMP (longer-chain polyphosphates can also be used) was added to each 100 cc of acid whey having a pH of 4.2. He then added hydrochloric acid to adjust the pH to 3.0. A precipitate of the protein-polyphosphate was thus formed, then isolated by centrifuging or filtering, and washed thoroughly with water. The polyphosphate was removed from its complex with the whey protein by treating the precipitate with a solution of calcium hydroxide containing a slight excess of calcium over that required to react with the polyphosphate. The pH was adjusted to 9.0. A precipitate and colloidal suspension of calcium hexametaphosphate formed after about 20 min holding time and was flocculated by acidification to pH 5.0. After the precipitate settled out, the pH was adjusted to 7.0. The calcium metaphosphate precipitate was removed by centrifuging or filtration, and the protein in the remaining filtrate was dried. The resulting protein was in an undenatured, water-soluble, heat-coagulable form.

McKee and Tucker[311] and Wingerd[312] described a second method in which long-chain sodium and potassium polyphosphates were used to precipitate a gummy sludge from hot whey, which was then separated by centrifuging and spray-dried. The polyphosphate was not separated from the protein in this case. Variations in the process and especially in the chain lengths and proportions of sodium and potassium polyphosphates resulted in a product containing between 40 and 70% protein and between 15 and 25% phosphate. In order to follow the common commercial terminology for isolated whey protein, the product was designated "lactalbumin phosphate."

While the patents on lactalbumin phosphate were being obtained, Hartman[313] and Hartman and Swanson[314] extended the investigation of

Gordon,[310] applying this work toward the isolation of whey protein from liquid whey. The investigation included the effects of pH, type of phosphate, phosphate chain length, quantity of phosphate required, and similar variables. The authors demonstrated that the optimum pH was approximately 2.5, that both SHMP with an average chain length of 12 and sodium polyphosphate with an average chain length of 50 to 60 caused heavy precipitation of the whey protein, and that the optimum level of polyphosphate was approximately 0.5% based on the weight of liquid whey. SHMP was found to remove 91% of the protein from the whey. Use of a sodium polyphosphate with a 50 to 60% unit chain length precipitated all the whey protein and a portion of the non-protein nitrogen compounds. Polyacrylamide disc electrophoresis of the precipitated proteins demonstrated that they were similar to the untreated whey proteins. The authors estimated the cost of precipitating the protein from liquid whey with SHMP at 3.5 to 4.0 cents/lb.

The work of McKee and Tucker[311] and of Wingerd[312] has led to the commercial production of sodium and calcium lactalbumin phosphate.[315] This product contains approximately 55% protein and 4% phosphorous, which is equivalent to approximately 18.6% phosphate as P_2O_5.

Lactalbumin phosphate has been found to be useful in numerous food-processing applications. McKee and Tucker[311] reported that lactalbumin phosphate would replace part or all of the nonfat, dried milk used in prepared cake mixes, biscuit mixes, self-rising flour, comminuted meats, pizza dough, refrigerated canned biscuits, and pancake, bread, and cookie mixes. Wingerd[316,317] described the ability of lactalbumin phosphate to form complexes and reaction products with gelatin for the preparation of "instant" cold-water, soluble gelatin products and with lipid-type emulsifiers, such as monoglycerides, lecithin, and similar recognized food emulsifiers. The complexes can replace protein such as sodium caseinate in whipped toppings and egg products in the cake mixes.

Lactalbumin phosphate has also been found to act as a direct functional and nutritional replacement for sodium caseinate in the preparation of imitation dairy products,[318] such as coffee whiteners, whipped toppings, artificial ice creams and frozen desserts, puddings, custards, frozen cream pies, imitation processed cheese, imitation sour cream, snack dips, baby foods, and "instant" breakfast and nutritional drinks. Lactalbumin phosphate has also been reported to serve as a partial or complete replacement for egg white in layer cakes, angel food cake, donuts, pancakes, brownies and cookies, and similar products that normally contain egg white.[319]

Other investigations have demonstrated that it is not necessary to isolate the whey proteins to utilize them in food applications. The ability of polyphosphates to cause complete dispersion of proteins is utilized in these applications. This is especially true where it is necessary to develop emulsions in which protein films are formed about fat droplets, thus forming charged barriers that prevent the coagulation of the fat particles. Stable, emulsified, mayonnaise-like products have been obtained when whey or other milk products were emulsified into oil with sodium polyphosphate and monoglycerides.[320] The proteins in liquid and dried whey, demineralized whey, and similar modified whey products are dispersed by the addition of polyphosphate so that these whey proteins form the proper charged layer about fat particles in imitation dairy products to form stable foods,[321] while the untreated liquid or dried whey usually would not function in such a manner.

Applications in Imitation Dairy Products

The high cost of production, processing, and delivery of dairy products has created intense interest in imitation dairy products; these have the taste and appearance characteristics, but do not contain the high-priced butterfat of real dairy products. Margarine was the first imitation dairy product to appear on the market. Its successful replacement of butter in a majority of homes and food-processing applications has stimulated dairies as well as other food manufacturers to consider marketing imitations of other dairy products.

Two types of products have appeared on the market. One is designated "filled" and is usually prepared by replacing the butterfat with vegetable fat and using all the other milk constituents, such as liquid skim milk or skim milk powder, in proper proportions to prepare a product with the desired characteristics. The second is labeled "imitation" and contains no milk fat or protein. Imitation dairy products are combinations of vegetable fat, sugars, proteins (such as soy protein, whey protein, or sodium caseinate), stabilizers, and emulsifiers as well as water to prepare a product resembling in appearance and other characteristics the dairy product being duplicated. Vitamins and minerals are added to some in order to duplicate the nutritional properties of the dairy product.[322-324] Most formulas for "filled" dairy products do not contain phosphates, although the ability of the phosphates, especially the polyphosphates, to aid in protein dispersion would be of value in products having such a problem.

The imitation dairy products usually contain a phosphate, which, in most cases, is either DSP or DKP. Although the DKP is more expensive, it is often preferred because it does not contribute as high a degree of astringent flavor as DSP might contribute. The function of the orthophosphates in imitation dairy products is usually as a buffering agent to maintain a pH optimum for proper protein dispersion. If the imitation dairy product had a pH below 6, there would be danger of the protein precipitating due to the proximity of the pH to its isoelectric point. If the pH of the product becomes higher than 7.5, off-flavors will develop.

Imitation Milk

Replacements for liquid milk have begun to appear in several of the larger markets on the east and west coasts of the U.S. These completely synthetic products are composed of vegetable fat, protein, lipid-type emulsifiers, buffering salts, stabilizers, body agents, sweeteners, and often also vitamins, minerals, and coloring agents in order to more closely duplicate the appearance, textural properties, and nutritional characteristics of cow's milk.[322-324]

Kozin and Rodionova[325] studied the changes in characteristics of an imitation milk emulsion as DSP was added. The pH of the emulsion decreased from 6.95 to 6.2 as the level of DSP was increased from 0.1 to 0.25 g/g protein. The isoelectric point of the protein was found to be at pH 6.2 because it apparently began to precipitate. However, as more phosphate was added, the pH rose and reached 7.30 at 0.5 g phosphate/g protein. The authors also reported that the viscosity of the emulsion and the degree of protein dispersion increased as the phosphate was increased. No fat globules with diameters greater than 10 μm were found at phosphate levels of 0.15 g/g protein, and the size of the fat globules decreased as more phosphate was added. The authors found that 76.6% of the fat globules had diameters of less than 2 μm when 0.1 g phosphate/g protein was present, and 90.7% when 0.5 g phosphate/g protein was added.

Sabharwal and Vakaleris[326] also studied the factors affecting stability of synthetic milk systems. These authors found that stable milk emulsions required the addition of proper levels in the proper sequence of calcium, phosphate, and citrate salts. The addition of phosphate, citrate and calcium, in that order, had an adverse effect, while the addition of 7.5 mM/l. citrate, 25 mM/l. calcium, and 25 mM/l. phosphate in that order gave stable emulsions containing 2% protein and 4% coconut fat. Separation of the fat from the aqueous solution was prevented by adding emulsifier systems of 3 to 6 HLB. They reported development of a heat and storage stable synthetic milk system containing all of the minerals of regular milk.

A number of commercial bases for use by dairies wishing to manufacture imitation milk products have been made available in recent years (see References 323 and 327). These require the addition of water and usually the vegetable fat to the base, pasteurization, homogenization, and subsequent packaging in order to prepare the completed imitation milk. Ingredient costs appear to be approximately 1/2 of the cost for similar unit volumes of cow's milk.[322-324]

Coffee Whiteners

Replacements for liquid coffee cream have been marketed in the U.S. for approximately 20 years. One of the first to appear was a "dried cream." It was prepared by treating whole milk, often fortified with additional cream, with DSP; the mixture was then given a heat treatment, concentrated and spray-dried to produce a powdered cream that could be used as a coffee whitener without prior reconstitution.[328] The joint action of the DSP and the heat treatment modified the casein in such a way that it resisted coagulation when the powder was stirred into coffee. A similar dried milk product was described by Brochner,[329] who found that the addition of alkali metal phosphates or citrates, or mixtures of the two at levels above those of the previous example, and additional lactose produced a powdered coffee whitener.

More recent technology for production of replacements for coffee cream, usually called "coffee whiteners," involves combinations of vegetable fats, either sodium caseinate or soy protein, gum stabilizers, stabilizing phosphate or citrate salts, corn syrup solids and sugar as bodying agents and sweeteners, lipid-type emulsifiers, flavor, and color to duplicate the characteristics of cream. Liquid products usually contain from 0.5 to 3.0% protein and 10% fat, while dried products can be prepared by spray drying homogenized emulsions containing similar ratios of ingredients. Typical ranges of the ingredient levels in formulas for coffee whiteners are shown in Table 20.

Several investigators have published the results of their studies of the functions of the various ingredients in coffee whiteners.[330-333] The fat used in coffee whiteners provides richness, a characteristic mouth-feel, and whiteness in the coffee in which it is used. Whiteness in coffee is inversely proportional to the size of the fat globules; the size of the fat globules is influenced by a number of factors. The type and efficiency of the emulsifying system influence the fat-globule size as much as do the processing conditions used in their manufacture.

The protein can also influence the whiteness in that it combines with the emulsifiers to form a protective barrier or charged layer about the surface of each fat droplet, preventing coalescence of the fat droplets and, therefore, formation of less stable larger globules. Thus, the protein can serve as an emulsifier as well as an improver of whitening power. It also imparts body and flavor to the whitener.[330-332] The phosphate buffering system contributes to the stability of the protein layer about the fat droplets by maintaining the optimum pH for formation of a uniform monomolecular film of the protein about the droplet. Thus, syneresis, or "wheying off," is prevented by the addition of the proper type and level of phosphate. Phosphates also prevent feathering or curdling and fat separation when the coffee whitener is added to an acidic coffee medium.

The proper type of phosphate is determined largely by the composition of the other ingredients in the system. Most coffee whiteners contain sodium orthophosphate salts, usually DSP. However, where flavor is a problem, potassium orthophosphate salts are used instead. Thus, DSP and DKP are the most commonly used,[330,332,333,335,337] and SALP has been reported to provide similar stabilization.[338]

The polyphosphates are especially effective where dispersion of protein is a serious problem,[336] as is the case with some soy proteins. This phenomenon suggests that the phosphates may have other functions than strictly buffering action, since SHMP is an extremely poor buffering compound. TSPP has also been suggested as a stabilizing salt in one formula.[334]

Because of their importance in coffee whitener and other imitation dairy product formulas, companies producing stabilizing mixtures for imitation dairy products are selling blends that

TABLE 20

Ranges of Ingredient Levels in Typical Non-Dairy Coffee Whitener Formulas[1]

Ingredients	Liquid whiteners, range in % (total wt)	Spray-dried whiteners, range in % (dry)
Vegetable fat	3.0 - 18.0	35 - 45
Protein	0.5 - 3.0	4.5 - 10.0
Corn syrup solids	1.5 - 10.0	40 - 60
Sucrose	0 - 3.0	–
Emulsifiers	0.1 - 1.0	0.2 - 3.0
Gum stabilizers	0 - 0.5	–
Phosphate stabilizing salts	0.1 - 1.0	1.0 - 2.0
Flavor	q.s	q.s.
Color	q.s	q.s.
Water	To 100%	–
Anti-caking agent	–	q.s.

[1] Compiled from References 330 to 338.

usually contain protein material, emulsifiers, buffering or stabilizing salts, color, gum stabilizers, and a portion or all of the sweetening agents required in the formulation. The producer furnishes the fat, which is purchased separately. The base is mixed with water, fat is added, and the mixture is pasteurized, homogenized, and packaged in order to produce the final dairy product.

Whipped Toppings

Imitation whipping cream is known in the trade as "whipped topping."[339] Whipped toppings represent substitutes for high-fat cream, while coffee whiteners represent low-fat cream substitutes. The shortage of dairy products during World War II stimulated considerable interest in producing whipped toppings. Mason and Justesen[340] patented a whipped topping preparation composed of vegetable fat, vegetable protein, ionizable inorganic stabilizing agents such as phosphates and/or citrates, vegetable gum stabilizers, emulsifiers, flavoring, and coloring agents.

Dry, powdered whipped topping compositions, such as the composition patented by Noznick and Tatter,[341] are also available. The phosphates were found to assist in stabilizing the protein films which are necessary in formation of foams and also to provide stiffening and flavor and to prevent weeping.

The whipped toppings have improved functional properties to those of whipping cream. These include the ability to incorporate a greater amount of air and water during whipping, greater overrun, superior stand-up, and, depending on the type of fat used, superior eating qualities.[339, 342-345] As is also the case with coffee whiteners, the quality and chemical characteristics of the protein can influence the syneresis, or "weeping," of a fluid or powdered whipped topping mix. Sodium and potassium phosphates and/or citrates prevent the separation of the proteins from the serum phase, which surrounds the air bubbles in the whipped topping foam (syneresis); they stabilize them so that uniform, monomolecular films are formed. These stabilizing salts also offset the adverse effects of calcium and magnesium ions introduced into the system through the water.[343-345]

Among the phosphates found useful in whipped topping formulations are DSP and DKP and TSPP at levels between 0.025 and 1.0% of the complete liquid base. The ranges of ingredients in typical liquid, non-dairy whipped topping formulas are presented in Table 21. Dry, whipped toppings are prepared by using similar ratios of the various ingredients to form concentrates containing 30 to 50% solids and by spray drying them, then instantizing them. The manufacturers of stabilizers and bases for whipped toppings also commonly add the required phosphate salts to their bases and stabilizer mixtures and provide proper instructions to the manufacturer for addition of other ingredients and for processing the finished product.

Imitation Sour Cream, Sour Cream Dressings, and Chip Dips

Replacements for fermented cream products are produced in considerable and growing quantities for the consumer market in the U.S. These non-dairy products contain vegetable fat, protein (either soy protein or sodium caseinate), gum stabilizers, stabilizing salts, flavorings, acids to provide tartness, and often coloring agents as well as imitation or real bacon chips, chives, spices and other natural flavoring agents.[331,339,346,347] Again, the stabilizing phosphate salts are used to cause the proteins to form the properly dispersed, protective films about the fat globules and to prevent syneresis.

Imitation Ice Cream or "Frozen Desserts"

Replacements for frozen dairy products are prepared by combining vegetable fat, either soy protein or sodium caseinate, emulsifiers, gum

TABLE 21

Ranges of Ingredient Levels in Typical Whipped Topping Formulas[1]

Ingredients	Range Based on % of total weight
Vegetable fat	24.0 - 35.0
Protein	1.0 - 6.0
Sucrose	6.0 - 15.0
Corn syrup solids	0 - 5.0
Gum stabilizers	0.1 - 0.8
Lipid-type emulsifiers	0.3 - 1.1
Stabilizing salts	0 - 0.15
Water	To make 100%

[1] From References 331, 339, and 342 to 345.

stabilizers, stabilizing salts, sugar, bodying agents, and the proper flavors and colors to produce products that resemble ice cream so closely that it is often difficult to distinguish between the natural and imitation products. Phosphate salts are used in most of these products to stabilize the protein system, as in other imitation or non-dairy products.[348,349]

Imitation Cheese, Cream Cheese, and Cheese Spreads

Replacements for all types of dairy cheese products can also be prepared from non-dairy products as well as from filled dairy products.[350,351] The proper formulation can result in products that, when properly processed, will resemble the natural product in all characteristics, including flavor and nutritional properties. The phosphates, including the mono-, di-, and tribasic orthophosphates, pyrophosphates, and polyphosphates, are useful in improving the physical characteristics of the proteins and in forming complexes with them to provide the proper dispersion so that the finished products will simulate the natural products.[351] The author is personally aware of experiments that have resulted in cheese products very similar to natural cheddar, natural Swiss, and natural cream cheeses. There is no reason to doubt that any type of natural cheese can be duplicated with synthetic, non-dairy products through proper formulation, including phosphate stabilizing salts.

Margarine

Even margarine has been found to be benefited by the addition of a phosphate. A Japanese patent[352] discloses that an oil-in-water-in-oil emulsion can be prepared by homogenizing a milk casein dispersion in a "trisodium polyphosphate solution" with hydrogenated oil to form a margarine product. The phosphate, assumed to be STP, most likely causes proper dispersion of the protein and inactivates any interfering calcium or magnesium ions in the system.

Applications in Cheese Products

The literature containing references to the applications of phosphates in cheese products is voluminous. The majority of the applications of the phosphates are in process cheeses, although they have also been found useful in the preparation of cottage and natural aged cheeses.

The initial steps in the manufacture of cheese products are similar for all types of cheeses including cottage cheese. Whole milk is usually used, although skim milk can also be used to produce low-fat cheeses. The pasteurized milk is treated with lactic acid bacteria or with lactic or phosphoric aicd, then usually with a protein-coagulating enzyme, causing the formation of a protein "curd," an insoluble paste or coagulum of the milk casein with the milk fat entrapped inside the curd in the form of emulsified droplets. The subsequent handling and processing of the curd determines the type of cheese that will be produced.[30]

Cottage Cheese

Merely washing the loosely packed, freshly coagulated curd produces "cottage cheese." Creamed cottage cheese is prepared by mixing the curd with whole milk often containing acids to provide tartness. Among the food-grade acids approved to be added to creamed cottage cheese is food-grade phosphoric acid.[353]

Bristol and Martin[354] reported that the use of H_3PO_4 or a mixture of citric acid and H_3PO_4 to preacidify skim milk along with an exponential-phase, noncoagulated culture of lactic acid bacteria reduced the setting time of cottage cheese curd by 44% as compared to the conventional procedure. Their method was faster than the conventional process, while it retained the natural cultured flavor.

A method of continuous processing of cottage cheese has recently been developed and is being promoted, particularly in European countries, because it requires less labor, provides greater uniformity, and increases the yield of cheese from the milk as well as the yield per hour of operation, resulting in a lower cost curd.[355] The process involves direct acidification rather than inoculation with lactic acid bacteria. The continual increase in operating and labor costs will probably result in the eventual replacement of all other processes for manufacture of cottage cheese by the continuous acidification process.

Natural Cheese

The product called "natural cheese" is produced by microbiological fermentation of milk curd. This involves a number of steps in preparation of the curd and a process requiring a considerable amount of time. The curd is finally compressed in forms into a compact mass fitting the shape of the form, then aged for various lengths of time to "ripen." In the ripening process the typical flavor of the cheese is developed by the fermentation of the curd, often by very specific cultures of bacteria or molds.

Phosphates have been found beneficial in a number of the steps involved in the manufacture of natural cheese. Kielsmeier[356] reported that the time required to produce natural cheese curd could be shortened through the use of calcium and magnesium sequestering salts, especially TSPP. In the conventional process a bacterial culture is added to the pasteurized milk for a period of time, then rennet is added and allowed to react. The cheese-milk is subsequently aged, during which time acid forms to lower the pH of the milk and coagulate the calcium or magnesium salt of the milk protein as curd. Kielsmeier found that by adding a calcium complexing agent, such as TSPP, to the milk, along with the bacterial culture and rennet, allowing it to react for 10 to 20 min, then adding a food-grade acid, such as lactic or phosphoric acid to lower the pH to 4.9 to 6.0 and preferably 5.4 to 5.5, he produced a cheese curd that could immediately be compressed into forms and could then be aged. The process was reported to lend itself to continuous cheese processing because of the short time involved in forming stable curd. Calcium chloride could be added along with the acid during the latter part of the process in order to provide firmer curd, if needed.

The addition of calcium salts to cheese curd has been known to form a more solid curd, a great advantage in natural cheese preparation. Vorob'ev[357] reported that the addition of 0.03% MCP to milk used in production of Edam-type cheese reduced the pH of the milk to a greater extent than calcium chloride and thus produced a higher yield and harder curd, leading to an improved grade of cheese. Cheese curd was also reported to be produced in less than 45 min by direct acidification of the milk, compared to 10 to 12 hr by the conventional process. Although food-grade hydrochloric acid could be used at lower cost, phosphoric acid was found to increase yield, decrease moisture, and increase the calcium level of the cheese.[358,359] Research is now being conducted to find a continuous process for the production of cheese curd in considerably less time than now required, resulting in lower production costs.[359]

Phosphates have been found to affect the rennet coagulation time of milk. Calcium and magnesium ions are required for coagulation of milk curd by rennet.[356] The addition of more calcium ion reduces the rennet coagulation time of milk,[360-364] while addition of citrate and/or polyphosphate salt stabilizes the rennet coagulation time, probably through formation of calcium and magnesium citrate and/or polyphosphate complexes. Orthophosphate salts, however, have no such effect on the rennet coagulation time.[360,361,363] Odagiri and Nickerson[219,362] reported results of their studies indicating that rennet coagulation time was prolonged logarithmically in proportion to the amount of the calcium-complexing agents added. Their studies included the effects of hexametaphosphate, oxalate, citrate, and EDTA as the complexing agents. Hexametaphosphate had the greatest effect, while citrate, oxalate, and EDTA, respectively, were less effective. The addition of more milk solids to the milk decreased the effect of hexametaphosphate on rennet coagulation time, indicating that rennet coagulation time was affected by factors other than the addition of complexing agents alone.[362]

deMan[363] reported that the addition of MSP to skim milk reduced the pH and slightly reduced the rennet coagulation time. Addition of DSP increased the pH and had no measurable effect on coagulation time. The addition of pyrophosphate or tripolyphosphate increased the pH, while tetrapolyphosphate and hexametaphosphate had no effect on pH; all polyphosphates greatly increased rennet coagulation time. The addition of 40 mg or more of polyphosphate phosphorus per 100 ml milk prevented any coagulation. deMan also reported finding less soluble calcium in the skim milk treated with DSP and polyphosphates; he found a higher level of soluble calcium after the addition of citrate. This seems to contradict an earlier work,[360,361] in which he and co-workers reported that orthophosphates did not affect calcium-ion concentration. Samuelsson[364] reported that the basic cheese-making characteristics of a milk depend on its ability to react with rennet and to form a stable curd and also on the ability of the curd to release whey. He reported that detrimental effects on these characteristics were produced by high pasteurization temperature, dilution with water, and the addition of sodium chloride, DSP, or TSP. Beneficial effects on curd formation were produced by increasing casein content, low pasteurization temperatures, ripening of the cheese milk, and adding acid, calcium chloride, or a combination of calcium chloride and DSP.

The phosphates have been found to affect the ripening of the cheese milk by the various bacterial cultures required to produce the typical flavor of the cheese being manufactured and the acidity required, along with rennet, to form the proper calcium and magnesium casein curd. Shew[365] and Potter and Nelson[366] discovered that bacteriophage, which could inhibit the growth of bacteria used in cheese processing, was dependent on calcium in the medium. Calcium-deficient milk media that would still grow the cheese culture bacteria were therefore developed. Babel[367] confirmed that the calcium-deficient media would prevent growth of the bacteriophage and still allow the necessary cheese-producing bacteria to grow. Hargrove[368] developed a calcium-deficient milk containing 2% of a 3:2 mixture of MSP and DSP. It had a pH of 6.5 that would prevent the growth of bacteriophage but allow growth of the necessary lactic *Streptococcus* in the medium. He added a sterilized solution of the concentrated phosphate mixture aseptically to sterilized milk to prepare the starter culture for addition to the cheese milk. Nine strains of the lactic *Streptococcus* having different bacteriophage sensitivities grew normally in the phosphated milk. Hargrove reported that the activity of 14 lactic strains, 4 commercial starter strains, and 12 *Leuconostoc* strains was not inhibited in the phosphated milk media. The starter cultures, when added to cheese milk, developed normal levels of acid, normal cheese texture, and normal flavor. Olson[369] reported that certain phosphate salts were not completely successful in preventing bacteriophage infection of cheese starter cultures. He found 0.5% ammonium oxalate precipitated the calcium more completely and allowed more successful propagation of phage-free cultures. In a detailed study of the effects of phosphates on bacteriophage inhibition, Hargrove and co-workers[370] reported that the orthophosphate salts were most effective in inhibiting bacteriophage growth under rather specific conditions. The minimum level found effective in inhibiting phage growth was 2% orthophosphate, but more resistant bacteriophage types required as high as 3%. They also reported that it was essential to heat the milk after the phosphate was added to bind most of the free calcium. The calcium content of the treated milks ranged between 10 and 30 ppm. The authors also reported that the optimum pH for maximum effectiveness ranged between 6.4 and 7.0. The most effective combination of conditions from the standpoint of economy, phage inhibition, and minimum precipitation of milk constituents was found to be 1.7% orthophosphate heated in milk at pH 6.6, followed by the addition of 0.3% pyrophosphate. This combination allowed the growth of 13 different strains of phage-contaminated lactic acid producing organisms within 3 to 4 subcultures in treated milk. The cultures were usually more active in the phosphated milk than in untreated milk.

Galesloot and Hassing[371] studied the growth of starter cultures for cheese production when added to skim milk, reconstituted decalcified milk, and skim milks containing 0.5% ammonium oxalate, 1% polyphosphate, 1.7% and 3.0% orthophosphate, or 0.3% pyrophosphate as calcium-binding agents. The various starter media were inoculated with 0.5% cultures of 3 *Streptococcus* and strains of *Leuconostoc* bacteria. The decalcified milk and those milks treated with calcium binding agents decreased the rate of citric acid fermentation and the number of cells of *Leuconostoc* strains, but they did not decrease the growth of at least one of the *Streptococcus* strains. Kadis and Babel[372] studied the effects of ammonium oxalate, several sodium salts of EDTA and STP as calcium binding agents in milk to prevent growth of bacteriophage. Although ammonium oxalate was effective at a level of 0.5%, it was not effective against the bacteriophage at a level of 0.3%. The ammonium oxalate allowed the growth of the *Streptococci*. The sodium salts of EDTA interfered with the growth of the *Streptococci* and did not prevent growth of the bacteriophage. A level of 1% of STP slowed the growth of the *Streptococci* and completely prevented growth of the bacteriophage in five of six cultures tested. Thus, 0.5% ammonium oxalate was more effective than 1% STP. The report[372] does not mention reheating the STP-treated milk as described by Hargrove et al.,[370] which may explain these results.

As a result of his work, Hargrove obtained a patent for the use of sodium, potassium, and ammonium phosphates and pyrophosphates as calcium-binding agents in milk.[373] The patent covers the use of these calcium-binding agents in milk that is subsequently heated to obtain maximum calcium binding. The preferred process calls for (1) the sterilization of milk and of solutions of the phosphates, (2) the aseptic addition of the necessary amount of phosphate solution to the milk in order to provide the desired level of phosphate, (3) reheating the milk-phosphate mixture, and (4) inoculation with the proper cheese-producing strain of bacteria. The invention was reported to prevent phage growth in bulk starters used for manufacture of various types of natural cheese, acidified butter, buttermilk, sour cream, yogurt, and similar dairy products.

Phosphates have also been found useful during the ripening or aging stage of cheese production. Natural cheeses normally are cured between 55 and 60°F and at the maximum possible humidity. According to Waters,[374] the time required to cure cheese can be reduced by maintaining maximum temperature and humidity conditions during curing. However, when blocks or wheels of natural cheese are aged at humidities greater than 95%, the surface of the cheese is subject to molds, rind rot, and excessive growth of cheese mites. Waters reported that it was possible to prevent growth of mold, rind rot, bacteria, and cheese mites at humidities of 96 to 97% if the cheese was periodically turned and the rinds were washed with a saturated solution of DSP.

Schulz[375] reported the development of a protective surface about the cheese. His process involved the development of a layer of process cheese on the outside of the natural cheese. This was accomplished by either immersing the block or wheel of cheese in a 20% aqueous solution of SAPP heated to 90°C or rubbing the surface with SAPP and subsequently heating the surface of the cheese at 100°C in a steam tunnel for a short time. The process was reported to produce a stable, airtight, edible rind that was similar to process cheese and increased the keeping quality of the cheese. Conochie and Sutherland[376] also reported methods of improved protection for cheese. Their process involved soaking cheese press cloths in a 2% solution of sodium tetrapolyphosphate at pH 6.26 to prevent adhesion of the press cloth to rindless cheese, as occurred during the usual practice in which the press cloth was soaked in sodium chloride solution.

Pasteurized Process Cheese Products

Pasteurized process cheese, pasteurized process cheese food, and pasteurized process cheese spread are commonly sold in the U.S. and in European countries. They are prepared by blending cheeses of various ages (e.g., large amounts of low-cost, fresh cheese curd with small quantities of the more expensive, highly aged cheeses) with water, salt, and emulsifier or plasticizer salts. The process is carried on in steam-jacketed kettles or in special processing units into which steam is injected directly to produce temperatures of 160 to 170°F to melt the cheese into a smooth, plasticized, fluid mass. The molten mixture can then be poured into containers for preparing process cheese loaves or allowed to flow over chilled drums or chilled, continuous, stainless-steel belts for solidification and cutting to produce sliced, process cheese.[30] J. L. Kraft obtained the initial patent for the preparation of a "sterilized cheese" in 1916.[377-380] Sommer and Templeton[378] attribute the development of pasteurized, process cheese by Kraft to the increased demand for food products, the necessity for maximum food preservation, the avoidance of waste, and the need for efficient distribution of foods during World War I. An excellent history of the development of pasteurized process cheese was published in their paper. According to these authors, one of the earliest patents claiming the emulsifying properties of sodium phosphate was that of Garstin, also mentioned in Irvine and Price.[380] Garstin[381] claimed the use of 5% sodium phosphate in cheddar and English dairy cheeses as well as other varieties to prevent fat separation in cheese processing. Irvine and Price[380] have provided an excellent summary of the major patents covering use of cheese-emulsifying salts prior to 1955.

The popularity of pasteurized process cheese has grown in direct proportion to the popularity of cheeseburgers, fondues, cheese sauces, and other food dishes requiring a cheese that will melt smoothly and uniformly. Most of the sliced, pasteurized, process cheeses now being marketed in the grocery store dairy case are prepared with either DSP or alkaline SALP emulsifying salts. These salts, in proper levels, and sometimes combined with TSP or sodium citrate, create the most acceptable melt and texture characteristics in the slice of process cheese. Recent reports indicate that cyclic sodium tetrametaphosphate, but not trimetaphosphate[382] and a starch phosphate,[383] would function as cheese emulsifiers. The adverse effects of rework (salvage) cheese can be overcome by the use of a combination of lipid-type surface active agents and phosphate emulsifying salts according to another recent report.[384] In addition to their functions as buffers that adjust the pH and maintain it in an optimum range, these phosphate salts have highly desirable and essential effects on the physical properties of the cheese protein, suggesting that their functions are more than just maintaining optimum pH values.[24,30,110,111,384-386]

The Federal Standards of Identity for pasteurized, process cheese[387] allow the addition of up to 3% of the following emulsifying salts or any combination based on the weight of the pasteurized, process cheese: MSP, DSP, DKP, TSP, sodium metaphosphate (SHMP), SAPP, TSPP, SALP, sodium citrate, potassium citrate, calcium citrate, sodium tartrate, and sodium potassium tartrate. They also allow the use of H_3PO_4, as well as other food-grade acids, as acidifying agents in pasteurized, process cheese.

The emulsifying salts may be added to the blended cheese at the time it is being ground in order to obtain an intimate mixture before the cheese goes to the cooking kettle.[379] The most common practice, however, is to add the emulsifying salts directly into the cooker along with the first portion of the ground cheese. Salt and other dry additives are usually added with the emulsifying salt. Color and any water also needed are then added, and the cheese is cooked by means of live-steam injection until the melt reaches the proper temperature, usually between 140 and 190°F. The bulk of the cheese processed in the U.S. is heated to a temperature between 160 and 170°F, the optimum temperature for the soft cheeses preferred in this country. Once the cheese has reached the desired temperature, it is discharged from the cooker to the packaging line.[30,378,379]

Bolanowski and Ziemba[388] described a continuous process for the manufacture of pasteurized process cheese. The raw cheese mixture is ground, pumped to a continuous, swept-surface heat exchanger, where it is cooked while a solution of the phosphates, salt, color, and other additives is continuously metered into it. After cooking, the cheese is pumped to the packaging line.

The cheese "emulsifying salts" are commonly phosphates with some use of citrates and, very infrequently, tartrate salts. These salts do not function strictly as true emulsifiers, such as the lipid-type emulsifiers. Instead they react with and change the physical properties of the protein so that it forms a very smooth, uniform film about each fat droplet in the processed cheese, thus preventing the separation and bleeding of fat from the cheese. It is possible to select and blend cheeses in such a way that fat separation is a minor problem without the addition of emulsifying salts. However, a manufacturer must have a large supply of raw cheese in order to make these selections, and it is very difficult to predict whether the blend will require assistance of emulsifying salts or not until it is actually processed. Therefore, it is common practice to add emulsifying salts to all batches of process cheese. In addition to preventing the bleeding of fat from the cheese, these salts also provide smoother flow properties so that the cheese is easier to process as loaves or as slices, particularly in ribbon slice operations. The cheese also acquires more desirable melting and slicing properties when used by the consumer than it would have without the assistance of the emulsifying salts. The type of salt used can also affect the "body" or firmness of the final cheese. This is very important when loaves of cheese are to be sliced because a cheese that is too soft would be very difficult to slice. A cheese that is too firm may also have an undesirable hardness when eaten.

Many investigations of the functions of the emulsifying salts in cheese have been reported in the literature. The most extensive investigations appear to be those reported by Templeton and Sommer.[378,389-396] These authors preferred the body and characteristics of the cheese when emulsified with sodium citrate over that of any other emulsifiers. However, when the properties of cheese emulsified with DSP and those prepared with sodium citrate were compared, the authors apparently concluded that the additional cost of the citrate was not warranted. They reported that DSP produced a weaker body in the cheese that they felt was undesirable.[389,390] The tartrate salts were found to produce "sandiness" in processed cheese; this sandiness was identified as crystals of calcium tartrate large enough to be detectable.[394-396] In addition, the tartrates were reported to produce brittle, mealy cheese that had poor slicing properties. The sodium polyphosphates produce acidic-flavored cheese and also produce cheese with greater firmness than when sodium citrate or DSP was used. Sodium citrate, tartrates, and DSP and TSPP, added at a 2% level, showed the most satisfactory results and were the most tolerant to variations in the blends of the cheeses used, in their judgment.[378] Templeton and Sommer[389,393] also found that phosphate emulsifying salts accelerated the darkening of tin foil if the finished processed cheese was more alkaline than pH 5.8 to 6.3. A study of longer-chain polyphosphates, such as tetrapolyphosphate and longer, showed that the cheese first had a firm texture that became mealy on aging.[391] The polyphosphates were found to dissolve incompletely in the cheese and resulted in lower pH values and higher levels of titratable acidity that caused fat leakage. Furthermore, cheeses produced with the polyphosphates did not melt, even at temperatures as high as 41°C., although such cheese is preferred in Europe.

Cheeses processed with different emulsifying salts were studied under the microscope by Templeton and Sommer[396] who reported that the fat globules of polyphosphate-treated cheese remained large and irregular in shape. They also reported that cheese processed with DSP had irregular textures and rapidly lost fat.[396] Cheeses emulsified with sodium citrate and TSPP were found to have small fat globules and a fine texture,[394] not corresponding with present judgments.

Palmer and Sly[397] also reviewed the use of various emulsifying salts in cheese processing. These authors preferred sodium citrate too, and recommended 3% of a mixture of 9 parts citrate to 1 part DSP, while maintaining the pH of the cheese between 5.3 and 5.4. They proposed some theories for cheese emulsification based on the fact that (1) process cheese is an oil-in-water suspensoid that becomes a stabilized emulsion consisting of hydrated casein and the emulsifying salt; (2) the fat is present as oil droplets that carry a negative charge, and the emulsifying salts and protein are adsorbed at the oil-water surfaces, thereby forming charged surfaces that prevent coagulation of the oil droplets; and (3) the functions of the emulsifying salts are to disperse or solubilize the protein (casein) and to provide hydrophilic cations and a lipophilic anion so that the emulsifier molecule orients itself at the oil and water surfaces along with the casein. As a result of their studies on cheese emulsifiers, Holtstorff and co-workers[398] demonstrated that all good emulsifiers have polyvalent anions, form alkaline solutions, and are calcium-complexing agents. Not all compounds that have these characteristics are good cheese emulsifiers, however. Conventional lipid-type emulsifiers, such as the polysorbates, actually act as de-emulsifiers in many cases. These authors also treated cheese with proteolytic enzymes in an attempt to determine whether cheese emulsification depended on hydrolysis or partial hydrolysis of protein. The action of proteolytic enzymes did not produce emulsification. The fact that the quantity of water-soluble nitrogen did not change between processed and unprocessed cheese provided further evidence that protein breakdown was not a factor.

Ruff[25] also studied the emulsification of processed cheese. He reported that, when natural cheese is prepared, the casein system of milk is converted to paracasein by the action of the enzyme rennet. A gel that is formed by the reaction between paracasein and calcium ions becomes the cheese curd. The calcium ions form cross-links between the molecules of paracasein by forming salt bridges between free H_3PO_4 and carboxyl groups in the paracasein molecules. As the natural cheese is aged, the large paracasein molecule aggregates, linked to each other by calcium, disintegrate by peptization to form a finely dispersed system.

When natural cheese is used to prepare process cheese, the emulsifying salts must convert the calcium paracaseinate system to a sodium paracaseinate system by forming colloidal, insoluble calcium phosphate salts and soluble, highly dispersed sodium paracasein salts. Such a system forms a highly stable, smooth emulsion that will not easily release its fat. Ruff reported that, although citrates are being used for this purpose, they are not as effective against the strongly cross-linked paracasein molecules. He reported that polyphosphates gave the optimum characteristics to the cheese, a reflection of the European preference for very firm cheese products with little or no melt properties.

Kiermeier and Mohler[399] and Kiermeier[400] also preferred the polyphosphates as cheese-emulsifying salts. Their studies demonstrated that polyphosphates, even though they changed the pH of the original cheese system very little, caused swelling of the casein that had been precipitated by rennin similar to that obtained when the casein system is treated with strong alkali. Sodium citrate had no such swelling effect on the casein molecule. These authors also found that citrate-treated cheese was more susceptible to microbiological spoilage than cheese treated with the polyphosphates; they speculated that the polyphosphates may have some type of bacteriostatic action.

Roesler[401] also preferred the polyphosphates as cheese-emulsifying salts. He studied the hydrolysis of the polyphasphates in the cheese with time, and found that they were rapidly hydrolyzed to ortho-, pyro-, and tripolyphosphates during the emulsification process. Upon storage, he found that the polyphosphates no matter what chain length was used, hydrolyzed completely to orthophosphate within two months in high-moisture cheese spreads and soft cheeses, while tripoly- and tetrapolyphosphates could be isolated from loaf cheeses after 80 days' storage.

Scharpf and Kichline[402] investigated the stability of long-chain sodium polyphosphates (Graham's salt) in aqueous cheese extracts. When the mixture was stored at 3 to 7°C for four weeks, the P_2O_5 remained unchanged. Similar mixtures stored at 20°C showed some hydrolysis of the long-chain polyphosphates, which decreased from 89% to 64% during a four-week period, while orthophosphates increased from 4% to 27%. As a result, commercial mixtures of IMP (insoluble metaphosphate) and DSP or TSP are being offered for use in pasteurized, process cheese. These mixtures contain between 40 and 70% IMP and between 30 and 60% DSP or TSP. When 3% is added to cheese, the mixture contributes between 1.2 and 2.1% of the DSP or TSP. The IMP acts as an inert diluent, as it can be recovered from the process cheese as insoluble particles that have the chemical and physical characteristics of the original IMP.

The appearance of inorganic salt crystals in pasteurized, process cheese has been investigated by several workers. Rank and Siebenlist[403] reported that sandiness in pasteurized, process cheese was due to crystals of calcium phosphate. These crystals do not appear in process cheese under normal circumstances, because the calcium phosphate forms amorphous, crystalline particles unless the cheese is processed at high temperatures or stored for prolonged periods of time. Blanchard[404] found crystals which appeared in process cheese to be those of calcium tartrate. It is assumed that this cheese had been emulsified with tartrate salts. This was confirmed by Russian workers[405] who observed that fairly large, visible crystals of tartaric acid appeared on the surface of processed cheese 3 to 5 days after manufacturing when the cheese was held at refrigerated temperatures. Storing the cheese at 25 to 30°C caused the crystals to vanish, but they again reappeared when the cheese was cooled. The authors recommended that tartrates should not be used for emulsifying process cheese.

Morris and co-workers[406] were able to identify calcium citrate crystals as a white, powdery deposit on the surface of processed cheese slices in cheeses emulsified with 0.4% sodium citrate and 1.6% DSP. Elimination of citrate from the processed cheese formulas prevented appearance of the crystals.

The quantity of DSP normally used in pasteurized, process cheese ranges between 1.5 and 2.25% of the anhydrous salt based on the weight of the finished cheese. TSP is usually added in mixtures with other emulsifying salts to raise the pH of acidic cheese blends and, in those few instances in which it is the only emulsifier, is seldom added at levels above 1.5% of the anhydrous salt. The use of orthophosphates at levels above this range leads to the formation of small clusters of calcium phosphate crystals that often form on the surfaces of the cheese. These are especially noticeable on the surface of sliced, pasteurized, processed cheese and give the appearance of small spots of mold. Although the crystals are completely harmless, consumers object to them, thinking that they truly are mold spoilage of the cheese.

This author has had personal communications with investigators in research laboratories of several cheese processors that confirm that crystals of DSP can appear on the surface of sliced, process cheese. It is, therefore, customary not to use levels of DSP above 2.25, and this may even be too high in some types of cheese containing high levels of calcium. These levels also are limits from the standpoint of flavors, since higher levels would cause an astringent flavor in the cheese.

Crystals of phosphate are completely eliminated by the addition of 3% of alkaline SALP to process cheese,[385] a fact confirmed by numerous production studies. In addition to preventing the development of phosphate crystals, the SALP-emulsifying system eliminated the development of astringent flavors during the normal storage period of the process cheese. Comparison of cheeses emulsified with DSP and SALP has demonstrated that, after two months, the DSP-emulsified cheese has a less acceptable flavor than that containing SALP.[29,110] The SALP-containing cheese has a richer, "cheesy" flavor that had not changed from its flavor when first processed; the cheese prepared with DSP or other orthophosphate salts had distinctly lost some of its original rich cheese flavor and developed an astringent background flavor. As a result, SALP has replaced most of the orthophosphates as the emulsifying system in sliced processed cheese, particularly in the ribbon slicing and individually wrapped, sliced, process cheeses.

Pasteurized, process cheese foods and spreads, as defined by Parts 19.765 through 19.781 of Title 21, *Code of Federal Regulations*, contain higher moisture (above 40% but not more than 44% moisture in cheese foods, and above 44% but not more than 60% moisture in cheese spreads) than pasteurized cheese defined in Part 19.750 of the same reference. The same emulsifying salts that are approved for use in these products are also approved for the pasteurized process cheese. A very brief discussion of cheese spreads and their preparation is found in a paper by Sommer and Templeton.[407]

Fat-free and low-fat, pasteurized, process cheeses also depend on emulsifying salts for their processing. Russian researchers[408] found that 0.1 to 0.15 g DSP or 0.08 to 0.09 sodium citrate was required to dissolve 1 g of fat-free cheese curd. The optimum pH range for solubility in DSP was found to be 6.23 to 6.53, and in sodium citrate, 5.74 to 5.83. Calcium phosphate compounds formed in the gel structure of the cheese curd when DSP was used and gave the appearance of incomplete solubility, although the cheese protein must have been completely dispersed. Allowing the cheese curd to swell in a solution of the emulsifying salt prior to production of the cheese promoted complete solubility. The viscosity of the gel formed during processing depended on the duration of the swelling, the temperature of processing, and the ratio of soluble to insoluble protein in the cheese treated with DSP. A low-fat, pasteurized, process cheese spread was reported by Hayter and co-workers[409] that contained 7.0% fat, 32.8% solids-not-fat, 1.7% sodium chloride, and 58.5% moisture. This low-fat, pasteurized, process cheese spread was prepared by using 40% skim milk Dairiworld cheese, 20% aged cheddar cheese, 4% nonfat dry milk, 4% dried sweet whey, and 2 to 2.5% DSP. Optimum body in the product was obtained when Dairiworld cheese was cured for 10 days before use. Longer curing resulted in decreased firmness and increased stickiness. The use of 1.5% DSP produced a mealy, gummy body; combinations of SHMP or TSPP with the DSP at a total of 2% emulsifying salt mixture increased the firmness and maintained quality. However, when 0.5% TSPP was added as a portion of the emulsifying salt, the product developed an objectionable, emulsifier flavor. The recent emphasis on the effects of high-fat diets in heart disease has created intense interest in the development of low-fat cheeses. The phosphate emulsifying salts, especially the SALP salts, promise to enable processors to develop numerous formulas for low-fat or no-fat, pasteurized, process cheese.

Phosphate Applications in Egg Products

Phosphates have been found to be useful in processing and improving the functional properties of eggs for use in food products. Their functions can be summarized to include complexing unwanted metal ions, buffering an egg-containing system to a pH optimum for some technical result, inhibiting proteolytic activity in the egg product, improving foam volume and stability, sensitizing *Salmonella* to heat pasteurization, and inhibiting oxidation of yolk fats.

Applications in Shell Eggs

Shell eggs, when collected from the nest in which they have been laid, often are contaminated on the surface of the shell by soil, feathers, nest materials, broken egg residue, and other similar substances. Farmers and egg handlers normally wash the soiled eggs to increase their acceptability to the public and to decrease spoilage. During a study of some of the problems involved in cleaning eggs and preventing an increase in the rate of spoilage, Pikarr[410] discovered that iron in the wash water contributed to more rapid spoilage, even though it was as low as 5 to 10 ppm. The spoilage results from the formation of iron complexes with the egg's conalbumin, the protein that acts as a protective mechanism against contamination entering through the shell. Pikarr discovered that the addition of iron-complexing acids to the wash solution would prevent the formation of the iron-conalbumin complex and thus increase the stability of the washed egg. Among the available food acids he found that citric, phosphoric, glutamic, acetic, and tartaric acids were excellent iron-complexing agents; a combination of any two of these acids was even better in this application. (The acids are listed in the order of descending preference.) By including a small amount of a soluble, acidic phosphate salt in the wash solution in addition to the acid, Pikarr improved its ability to complex the iron and protect the egg. Among the preferred soluble acid phosphates named were the monoalkali orthophosphates and dialkali metal dihydrogen pyrophosphates. The wash solution was compounded to contain between 0.1 and 5%, usually between 0.1 and 2% by weight, of the acid and sufficient quantities of the alkali or ammonium salt of the phosphate to increase the initial pH to approximately 3.5 but not above pH 5.

Applications in Whole-Egg Products

When whole eggs or mixtures of egg white and egg yolk in any proportion are allowed to stand at room temperature or are refrigerated for a period of time, the characteristic golden-yellow color of the egg composition gradually turns to a muddy brown. This muddy-brown color can even be transferred into scrambled eggs made from the yolk and white composition. Chin and Redfern[411] discovered that the addition of H_3PO_4 or polyphosphoric acid, or any one of the alkali metal or ammonium salts of these acids, would prevent the formation of this disagreeable color and would restore the golden-yellow of liquid eggs that had turned muddy brown. The quantity of the phosphate compounds needed is such that it will reduce the pH to a range between 5.5 and 7.0. According to the authors the most desirable color is noted between pH 6.4 and 6.6 when using H_3PO_4. Between 0.3 and 0.7% of the phosphate salt appears to be optimum for preservation or restoration of the golden-yellow color. As a result of this work, the Federal Standards for whole-egg product, whether liquid, frozen, or dried, were revised in order to allow the addition of up to 0.5% MSP for color preservation. The standards were again modified in 1968 to allow the addition of up to 0.5% MKP for a similar purpose.[412]

Liquid egg products, egg whites, egg yolks, or whole eggs, after removal from the shell, will undergo proteolytic action and coagulation, especially if frozen eggs are accidentally thawed, or upon heating eggs to dry them. Hall[413] reported that the addition of small quantities of a polyphosphate with a chain length of two or greater, and preferably three or more, stabilized the eggs to this proteolytic action and subsequent coagulation even though the eggs were heated to between 160 and 180°F. The effective ratios of phosphate to egg product were reported to be very wide, ranging between 1:20 and 2½:1.

Bellamy[414] obtained a patent on the use of SAPP to be added to egg whites separated from the yolks of whole eggs, allowing the egg whites to ferment by the natural organisms in them until the pH has been lowered to between 6.3 and 6.5, evaporating them to about one fourth of their original volume, then reconstituting them with the equivalent amount of egg yolk that had been removed from them when they were broken out of the shell. This reconstituted and concentrated whole-egg product was reported to have superior whipping properties and foam stability.

Lewis and co-workers[294] discovered that the addition of tripolyphosphate or hexametaphosphate to foams made from dried, whole eggs improved their whipping ability and stabilized them to deterioration during storage. These effects are shown in Tables 22 and 23. The effective levels were found to be between 1.3 and 2.0% STP and 2.3 to 3.0% SHMP. STP gave the greatest improvement and was especially noticeable in improving the foaming properties of dried, whole eggs stored at room temperature for three months, while the controls would not foam at all.

Tongur[415] reported that oxidation of the fats in dried, whole eggs was inhibited by intimate blending of the powdered eggs with 0.01 to 0.03% ascorbic acid and that the effect of the ascorbic acid was significantly improved by the further addition of 0.01 to 0.05% TSPP.

Salmonella in whole-egg products has been reported to be sensitized to heat and destroyed at lower pasteurization temperatures by the addition of 0.2 to 2% SHMP prior to pasteurization.[416] However, when the eggs were heated to the pasteurization temperatures of 51.5 to 63°C (125 to 145°F), they became discolored and, in some cases, coagulated. It was discovered that the addition of between 0.075 and 0.75% of soluble calcium or zinc salts, or mixtures thereof, would overcome this defect. A pasteurization process involving the use of 0.5% SHMP, 0.5% calcium chloride, and a temperature of 130 to 135°F was proposed. The pasteurized, whole-egg products were reported to have superior functional pro-

TABLE 22

Effect of Polyphosphates and Storage on Foam Volume of Whole Egg Meringues[1]

Kind of egg	Storage time (weeks)	Storage temperature	Polyphosphate used	Optimum phosphate in dried egg	Maximum volume increase
		°C		%	%
Fresh	None	–	–	–	560
Dried No. 1	None	–	None	–	200
	None	–	Hexametaphosphate	2.3 - 3.0	330
	None	–	Tripolyphosphate	1.3 - 2.0	360
Dried No. 2	None	–	None	–	220
	None	–	Tripolyphosphate	1.5	380
	13	4	None	–	220
	13	4	Tripolyphosphate	1.5	380
	13	25	None	–	0
	13	25	Tripolyphosphate	1.5	160

[1] Source: Lewis, M. A., Marcelli, V., and Watts, B. M., *Food Technol.*, 7, 261, 1953, Table 3, with permission.

TABLE 23

The Effect of Polyphosphates on Whipping Ability of Whole Dried Egg in Sponge Cake[1]

Kind of egg	Polyphosphate in dried, whole egg	Weight of one cup of batter, g	Volume of cake from 175 g of batter, ml
Fresh		62	980
Dried – 4% Moisture	None	117	435
	2.9% Hexametaphosphate	90	660
	4.6% Hexametaphosphate	97	645
	1.5% Tripolyphosphate	84	710
Dried – 4.5% Moisture	None	110	525
	2.3% Hexametaphosphate	81	815
	1.5% Tripolyphosphate	74	895

[1] Source: Lewis, M. A., Marcelli, V., and Watts, B. M., *Food Technol.*, 7, 261, 1953, Table 4, with permission.

perties over those of whole eggs pasteurized by other available processes.[416] The peroxyhydrates of alkali metal salts of phosphates, sulfates, or carbonates are reported to be even more effective in the inhibition of *Salmonella* and other microorganisms in pasteurized egg products.[417]

Applications in Egg Whites

By Federal regulation all egg products, including liquid and dried whole eggs, liquid and dried egg whites, and liquid and dried egg yolks, must be free of *Salmonella* organisms. The federal standards for egg products have been amended to require the elimination of *Salmonella* organisms. Few eggs are used in food preparation that have not been treated by pasteurization or some other manner to eliminate *Salmonella*.

Pasteurization processes invariably affect essential properties of the eggs, such as whipping time, foam stability, and viscosity. Numerous studies have reported that these changes cause considerable difficulty in baked products, particularly those requiring light, stable foams. Egg whites are most affected. Since the egg whites used in angelfood cakes are essential to the cake structure, they are usually treated with some type of foam stabilizer or whipping agent. Kothe[418] discovered that the addition to egg whites of between 0.02 and 0.05% of an edible, unsulfonated ester of an aliphatic alcohol and an aliphatic polybasic acid significantly reduced whipping time and increased foam height and stability. The most commonly used ester is triethyl citrate; many of the pasteurized egg whites now commercially available contain this whipping agent. Kothe found that he could increase the tolerance of the treated egg whites to overbeating by adding sufficient MSP to reduce the egg white pH to approximately 8.0.

The whipping properties of dried egg whites are also affected adversely by the drying process. Finucane and Mitchell[419] discovered that alkali metal and ammonium hexametaphosphate salts could be added directly to the dried egg whites to provide highly improved foam properties. Not only were foam heights increased, but the foam stabilities were also increased. A level of approximately 2½% SHMP added to the weight of the dried egg white produced the desired effect. Finucane[420] extended this work, finding that the addition of between 7 and 15% of a partially hydrolyzed soy protein and between 2 and 7% SHMP improved the whipping properties of the egg white to such an extent that all of the ingredients of an angelfood cake mix could be included in a single mixing stage instead of the usual two stages. This innovation eliminated the need to whip the eggs, acid, and part of the sugar to a stiff meringue before adding the balance of the ingredients, including the flour and flavorings. Finucane also reported that the soy protein and hexametaphosphate additives reduced whipping time, increased foam height and stability, and increased tolerance to overbeating and to variations in the characteristics of the other ingredients.

Chang[421] and Chang et al.[422] reported that SHMP stabilized the heat-labile protein, conalbumin, when egg white was heated to 61.7°C (143°F) for 3 min. This stabilization prevented the development of turbidity in the pasteurized egg white. Although the foaming ability of the hexametaphosphate-treated egg white was somewhat reduced, the addition of a whipping agent, triacetin, restored the foaming power to that of the native albumin.

The foaming ability of egg white is sharply reduced when it is contaminated by egg yolk. Quantities of egg yolk as low as 0.5% severely reduce the foam height and stability of the egg white. Chang and co-workers[422] demonstrated that the addition of 2% SHMP would reverse this reduction in foam height and would stabilize the egg white to the presence of contaminating egg yolk. This discovery could lead to significant improvements in foamability of commercial egg whites, since it is very difficult to prevent contamination with the yolk during commercial egg-breaking processes.

Kohl et al.[423,424] reported that, by adding between 0.2 and 2.0% SHMP, based on the weight of liquid egg whites, and by adjusting the pH of the egg whites to between 8.0 and 10.0, up to 10^6 *Salmonella* were destroyed at pasteurization temperatures of 125° to 130°F with 0.5 to 10 min holding time. The process was reported to be especially effective against the most heat-resistant types of *Salmonella* and to completely destroy high levels of coliforms in the egg whites. The treatment also was reported to contribute residual microbiological inhibition in the pasteurized eggs so that bacteria surviving pasteurization and/or introduced by recontamination decreased in numbers even after thawing and holding at 75°F for 72 hr. The hexametaphosphate was believed to have sensitized the *Salmonella* and other bacteria so that they were destroyed at lower pasteurization temperatures than in egg whites pasteurized without the added phosphate.

Phosphate Applications in Fats and Oils

Various phosphates have been found useful in the processing and applications of fats and oils in foods. The phosphates have been found to aid in the extraction of fats and oils and their refining, in fat stabilization, and in obtaining stable emulsions.

Applications in Oil Extraction

Rousseau[425] reported that use of the Drupex process for extraction of olive oil from pastes, made by grinding and working the raw olives, increased the yields of olive oil from 1.5 to as high as 3% over those of untreated controls. Prior to grinding, the olives were treated with 2% of the Drupex mixture, which consisted of three parts TSP and one part sodium bicarbonate. The function of the phosphate was to increase protein dispersion or solubility in the aqueous phase and in this way to break up lipoprotein complexes that decrease oil yields. The ability of phosphates, particularly the polyphosphates, to solubilize protein should be highly useful in fat and oil extraction.

Applications in Processing Fats and Oils

Crude fats and oils, as they are extracted or expressed from vegetable or animal matter, contain a considerable number of impurities, including free fatty acids, phosphatides of various types, mucilaginous materials, protein from plant or animal tissues as well as soluble proteins, and similar contaminants. These impurities must be removed by refining the crude extracts. There are two general processes used in the refining of fats and oils – alkali refining and acid refining.

Alkali Refining

The alkali-refining process involves agitating the melted fat or liquid oil with a solution or suspension of an alkali at temperatures and times that vary with the desired results. The mixture of aqueous alkali solution and oil is then filtered or centrifuged and dried. After the dried oil is treated with a decolorizing agent, such as Fuller's earth or another decolorizing clay, it is filtered or centrifuged to remove the decolorizing agent. The oil is finally given a vacuum treatment to deodorize it and to make it a bland-flavored, odorless oil; it may also be hydrogenated during this process. Because fats and oils refined by the alkali process are neutral, very bland, and practically tasteless, they are suitable for use in foods. The alkali treatment forms soaps of any free fatty acids present and from the lecithin and other phosphatides and saponifies the sterols to make them soluble in the aqueous solution. They are, therefore, removed with the aqueous phase.

A disadvantage to the alkali-refining process is that it may produce refining losses of 2% or more. A portion of the refining loss is felt to be due to saponification of the desirable glycerides to form soaps and glycerine, which are soluble in the aqueous solution. The addition of inorganic phosphate salts to the oil during the alkali-refining process or the formation of soluble inorganic phosphates by use of phosphoric acid and subsequent neutralization with alkali has been reported to overcome these losses. Very low levels of phosphates in the range of 0.05 to 1% are necessary to reduce refining losses from 2% or more to less than 1%.[426-429]

Acid Refining

Sulfuric acid is most commonly used in the acid-refining process, although food-grade hydrochloric and chromic acids can also be used. Acidic salts of these acids have been used in some applications. By varying the proportion of the acid to oil, the concentration of the acid by dilution, and the temperature of the reaction, it is possible to obtain oils with levels of desired fatty acids within narrow ranges. This process has been used in the preparation of certain commercial shortenings in which some level of free fatty acids is sometimes desirable. When using this treatment, the oil is treated with the desired concentration of acid at a temperature that varies with the acid concentration. The oil is filtered in order to remove the aqueous phase, which contains hydrolyzed impurities, acid, etc. The oil must be thoroughly washed to remove every trace of the acid. After the acid is completely removed, the oil is bleached with one of the bleaching clays and deodorized as in the alkali-refining process.

Appleton[430] reported that the use of sulfuric acid, as is normal in the acid-refining process, resulted in a substantial amount of sulfonated triglycerides; these discolored readily upon heating and, therefore, rendered acid-refined oils inadequate for use as frying fats. Appleton found that he was able to use approximately 0.75% of 75% H_3PO_4, based on the weight of the oil at about 80°F, to accomplish an acid-refining process. The acid was easily removed from the oil by washing or could be precipitated by the addition of lime to form TCP, which could then be filtered from the oil. Eckey[431] reported that he was able to prepare fats with improved flavor and odor, with increased resistance to darkening, and with improved resistance to oxidative rancidity. He treated the oil with approximately 0.01% H_3PO_4 or less or with an acidic phosphoric acid salt or ester that provided an equivalent proportion of the P_2O_5. Others reported that treatment of the oils with H_3PO_4,[432] H_3PO_4 followed by an aqueous solution of DSP,[433] MSP and TSP solutions,[434] or small quantities of concentrated H_3PO_4 at high temperatures followed by neutralization with alkaline materials[435,436] would solubilize the protein, mucilaginous materials, phosphatides, and fatty acid contaminants and greatly improve the odor, flavor, and stability to rancidity of the treated fats and oils.

Where low levels of free fatty acids are desirable, the other contaminants can be removed by treating the fats and oils during refining with TSP[437-439] or with ammonium phosphate,[440] leaving small quantities of fatty acids. Another case in which some free fatty acids were desired was reported by Coith and Votaw.[441] Free fatty acids were formed by deliberate hydrolysis under controlled conditions with alkali and subsequent neutralization with sulfuric acid or H_3PO_4; this produced 2 to 10% free fatty acids having no less than 16 carbon atoms in their molecules. The acid-treated fat was washed with water, then dried and used in preparing a common commercial bakery shortening known as Sweetex®.

Other Refining Applications

Irmen[442] reported that he was able to prepare a fat from soft fats or liquid oils by treating the fat or oil with one of the phosphoric acid salts, such as TKP, TSP, or TSPP, along with higher alcohols and esters of high-molecular-weight fatty acids, such as waxes. The treated fat or oil separated to the surface of the mixture and hardened into a cake, which then could be removed. Irmen's process eliminated the need for hydrogenation to harden fats and oils for use as shortenings.

A treatment reported to have numerous advantages was developed by Beck and Klein.[443] It involved percolating or filtering the crude fat or oil through a bed of granular TCP. The treatment was reported to remove nearly all of the phosphatides, free fatty acids, colored contaminants, and the odors and flavors typical of the fat or oil source. By this process even deeply discolored, highly flavored crude oils could be refined to provide colorless, bland, flavorless fats and oils. In addition frying fats were found to have higher "smoke points" when refined in this manner.

Toxic alkaloids were reported to be removed by use of 1% manganese dioxide, aeration, then treatment with 1 to 1.5% H_3PO_4.[444] Alkali metal pyrophosphates, such as TSPP and similar alkali metal salts, were also reported to aid in the removal of sterol-containing contaminants.[445] The presence of heavy metal ions can cause considerable difficulty through their ability to catalyze oxidative rancidity in the unsaturated fatty-acid constituents of the triglycerides. Copper, iron, nickel, and tin, in lard, were effectively deactivated as oxidation catalysts by phosphoric, citric, tartaric, and ascorbic acids and several fatty-acid esters of ascorbic acid.[446,447]

Bleaching

The phosphates also have applications in bleaching or decolorizing fats or oils. Orthophosphoric acid (H_3PO_4) was preferred, although sulfuric and metaphosphoric acid could also be added to bleaching clays, according to Eckey.[448] Other bleaching processes utilize the following treatments:

1. treating with H_3PO_4 and hydrogen peroxide at 60 to 70°C for 0.5 to 1.5 hr,[449]
2. treating the fat or oil with H_3PO_4, $H_4P_2O_7$, or any of their acidic salts plus an inorganic silicone acid or its salt while air is bubbled through it,[450]
3. bleaching the fat or oil with TSPP in a solution of hydrogen peroxide,[451]
4. adding acids, especially H_3PO_4, to acid clays, which are then used to decolorize the fats or oils,[452]
5. treating the fat or oil with an alkali metal carbonate and H_3PO_4, and later adding a solution of hydrogen peroxide.[453]

Loury[454] reported that it was necessary to

remove metallic salts and hydrolyze mucilages in fats or oils prior to bleaching them. He was able to precipitate the metallic salts and hydrolyze mucilages by use of dilute H_3PO_4; these could then be removed prior to bleaching the fats and oils.

Rearrangement

Rearrangement, or interesterification, requires the use of sodium methylate as a catalyst, which must be destroyed before further processing of the fat or oil can be accomplished. Phosphoric acid (H_3PO_4) was found useful for this purpose, with subsequent removal of the acid and inactivated catalyst by neutralization with sodium compounds and washing.[455,456]

Hydrogenation

Phosphoric acid has been used to remove nickel catalysts from hydrogenated fats and oils.[457] Hydrogenated vegetable oils were found to lose the hydrogenation odors and to be stabilized against odor and flavor reversion by treatment of the hydrogenated fats with H_3PO_4 and a neutral bleaching earth and bubbling air through them at elevated temperatures.[458] H_3PO_4 has also been used in nonselective hydrogenation processes of fats and oils.[459]

Monoglyceride Preparation

Monoglycerides are prepared by (1) treating a fat or oil with additional glycerine dissolved in caustic soda, (2) heating, and (3) agitating the mixture until all the glycerine has been esterified by transfer of fatty acids from the triglycerides of the fat. Such compositions, containing mono- and diglycerides, must again be bleached to remove undesirable colored compounds. This can be accomplished by treating the mixture with filter clays and H_3PO_4.[460] It is also necessary to neutralize the sodium hydroxide used as a catalyst; a combination of H_3PO_4 and glycerol has been used to accomplish this neutralization.[461]

Other Processing Applications

The presence of glyceride polymers causes foaming, discoloration, and development of off-odors in frying fats. The polymers can be removed by treating a fat or oil with a combination of phosphoric acid and bleaching clay.[462] Mixtures of filter aids and H_3PO_4 could be used by processors of fried foods to filter their frying fats at the end of each day's use and thus provide additional protection against these polymers in the frying oils.

Applications in Fat Antioxidant Systems

Dry Fat Systems

Eckey[431] first reported that combinations of tocopherol and H_3PO_4 and/or its acidic salts exhibited antioxidant activity in dry fats. He also reported increased stabilization when H_3PO_4 was used in bleaching fats or oils.[448] Richardson et al.[463] confirmed that dry lard and cottonseed oils could be stabilized against oxidative rancidity by saturating them with solutions of H_3PO_4, sulfuric acid, or their acidic salts. Musher[464] stabilized fats containing high levels of unsaturated fatty acids by heating them to temperatures between 375 and 450°F for 15 to 30 sec in the presence of a phosphorus compound and a poly-substituted benzene-ring compound. The addition of 0.005 to 5% of a mixture of equal parts of the phosphorus and aromatic compounds was sufficient to increase the shelf-life of lard from 2 to 50 times greater than that of untreated lard. The phosphorus compounds used were H_3PO_4 and its salts, as well as organic phosphorus compounds. Hydroquinone was the preferred aromatic compound.

Calkins[465] studied the synergistic action of H_3PO_4 with tocopherols and other quinone-like phenolic compounds as antioxidant systems in fats and oils. He reported that a maximum of 0.0002% H_3PO_4 dissolved in the lard ethyl esters used as the test system in his experiments. This was an insufficient quantity of phosphoric acid to provide synergistic activity with 0.02% quinone compounds. However, the stability of his test system was dramatically increased as the level of H_3PO_4 increased from 0.009% to 0.09% in the presence of the 0.02% quinone. Calkins proposed a mechanism for the antioxidant activity that involved absorption and exchange reactions between the H_3PO_4 and the fat system containing unsaturated carbon-to-carbon bonds. The H_3PO_4, according to Calkins, forms an addition compound with the quinone that can then diffuse throughout the fat system. The phosphorylated quinone is thought to enter exchange reactions with the free radicals formed during the initial stages of formation of peroxide compounds in the unsaturated fatty acids. The phosphorylated-fatty-acid-peroxide compound then splits off H_3PO_4 to regenerate the unsaturated bond instead of undergoing the usual breakage of the unsaturated bond to form aldehydes and other oxidative rancidity by-products.

Investigators at the American Meat Institute Foundation[466] reported that H_3PO_4 and other acidic phosphates are synergistic with BHA in fat-antioxidant systems, increasing stability of the fats by 8 to 40 times greater than controls. This work led to patents for combinations of 0.001 to 0.01% BHA with up to 0.02% H_3PO_4,[467] or for butylated hydroquinone ethers, propyl gallate, and acidic compounds such as H_3PO_4.[468]

Other investigators also studied the synergistic effects of phosphoric acid with phenolic antioxidants.[469-472] Lips[471] reported that metaphosphoric acid, among other preservatives used, stabilized linseed oil against oxidative rancidity but not against development of reversion flavors and odors. He reported that citric acid exhibited no synergistic or additive effects alone or in combination with other stabilizers. Privett and Quackenbush[472] demonstrated that H_3PO_4 prevented the estimation of fatty-acid peroxides through its reaction with the peroxides at the temperatures used in accelerated stability tests. A colored, polymeric, phosphorus-containing precipitate was formed during the reaction. These authors also reported that H_3PO_4 formed an undissociated, fat-soluble complex with oxidized fat, but it would not react with fresh, unsaturated fat. H_3PO_4 was reported to have a "sparing action" on tocopherols and quinones used as antioxidants; the presence of the aromatic compounds also had a "sparing action" on the H_3PO_4.

The synergistic effect of H_3PO_4 with organic antioxidants in dry fat systems has led to several patents. Hall[473] patented the use of antioxidant flakes composed of hydrogenated fat, an acid, such as H_3PO_4, and a gallic acid ester. Magoffin[474] patented combinations of hydroxyphenyl ethers with acids, including H_3PO_4, and propyl gallate. Other examples of combinations of synergistic acid, including H_3PO_4, with aromatic antioxidants and sometimes with derivatives of gallic acid or its esters were patented in several countries.[475-482]

Kuhrt[483,484] obtained a series of patents for increasing the stability of dry fats, oils, and monoglyceride emulsifiers and their mixtures with fats and oils. He washed the liquid fat at 40 to 95°C for 3 to 30 min with a solution containing between 0.2 and 1% by weight of an equal mixture of the amino acid, glycine, and H_3PO_4 or their salts. Stability of the dry-fat system was reported to be increased from 3 to 30 times that of the untreated controls.

Aqueous Fat Systems

Watts and her co-workers have studied the antioxidant activity of phosphates in food and pharmaceutical systems that have combined fat and aqueous phases. Baked goods leavened with baking powders containing SAPP were found to be more stable against development of rancid flavors and odors than those leavened with baking powders containing tartrates, calcium orthophosphates, or a combination of calcium orthophosphate and sodium aluminum sulfate.[485] The SAPP was shown to exhibit synergism with tocopherols natural to the lard used in these tests. The antioxidant activity exhibited by the SAPP was not due to copper binding alone, according to these authors, as had been suggested by Lea.[486] Watts extended these studies[487] to include a number of antioxidant-system synergists. She reported that the orthophosphates were ineffective as synergists but that the antioxidant activity of polyphosphates increased as the chain length was increased from pyrophosphate through hexametaphosphate to long-chain phosphates, such as Maddrell's salt. Citrate was effective as an antioxidant synergist and compared with STP and the pyrophosphates in activity. The antioxidant activity of citrate and the polyphosphates was inhibited by large quantities of orthophosphate. The stability of lard in contact with an aqueous phase containing phosphates of various lengths was increased from approximately two times for pyrophosphate, tripolyphosphate, and citrate to as much as six to twelve times for hexametaphosphate and Maddrell's salt. Similar lard systems containing tocopherols that were in contact with aqueous phases containing phosphates were increased in stability against oxidative rancidity by two to eight times over controls. The polyphosphates were less effective in dry-fat systems than in aqueous-fat systems.

Lehmann and Watts[488] studied the effects of primary phenolic rancidity inhibitors and their combinations with synergistic compounds on a dry-lard system and an aqueous system containing the same lard. They reported that the effectiveness of the primary phenolic antioxidants tested in the dry-lard system, in order of decreasing effectiveness, was NDGA, BHA, propyl gallate, lauryl gallate, and α-tocopherol. When the same lard containing the phenolic inhibitors was in contact with an aqueous solution buffered at pH 7.5, BHA was most effective, while NDGA and propyl gallate were only about one half as effective; the others, even less so. Hexametaphosphate and Maddrell's salt were the most effective synergists when included in an aqueous solution that was in contact with either the plain lard or the lard containing 0.005% of the phenolic inhibitors. The presence of ascorbic acid in contact with either the plain lard or the lard containing the phenolic inhibitors generally increased the development of rancidity (reduced stability of the lard); in the presence of NDGA, ascorbic acid increased the stability of the system by about ten times over the control containing borate buffer in the aqueous system. Polyphosphates were reported to counteract the destabilizing effect of ascorbate, while combinations of polyphosphate with ascorbate and the primary inhibitors further increased the stability of the lard. Orthophosphate, tripolyphosphate, and citric acid had about the same activity and were less effective than hexametaphosphate and Maddrell's salt. Table 24 summarizes the results of this study. The work of Watts and her co-workers led to patented applications for phosphates as antioxidants and synergists for antioxidants in aqueous fat systems, food products, and pharmaceutical preparations.[489,490]

Ozawa and Ota[491] demonstrated that condensed phosphates increased the preservative effects of a number of "antiseptics" they added to boiled fish meat. Hexametaphosphate was most effective. The effect, however, could be eliminated by the addition of calcium.

Holman[492] patented the use of a thin film of MSP or MKP around the metal exterior of the pouring spouts of cans; this thin film prevented formation of oxidative rancidity by-products that could form on the spout and contaminate the fat or oil poured from the can.

Mahon and Chapman[493] reported that citric acid, H_3PO_4, glycine, and EDTA were ineffective as acidic synergists in increasing the shelf-life of pie crusts. These authors apparently did not test the longer-chain polyphosphates or polyphosphoric acids in this application. Further evidence of practical applications of the phosphates as synergists and antioxidants in food systems will be

TABLE 24

Some Synergistic Effects of Phosphates with Various Primary Antioxidants in Aqueous Lard Systems[1]

	Days to turn rancid at 45°C[2]					
Synergist used[3]	Plain lard	Lard and 0.005% tocopherol	Lard and 0.005% NDGA	Lard and 0.005% BHA	Lard and 0.005% propyl gallate	Lard and 0.005% lauryl gallate
Buffer	2	4	5	22	11	6
Ascorbic acid	<1	<1	23	1	<1	<2
Citric Acid	3	12	20	66	15	16
DSP	2	7	11	43	7	6
STP	3	12	21	48	18	16
SHMP	7	–	34	57	28	21
Maddrell's salt	8	–	31	64	28	27

[1] Source: Lehmann, B. T. and Watts, B. M., *J. Am. Oil Chem. Soc.*, 28, 475, 1951, Table 2, with permission.
[2] Time required for half bleaching of carotene.
[3] Concentration = 0.1% in pH 7.5 borate buffer.

reported in further sections in this book, particularly in the section on applications in meats.

Applications in Oil and Water Emulsions

Phosphates have been reported to increase the stability of emulsions against separation. van Hees[494] reported that a combination of polyphosphate and pyrophosphate stabilized dispersions of fats and oils with an aqueous phase for "a long time without separation of the fat from the water." A Spanish patent reported that the addition of a phosphate salt to mayonnaise increased the stability of its emulsion.[495] Mitchell[496] reported that the stability of an aqueous peanut butter suspension was improved by the addition of sufficient DSP to raise the pH of the aqueous phase to between 6.6 and 6.8. Vincent[497] reported a similar stabilizing effect of phosphates on comminuted peanuts when the pH was buffered so that it would remain between 6.5 and 7.0. These emulsion stabilization applications may depend on buffering capacity to produce the optimum pH for maximum protein film dispersion to stabilize the system against separation into aqueous and fat phases.

Phosphate Applications in Fruit and Vegetable Products

As shown in Table 1, fruits and vegetables contain significant levels of phosphorus. This element exists in the plant as phosphoric acid and phosphate anions. For example, Langendorf and Lang[498] reported that commercial lemon juice they analyzed contained 1 millimole of phosphoric acid per liter, the highest level of phosphoric acid in the fruit and berry juices analyzed. Benk[499] reported the phosphate P_2O_5 content of 59 samples of domestic and commercial lemon juices to range between 13 and 41 mg/100 ml. Diemair and Pfeifer[500] reported that, by their analyses, apple juice contained 1.7% dihydrogen phosphate anion and 0.08% phosphoric acid; currant juice contained 1.3% dihydrogen phosphate anion and 0.1% phosphoric acid; orange juice contained 2.6% dihydrogen phosphate anion and 0.13% phosphoric acid; and grapefruit juice contained 2.7% dihydrogen phosphate anion and 0.16% phosphoric acid. In spite of these high levels of phosphate and phosphoric acid, phosphates have significant functions as food additives in fruit and vegetable processing. For example, they provide microbiological stabilization, stabilization against oxidative rancidity, stabilization of vitamins and color, increase and optimization of thickness, tenderization, firming, and other functions.

Applications as Inhibitors of Microbiological Spoilage

Surface Applications

Phosphates provide microbiological stabilization of the surfaces of fruits, from their action as detergents or wetting agents and from other functions beyond the detergent action. Laucks et al.[501] removed spray materials and the natural waxy coating of fruits, especially of apples, by adding a strongly alkaline salt, particularly TSP, to wash solutions. They reported that it was essential to obtain a high pH but one not greater than pH 13. Concentrations between 1 and 10% TSP in the water were used, the concentration depending on the fruit or vegetable being cleaned and the temperature of the water being used. Bates[502] stabilized fruit against decay organisms for a significant time period by (1) immersing fruits in a solution of 1 lb TSP or DSP in 30 gal of water for 5 min at 75 to 80°F, and (2) drying them so that a dry film of the phosphate remained on the fruit and especially in any cuts, scratches, or abrasions on the surface. The treatment was reported to be most effective against blue molds, which cause surface decay of citrus fruits. Rippey[503] reported that fruit washed in a 5% solution of TSP or sodium carbonate, and preferably in a 50-50 combination of the two, suffered less decay due to mold on the surface of the fruit. He reported that the molds were removed from the fruit during treatment and that spread of mold from one fruit to another was therefore reduced. Concentrations between 2 and 10% were used.

Kalmar[504] combined the use of a buffer or regulator, such as TSPP, TSP, or sodium carbonate, and preferably combinations of these, to obtain solutions buffered at pH values above 10 but not higher than 11.5. These solutions were used to apply sodium orthophenylphenate to the surface of the fruit without obtaining substantial hydrolysis of this mold inhibitor. The combination was reported to have a significant value in preventing decay of citrus fruits. Kalmar[505] further demonstrated that he could prevent decay by blue and green molds and other spoilage organisms between the harvesting of fruits or vegetables and their consumption by disinfecting their surfaces with solutions of sodium hypochlorite, containing a buffer composed of DSP and sodium carbonate and a wetting agent. The buffer system, which could also contain TSPP, maintained the pH in the optimum range between 10.2 and 11.5 and prevented substantial hydrolysis of the disinfecting solution.

Smith and Krause[506] treated peeled, raw potatoes with an aqueous solution containing 0.5 to 4% phosphoric acid, 0.5 to 1.5% propylene glycol, 0.05 to 0.1% sodium benzoate, and 0.05 to 0.1% sodium metabisulfite; they reported that submerging the potatoes, as well as sweet potatoes and other peeled fruits and vegetables, for 10 to 100 sec in the solution at 50 to 115°F significantly extended the shelf-life and prevented bacterial growth and enzymatic browning. The phosphoric acid increased the effectiveness of the sodium metabisulfite in preventing browning and inhibited and/or destroyed enzyme-producing bacteria.

Brandt[507] reported that fruits and vegetables could be preserved by dipping them for short time periods into a solution containing between 0.1 and 1% pyrophosphoric acid. Brandt reported that the shelf-life of fruits and vegetables treated at ambient temperatures in such solutions was significantly extended, as microbiological spoilage was inhibited. In addition, the original colors of the fruits and vegetables were stabilized, and discolorations could be reversed, as in the after-cooking darkening of potatoes. The microbiological preservation was not completely due to the low pH of the acidic solution, because Brandt found that buffered solutions were equally as effective and often were required in cases where the food was adversely affected by low pH values. Increased protection against spoilage was accomplished by (1) making a solution of gelling agents such as cellulose derivatives, pectin, gelatin, or alginates, (2) dipping the fruit or vegetable into the solution and allowing it to gel as a film about the food, and (3) storing it under conditions that would prevent complete desiccation. Concentrations of pyrophosphoric acid under 0.5% did not prevent mold growth; however, combinations of pyrophosphoric acid at levels below 0.5% with benzoic acid were highly effective as were combinations with other common microbiological inhibitors.

Post et al.[508] reported that certain of the polyphosphate salts were effective as mold inhibitors on the surface of fresh cherries. Mature, ripe Montmorency sour cherries were treated by immersion for 2 min in solutions of 1%, 5%, or 10% concentrations of the phosphate compounds, thoroughly drained, then stored at 1.1°C. The most effective of the phosphate compounds tested was sodium tetrapolyphosphate, which more than doubled the shelf-life of 14 days for the controls to 30 days for the treated cherries. Three other compounds tested, in order of decreasing effectiveness, were SHMP, STP, and TSPP. A 10% solution of the polyphosphate, used as a dip, appeared to be the most satisfactory. The phosphates were found to be effective against the most common mold spoilage organisms – *Penicillium expansum, Rhizopus nigricans*, and *Botrytis* species. The authors believed that the germination of the fungal spore was prevented in some unknown way; resistant spores eventually did germinate, and weak mycelial growth eventually occurred. The authors explained the mechanism of action as most likely due to chelation of metabolically essential metallic ions; the eventual growth of the molds was probably due to degradation of the polyphosphate through the action of phosphatase enzymes produced by the mold cells.

Applications in Juices

Kohl and Ellinger[509] reported that SHMP alone, and in particular, when combined with either sodium benzoate or potassium sorbate, was effective in inhibiting spoilage of apple cider and other fruit and vegetable juices. Yeasts and molds were most effectively inhibited, although bacterial growth was also retarded. Taste tests indicated that the shelf-life of the phosphate and benzoate or phosphate and sorbate treated samples was at least doubled over the acceptable shelf-life of the samples treated with benzoate or sorbate alone. The ciders containing phosphates were significantly preferred for flavor over other treated samples during the entire study.

Applications as Antioxidants

Several investigators have reported that phosphates can effectively retard oxidative spoilage of fruits and vegetables, either alone or in combination with other approved antioxidants. Deobald et al.[510] evaluated combinations of BHA, BHT, a mixture of the two, α-tocopherol, citric acid, and SAPP, alone and in combinations, for preventing oxidative development of off-flavors in sweet-potato flakes. Combinations of citric acid and

SAPP with the various aromatic antioxidants were most effective, particularly when these acidic compounds were combined with a mixture of BHA and BHT. The authors reported that citric acid was more effective than SAPP as a synergist for BHA and BHT; however, an examination of their tables indicated that the control for the SAPP-treated samples spoiled in approximately one half the time as did the control for citric-acid-treated samples, indicating this may have been a less stable lot of sweet potatoes.

The antioxidant effects of TSPP were reported to extend shelf-life of dried vegetables when the vegetables were dipped into a solution of the phosphate before blanching and were heated at approximately 80°C for 2 to 10 min.[511] No discoloration or alteration of taste was noted. Evidence that STP failed to inhibit lipoxidase activity in model systems containing pea or soy bean lipoxidase was presented by Rhee and Watts.[512] Although the STP was relatively ineffective in the artificial systems, it and turnip-green extract were found to be very effective when used in slurries prepared from the peas themselves. Other antioxidants, such as BHA, gallic acid and its esters, propyl gallate, and quercetin, were less effective or ineffective in the latter systems but highly effective in the artificial systems. The authors were unable to explain the effect of the STP.

Molsberry[513] reported that the flavor and texture of frozen mushrooms were retained by treating them with a "sealing mixture" of approximately one part sodium sulfate, one part DSP, and two parts sodium bisulfite in water. The mushrooms were (1) immersed in a solution of 1.5 oz of the salt mixture in 6 gal of water for 1 to 1.5 min, (2) dipped into a bleaching solution, (3) sliced or cut into pieces, (4) again dipped into the sealer compound, and (5) dipped into the bleaching mixture. The mushrooms were frozen or refrigerated and found to contain levels of sulfur dioxide well below the levels allowed by Federal regulations. Upon being cooked, the mushrooms had excellent flavor, color, and texture. The DSP probably acted as a buffering system in this particular application and maintained a pH for optimum action of the bisulfate.

Phosphates were also reported to have a preservative action on ascorbic acid in fruit and vegetable products. Oxidized ascorbic acid no longer functions as a vitamin, and, therefore, ascorbic acid preservation in fruit and vegetable products is essential for proper human nutrition. Between 50 and 400 mg% of SHMP was reported to be an efficient preservative for ascorbic acid in ascorbic acid solutions having a pH of 3.5 or lower. SHMP was also found to exhibit similar preservative action in sugared and nonsugared orange juices.[514] The addition of 2% metaphosphoric acid to apple, lemon, and orange juices and to lemon and orange juice concentrates was reported to be highly effective in the stabilization of ascorbic acid for 10 days.[515]

Applications in Stabilizing Fruit and Vegetable Color

The phosphates have been reported to function as color stabilizers in a number of fruit and vegetable applications. Hall[516] reported that the natural, ripe colors of tomatoes, berry and cherry products, and the colors of jellies and preserves could be retained if the fruit products were treated with certain polyphosphates, such as the pyrophosphates, tripolyphosphates, and the longer-chain metaphosphates. STP and longer straight-chain phosphates, including the long-chain potassium and sodium polyphosphates, were more effective than pyrophosphates. Hall described tests in which he added 0.25% of a typical commercial SHMP to tomatoes prior to cooking them. The treated tomato product was then used to make catsup, which was reported to have, and to retain, a deeper red color. The catsup also seemed to have more body and cook more rapidly, probably due to a more complete extraction of pectin from the tomato cell walls due to the action of the polyphosphate. The same type of polyphosphate was reported to stabilize the bright red color of dehydrated tomato paste and to aid in its reconstitution with water. The addition of 0.1 to 0.2% polyphosphate to strawberries, based on their weight, for strawberry preserves prevented the brown discoloration that strawberries normally undergo during preparation of cooked products. Red-raspberry preserves were also brighter and more naturally colored with the addition of 0.1 to 0.4% polyphosphate. Cranberry sauce was reported to have brighter color, greater gel strength, increased yield, and none of a distasteful bitterness (especially noticed in the control sample) when the cranberries were treated with 0.2 to 0.8% polyphosphate, based on the weight of the raw fruit.

Nelson and Finkle[517] studied the oxidative darkening of apple juice and apple fruit sections on exposure to oxygen in the air. This discoloration, the development of a brown surface color, results from an enzyme-catalyzed reaction between *o*-diphenolic compounds in the fruit tissues and oxygen. These authors discovered that treatment of apple slices with a mildly alkaline solution resulted in methylation of the *o*-diphenolic compounds, which prevented them from being substrates for this enzymatic reaction. Bolin et al.[518] developed a commercial process based on this work, in which the sliced apples were dipped for 45 sec in a 0.25% solution of sodium bisulfate and then for 5 min in an 0.2% solution of DKP, maintained at pH 8.8. The apple slices were then held at 34°F until evaluated. The authors reported that the combination, bisulfate-phosphate buffer treatment, caused the apples to maintain excellent color and a very crisp texture for at least three times as long as those produced by present commercial processes. The flavor of the treated apple slices was preferred over the flavor of the untreated controls.

Phosphates have been shown to preserve the bright green color of vegetables dipped in a DSP buffer, designed to maintain a pH of 6.8 to 7.0 prior to canning or freezing.[110] Fleischman[519] reported that the green color of purêed vegetables changed to brown due to degradation of chlorophyll to the brown-colored pheophytin during normal cooking processes. Buffering the vegetable purées at pH 8.0 with sodium phosphates (assumed to be the necessary conbination of sodium orthophosphates to produce this pH) increased the production of green chlorophyllide in the vegetable product. A further increase in the stable green compound was found to occur when magnesium carbonate and sodium phosphate were added together. A taste panel preferred the sodium phosphate-magnesium carbonate processed product over controls. However, although this process appeared to stabilize the color of spinach, it would not stabilize the color of pea and string bean purées.

One of the most extensively investigated discoloration reactions in fruit and vegetable processing is that of blackening of raw and cooked potatoes. The blackening of raw potato tissue is due to the oxidation of *o*-diphenolic compounds in the presence of heavy metal ions, just as with apple tissue. In the potatoes iron is known to catalyze the reaction in the presence of oxygen to form a black discoloration on all exposed surfaces of raw potato tissue. Therefore, peeled potatoes have often been treated with sulfur dioxide or with sodium bisulfite in acidic solution (to provide sulfur dioxide) to inhibit the enzyme that catalyzes the reaction. A treatment to aid in preventing or inhibiting discoloration of exposed, raw potato tissue was patented by Michels.[520,521] He used polymerized phosphorus acids, probably metaphosphoric acids, and calcium or magnesium ions for this purpose. He also prevented after-cooking discoloration of potatoes by adding 0.25 or 0.3% of a 1:1 mixture of sodium metaphosphate and monocalcium citrate to their cooking water. Combinations of sulfites such as sodium sulfite with a mixture of the acid and alkaline sodium pyrophosphates have been reported to aid in the control of enzymatic discoloration in raw fruit and vegetable tissues.[29] The acid and alkaline sodium pyrophosphates are probably blended in the proper proportions to maintain the optimum pH for the sulfite reaction and to inactivate any of the iron and copper that might be required for the oxidation reaction. Levels of iron as low as 1 ppm and copper as low as 0.1 ppm in fruits or vegetables have been reported sufficient to assure the discoloration; most fruits and vegetables normally contain higher levels than this.[29]

The after-cooking darkening of potatoes is very effectively controlled by sequestering iron, thus preventing its involvement in the oxidative reaction responsible for this discoloration. Juul[522,523] apparently was the first to discover that SAPP was highly effective in preventing after-cooking darkening if the potatoes were either dipped or cooked in a solution of the phosphate. Smith[521,524] extensively studied after-cooking darkening of potatoes, discovering that dipping the potatoes into solutions of EDTA or SAPP would prevent discoloration when the potatoes were exposed to oxygen after having been cooked. Smith and Davis[525] further reported that if discolored, cooked potatoes, even after being frozen, were dipped for 2 min in a 2% solution of SAPP at 40°F, the color would be completely removed and the natural white color of the cooked potato would reappear. This activity of SAPP appears to be the complexing of ferrous ions in a colorless form.

SAPP is now commonly used in commercial

processes to prevent after-cooking darkening. It is either applied as a spray to cover the potato surface prior to cooking, or it is added to the blanch water.[29,30,110,111,521,525-527] The use of this process has been very thoroughly reviewed by Smith[521] and by Talburt and Smith.[526] A complete bibliography of the literature on this topic has been published very recently by Talley et al.[527]

SAPP was found to be highly effective in inhibiting discoloration and preserving the original orange color of sweet-potato flakes[528] and of sweet potatoes for canning and freezing.[523] TSPP was also effective but less so than the acidic compound. STP had some effect, but less than either of the pyrophosphates, while hexametaphosphate and MSP had very slight effects. Optimum flavor and acceptable color stability were obtained when SAPP and TSPP were blended to produce a mixture that, in water solution, had a pH equivalent to that of the sweet potatoes. Between 0.3 and 0.5% of the pyrophosphate mixture was needed to obtain color stabilization.

Applications in Obtaining Optimum Texture

Just as calcium has dramatic effects on the physical and chemical properties of milk proteins, calcium also has been shown to have dramatic effects on the physical characteristics of fruits and vegetables. Kertesz[529] reported that fruit or vegetable tissues could be made firmer by providing soluble calcium salts when the fruit or vegetable tissues were damaged; thus the calcium would be available to react with the pectic acid (polygalacturonic acid) formed when the natural pectin in the cell walls was demethoxylated. This reaction has been studied to a considerable extent and is the normal reaction that occurs during gelling of fruit juices to prepare preserves, jams, and jellies. Proper control of the calcium ion can be highly beneficial in processing of fruits and vegetables. The addition of soluble calcium can introduce firmness to the structure of the tissue such as canned, sliced fruits or vegetables. Removal of the calcium naturally present can soften tissue that is too firm. Changes in viscosity or texture can also be introduced by proper manipulation of calcium levels in the food. This section will attempt to review briefly the ability to control calcium in order to thicken, tenderize, and firm fruit and vegetable products.

Pectin Gels

Pectin, which is necessary for proper preparation of fruit preserves, is normally extracted from apple pomace or from the peel of citrus fruits by heating with acid. Natural pectins in plant materials were thought by early investigators to be in the form of insoluble calcium salts. It was, however, discovered that pectins were methyloxylated galacturonic acid polymers; these became demethyloxylated by natural enzymes in the tissues when the cell walls were broken. These facts were reviewed by Kertesz.[529] Baker and Woodmansee[530] revealed that the addition of approximately 2.5% SHMP, based on the weight of the fruit pulp, would increase the yield of pectin as much as 25% and decrease extraction time up to 50%. The polyphosphate was believed to sequester calcium ions that would otherwise precipitate the pectins and slow or prevent extraction of the maximum amount. Sodium tetrapolyphosphate was approximately as effective as SHMP, but TSPP was considerably less effective, since lower yields were obtained when it was used. The addition of sodium orthophosphates decreased the yield of the pectin that could be extracted by the acid process at pH 3.5. When pectins, extracted with SHMP, were used in preparation of fruit jellies, they produced slow-setting jellies of greater clarity than the controls. The jellies contained approximately 0.09% SHMP.

Baker and Gilligan[531] reported that they obtained highly significant increases in gelling strength of the juices expressed from the pulp of apple, orange, blackberry, and plum fruits for use in jelly-making in the presence of SHMP. The increased gelling capability ranged from 62% increase with apples and 82% increase with blackberries, to as high as 184% increase with oranges when 1% polyphosphate was added to the fruit prior to cooking.

Baker and Murray[532] described a method for extracting the pectin from dehydrated fruits and vegetables using a solution of SHMP. The pectin was extracted by soaking 100 g of the dried product in 900 ml water containing 2.5 g SHMP at 79°F for 1 hr. After comminuting in a Waring blender for 5 min, boiling for 10 min, filtering, then clarifying the filtrate, the pectin was precipitated from the extract with 2 volumes of 90% isopropyl alcohol and then dried. The Ostwald viscosity of a 0.5% solution of the pectin was found to correlate directly with the quality of the pectin and also with the appearance, flavor, and eating quality of the rehydrated food product.

McCready and co-workers[533] confirmed that polyphosphates were effective aids in extracting enzymatically demethoxylated pectinic acids (pectins) from ground citrus peel. SHMP was more effective than sodium tetrapolyphosphate in most cases, and its effectiveness increased as the pH increased from 2.0 to 3.5; neither polyphosphate was effective below pH 2.0. The quantity of SHMP required to disperse calcium pectinates has been found to be directly proportional to the calcium content.[534] During a study of formation of calcium pectinate films, Waller and Baker[535] found that the polyphosphates hydrolyzed to orthophosphate at a slightly faster rate in the presence of calcium pectinate than in the presence of calcium chloride. Furthermore, they found evidence of chemical phosphorylation of the pectin by its reaction with the orthophosphate ion removed from the polyphosphate chain during this hydrolysis. Evidence was also found for cross-linking within and between chains of the pectinic acid, either through phosphorylation of terminal glucosidic bonds or by phosphate linkages between calcium ions attached to the carboxyl groups of the pectinic acid chains. SHMP was again found most effective, and sodium tetrapolyphosphate slightly less so, while shorter-chain polyphosphates were decreasingly effective. An excellent review of the chemistry and uses of pectin was published by Woodmansee[536] and contains a bibliography for further study.

The patent literature demonstrates the applications of phosphates in obtaining optimum characteristics in pectin gels, such as fruit jellies and preserves. Olsen and Fehlberg[537] patented the use of calcium phosphate in preparation of pectin jellies. Leo and Taylor[538] patented a method of pectin jelly preparation in which a calcium phosphate (or calcium carbonate or citrate) could be used to form the proper gel. TCP was preferred because it is converted to the soluble MCP at the low pH required, and thus provides the required calcium ion for formation of calcium pectinate jellies. The authors found greater quantities of fruit juice, or water, could be used at wide variations of sugar levels (from 0 to 75% soluble solids). The process could be used for numerous other food applications, such as the preparation of confections and the manufacture of ice cream. Leo and co-workers[539] extended this work by using SHMP to provide a controlled release of calcium ions to provide a slow, uniform "set" of the jelly, due to formation of the calcium pectinate network uniformly throughout the jelly.

Pedersen[540] used SAPP or TSPP to prevent the precipitation or formation of insoluble salts of pectinic acid in liquids containing polyvalent metal ions. An Italian patent reported a method for preparing low-methoxyl pectins suitable for making low-sugar jellies by use of a calcium phosphate at low pH values and SAPP at pH values above 3.5.[541]

A British patent revealed that dry, cold-water-soluble, dessert gel mixes could be prepared from a gelling mixture containing low-methoxyl pectin, SHMP, and TCP.[542] A method for the preparation of a low-methoxyl pectin has been described by Leo and Taylor[543] for the preparation of gels with and without the presence of calcium. The gelling mixture is composed of the pectin, SHMP, and either sodium or potassium citrate. The pectin composition can be used to form fruit jellies by combination with fruit juice, and, if desired, with sugar; it also can be used to prepare milk custards, depending on formation of calcium pectinate for the gelling action.

H_3PO_4 has also been used in preparation of jellies in order to produce tartness and the optimum pH for the setting of pectin gels through reduction of the hydrogen bonding in the pectin. It has the added advantage that it will not be driven off by heat.[29,110,111]

Tomato Products

Baker and Gilligan[544] studied the effects of phosphates on the extraction of pectin and the influence of these factors on viscosity and texture of fruit and vegetable products. They reported that more viscous tomato sauces could be produced by the addition of approximately 1% SHMP to the tomato fruit prior to cooking. The addition of the sodium hexametaphosphate extracted a greater amount of pectin from the tomato pulp, which accounted for the greater thickness of the sauce. The addition of soluble calcium salts in sufficient quantities to provide free calcium ion thickened the sauce to an even greater extent. It thus appeared possible to control the viscosity of a tomato sauce to any desired thickness.

Morse[545] investigated the addition of various levels of SHMP to tomatoes in the extraction of tomato juice. He reported that he obtained a 3.7%

increase in the yield of juice with 0.25% SHMP, 6.9% increase with 0.5% SHMP, and 9.8% increase with 1% SHMP based on the weight of the fruit. The addition of the SHMP increased the pH, and it was, therefore, necessary to use phosphoric acid to reduce the pH to the optimum of approximately 3.35. Morse reported that he also observed the increased viscosity of the juice reported by Baker and Gilligan[531] and the enhanced color reported by Hall.[516] Furthermore, he reported that the ascorbic acid in the tomato juice was preserved to a greater extent in the presence of SHMP than in its absence, which confirmed the reports of Japanese[514] and German[515] investigators.

Peters et al.[546] found that hot-break tomato purées with and without SHMP contained significantly higher levels of pectin compounds than the purées prepared by the cold-break process, whether they contained SHMP or not. The hot-break purées containing no hexametaphosphate had higher relative viscosities, as measured by the Ostwald-Fenske viscosimeter, than the hot-break process purée containing hexametaphosphate and greater than cold-break purées prepared with or without SHMP. The addition of calcium ion in increments from 0 to 30 ppm resulted in a decrease in the viscosity with the conventional hot-break process purée; it caused a significant increase in viscosity in proportion to the level of added calcium ion in the hot-break purée containing hexametaphosphate. The maximum viscosity was obtained with the hot-break purée containing hexametaphosphate on addition of 200 ppm of calcium ion.

Tomatoes used in preparation of tomato juice, tomato purée, and other similar products must be heated as quickly as possible to approximately 190°F or higher, after they have been chopped or crushed; this inactivates the enzymes that cause hydrolysis of the tomato pectin compounds and thus leads to thinner, lower viscosity products than may be desirable. However, if strong acids, such as H_3PO_4, hydrochloric, sulfuric, or tartaric, are added to the tomatoes prior to heating, higher consistency tomato products are produced.[547] The acid must produce pH values of 2.5 or lower in order to inhibit this enzymatic action. These investigations have led to an improved hot-break process for preparation of tomato juice and purée.

Potato Texture

Phosphates have been shown to improve the texture and consistency of potato products. Smith and Davis[548] demonstrated that the addition of 2% of SAPP increases mealiness of processed potato products, especially of dehydrated potato flakes. Microscopic examination demonstrated that the cells of the potato product treated with SAPP did not lose their contents when they were reputured during the cooking process. Cole[549] reported that buffering the cooking water in which potatoes were cooked prior to preparing the dehydrated products would improve the textural characteristics of the dehydrated potato product. The optimum pH appeared to be between 6 and 8 and was obtained by adding a mixture of MSP and DSP, or the potassium equivalents, to the cooking water at a concentration of at least 0.01 *M*. The addition of approximately 0.2 *M* of either sodium or potassium chloride to the buffered solution gave further improvement in the characteristics of the dehydrated potato products prepared from the treated potatoes. The improved process also eliminated a cooling step (which had been essential in previous processes) between a precooking step and the final cooking of the potatoes until they were soft enough to mash prior to dehydration.

The effect of phosphates on improving potato mealiness was reviewed by Kintner and Tweedy[550] and, in considerable detail, by Smith.[521] Smith reported that the phosphorus content of dehydrated potato flakes treated with SAPP was increased in proportion to the increase in textural characteristics; this indicated to Smith that a small amount of phosphate was esterified with the amylopectin in the potatoes. Starch grains were retained within the starch granules after cooking and did not extrude upon rupture of the granule. Smith believed this was due to increased hydrogen bonding among the starch molecules with their esterified phosphate groups.

Tenderization

In the preceding examples the phosphates have been shown to have significant effects on thickening and viscosity control of fruit and vegetable products. They can also be utilized in tenderizing or firming fruit and vegetable tissues. The following discussion will treat the ability of phosphates to aid in tenderizing fruit and vegetable products, particularly the skins of the fruits or vegetables to attain better eating characteristics.

Frozen vegetables, particularly peas, lima beans, snap beans, sweet corn, and similar seed types, often have tough skins. The use of hard water in blanching the fruits or vegetables and the use of fruits or vegetables that are too mature have been shown to result in high calcium content in the skins of the frozen products. Nielsen and coworkers[551] investigated the use of SHMP in blanch water as a means of reducing some of the calcium content that toughens the skins. Their investigations showed a definite increase in tenderness of the skins of peas, lima beans, and snap beans when blanched in water containing between 0.0005 and 0.4% SHMP. Levels between 0.1 and 0.4% were more effective. No improvement was noted when corn or soy beans were similarly treated. The authors also reported that the green color and fresh flavor were stabilized, particularly in the treated snap beans, and that the hexametaphosphate blanch did not have any effect on the content of vitamin C or sugar in the vegetables. Although this method of blanching improved the skin texture and color of frozen vegetables, it could not overcome other undesirable quality factors of poor-grade products. The quality of the product resulting from the hexametaphosphate blanch was directly proportional to the quality before blanching, and the hexametaphosphate could not be used to upgrade poor quality vegetables. Holmquist et al.[552] reported that the addition of quantities of SHMP equivalent to the amounts of calcium and magnesium in hard blanch waters resulted in significant and noticeable softening of the skins of peas blanched in this water. Additional softening of the skins was noticed with the addition of SHMP in excess of that required to soften the water. Adding excessive amounts of SHMP resulted in oversoftness of the peas and cloudy, viscous or gelled brines.

Neubert and Carter[553] reported that the calcium content of the skins of Italian prune plums determined the toughness of the skins in the canned fruit. They found that the addition of SHMP to the canning syrup had a significant tenderizing effect, as shown in Figure 3. The addition of increasing quantities of SHMP to the canning syrup decreased the toughness of the skins of the prunes until measurement of their skin textures on a tenderometer resulted in values equivalent to that for peeled prunes. The calcium content of the water used in preparing the canning syrup, as well as the calcium content of the skins, contributed to increasing toughness. Although one might suspect that the SHMP would contribute to the softening of the flesh of the plums, as well as the skins, this was not noticeable at levels of 0.5%

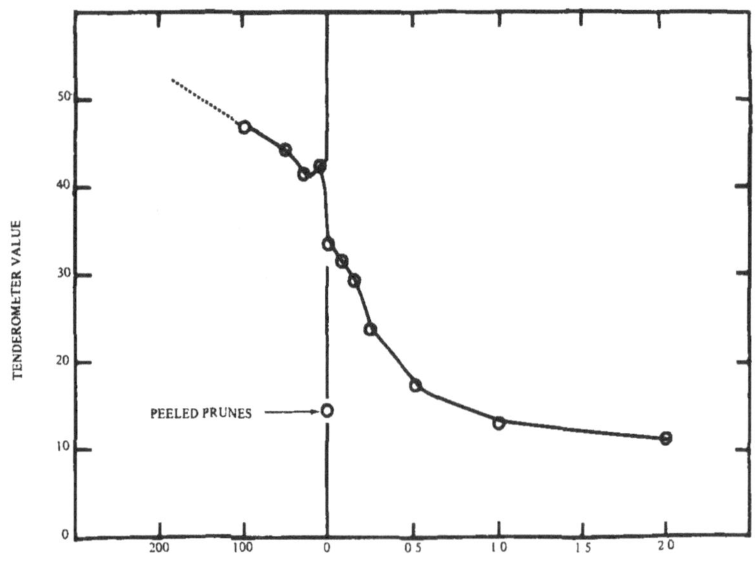

FIGURE 3. Effect of addition of calcium and sodium hexametaphosphate on texture of canned Italian prunes. (Source: Neubert, A. M. and Carter, G. H., *Food Technol.*, 8, 518, 1954, with permission.)

sodium hexametaphosphate and could be noticed only slightly at 1% levels, as confirmed by organoleptic tests. Storage studies of the canned plums also indicated that the skin texture of the plums treated with SHMP was more tender within one day after canning than that of the distilled water controls and increased in toughness only slightly during the next eight months of storage. The control plums canned with distilled water, which were initially tougher than the SHMP-treated plums, increased in toughness with storage during the first week and then remained relatively constant for the next seven months of storage. As would be expected, the addition of calcium chloride to the canning syrup increased the toughness of the skins; however, the addition of magnesium chloride had no noticeable effect. Since the effectiveness of the SHMP was thought to be due to its ability to sequester calcium from the skins of the canned Italian prune plums, it was thought that EDTA should have a similar effect. The use of trisodium EDTA in the canning syrup had no effect on increasing the tenderness of the canned plum skins. A brief report has also appeared indicating that SHMP would increase tenderness of the skins of Damson plums.[554]

Both STP and SHMP have been reported to tenderize skins of peas, corn, and beans.[29,110,111] SHMP in a salt brine affects the shells of peanuts in such a way that the salt brine penetrates the shell, which permits processors to salt peanuts inside the shell.[111]

Blakemore[555] reported that fruit that was to be brined and sweetened would absorb greater quantities of the sweetening syrup faster if the fruit was first brined in a solution containing a combination of SHMP and papain, a proteolytic enzyme. Blakemore reported that after being soaked in brine for 16 days, then washed 22 hr in running tap water, and finally sweetened in a high-sugar-content syrup overnight, the cherries were bright, clear-colored, firm, and plump.

Firming Tissues

The addition of soluble calcium salts to fruit slices firms them because of the effects of calcium ion on the precipitation of calcium pectate in fruit and vegetable tissues. MCP, calcium chloride, and other soluble calcium salts have been proposed for this purpose. Loconti and Kertesz[556] have studied the effects of the calcium salts and definitely identified the formation of calcium pectate as the tissue-firming compound in tomatoes. Calcium salts have been effective in firming peeled, sliced apples used in freezing and canning and as fresh sliced apples[557] and in the processing of snap beans.[558] It was found more effective if the calcium was added to the beans before blanching than when added after blanching.

The use of calcium phosphate to firm vegetables has been approved by FDA for use in some canned vegetables under the federal standards of identity, but not for all vegetables. Only canned tomatoes, potatoes, green or red sweet peppers, lima beans, and carrots may be treated with calcium salts for this purpose. The approved calcium salts are MCP, calcium chloride, calcium sulfate, calcium citrate, or any mixture of these. The levels of calcium that can be applied by this means are established for each of the vegetables under the standards of identity described in the *Code of Federal Regulations.*[559] Anyone who desires to use calcium salts to treat fruits or vegetables with phosphates for firming or tenderization, or for other textural characteristics, is advised to refer to the regulations for use of such additives. The fact that some of these additives are not approved for use in standardized products may only be because no applications have been made for their use, since all are considered to be GRAS by FDA; however, new applications are constantly appearing. For example, an application has recently been made and a temporary permit has been granted to one vegetable processor for marketing canned, cut sweet potatoes, containing calcium salts and pectin as firming agents. One of the approved calcium salts is MCP.[560]

Miscellaneous Applications

Polyphosphates are reported to have beneficial effects on other constituents of vegetable tissue. Dakin[561] reported that pyrophosphates, as well as sulfur dioxide, iodine, EDTA, ethanol, and formaldehyde, retarded softening of cabbage tissue, which he reported was due to the oxidation of cellulose to form pectic materials. Takeda[562] reported the use of EDTA or polyphosphate to soften salted pickles prior to brining. DSP or TSP has also been reported necessary in the production of imitation, or fabricated, potato chips. One formula was composed of potato powder, potato starch, sodium alginate, and TSP.[563] The paste prepared from these ingredients was extruded and cut into rod-like pieces that were boiled in water until the starch was completely gelatinized; they were then dried and cut into thin slices that were further dried. These were stored until needed for use in preparing the imitation potato chips. They were reported to have a long shelf-life.

Phosphate Applications in Gums and Gels

A number of vegetable-derived gums, in addition to pectin, are used as stabilizers and thickeners in the food industry. Most of these vegetable gums, like pectin, are long-chain, high-molecular-weight polysaccharides, or mixtures of polysaccharides. Each specific gum has its own advantages and disadvantages for uses in food systems. Some only swell when hydrated and merely increase viscosity of the food system; others form thick rubbery gels and are used to form jellies similar to those prepared from gelatin or pectin. Glicksman has published an excellent review of the properties and uses of the various vegetable gums.[564]

The gums are long-chain polyelectrolytes and, therefore, interact with the various phosphates, particularly with the longer-chain polyphosphates, in many useful food applications. The phosphates modify the physical characteristics of the gums to provide increased viscosity, thickening, or gelling of food systems in which they are used. Among the gums beneficially modified by use of phosphates are agar, alginates, carrageenan, guar gum, and locust bean gum.

Applications in Agar Gel Systems

Agar, which has been used as a gelling component in food systems for centuries, is still commonly used as a stabilizer in numerous baking applications, such as chiffon pies, meringues, icings, and toppings. Agar has also been commonly used in confections and was probably the gelled constituent of sweet gums prepared by the early Chinese and Japanese. The effect of DSP on a gel composed of agar, sugar, and molasses for use in preparing orange and lemon slices was studied by Sergeeva.[565] The strength of the agar, sugar, and molasses gel was significantly increased by the addition of 0.12 to 0.3% DSP.

Few other investigators appear to have studied the effects of phosphates on agar gels. However, because agar is composed of at least two polymeric electrolytes, the polyphosphates should have useful effects on its gels.

Applications in Alginate Gel Systems

Alginic acid, believed to be the basic component of alginate gels, is a polymer of anhydro-1,4-β-D-mannuronic acid. Therefore, it is a powerful anionic polyelectrolyte and reacts very readily with cationic polyelectrolytes. Although alginic acid is insoluble, its sodium, potassium, and ammonium salts are very soluble and are, therefore, used in the food applications to be discussed.

When solutions or suspensions of the soluble alginic acid salts are treated with di- or trivalent metal ions, an insoluble salt of the di- or trivalent metal alginate is formed. If the rate of reaction is too rapid, grainy, discontinuous gels, or even precipitates, are formed. However, if a slow, uniform rate of formation of the gels is obtained, gels having the desired physical characteristics can be formed. If the rate is too slow, however, the gel formation will take too long for practical use, and the resulting gel also may be softer than desired.

The calcium ion is preferred for reaction with the soluble alginates to form gels, since calcium alginate gels are much easier to control than some of the others and also have the proper gel characteristics for most food applications. A proper calcium salt must be carefully selected. It is most desirable to have a sparingly soluble calcium salt, especially one that will release calcium ions slowly at proper pH values. The rate of release of calcium ions is commonly retarded by use of sequestering agents. Glicksman[564] has prepared an excellent discussion of the conditions required for obtaining optimum food gels with alginates.

Early methods of preparing alginate gels involved mixing the dry, powdered alginic acid with sugar so that it would disperse properly. Considerable difficulty was experienced in obtaining uniformly dispersed gels when such mixtures were added to milk. They formed stringy or lumpy suspensions unless the alginate was first carefully dissolved in water prior to being added to the milk product. Since the addition of water to a milk product is undesirable, Lucas[566] developed an improved product composed of alginic acid solubilized by mixing it as a paste with soda ash to form sodium alginate. About 0.125% of TSP was then added, based on the weight of the anhydrous alginic acid. The TSP assured complete, uniform dispersion of the sodium alginate throughout the milk, after which the calcium in the milk would interact to form the gel network of insoluble calcium alginate.

Steiner[567] provided a further improvement by (1) preparing soluble sodium, potassium magnesium, or ammonium alginate, and (2) adding to it a slowly or difficultly soluble calcium salt, such as TCP, DCP, calcium sulfate, calcium oxalate, or calcium alginate. TCP, calcium tartrate, or calcium alginate were preferred, as they formed the best gelatinized food products. Steiner added a weakly acidic material to this mixture, such as glucono-delta-lactone or similar compounds, which slowly released acidity upon hydration. The acid was required to free calcium from the calcium phosphate or other difficultly soluble salt. A gel-inhibiting or retarding salt was also added. TCP could be used for this purpose because it raised the pH to a point where calcium ions were only slowly released from the calcium salt. However, the most successful gel-retarding agent was found to be SHMP and similar polyphosphates. In some cases, mixtures of the orthophosphates and polyphosphates were found to give optimum results. Steiner reported that his four-component system was useful in preparing fruit jams and jellies, jellied salads and broths, water-jellied desserts and candy jellies. He also mentioned its use in making dental impressions, gelled lubricating and insecticidal compositions, and similar non-edible applications.

Steiner[568] later extended this work to include compositions useful in all of the previous applications, as well as in controlling body and texture of ice cream and other frozen desserts, viscosity control in chocolate milk to prevent settling of cocoa or chocolate, and processed cheese products. By using alginates normally producing low viscosities when hydrated, Steiner was able to stabilize milk products without causing the undesirable thickening after they were cooled, which prevented their flowing over cooling surfaces, a common complaint of dairies that used the alginates at that time. In addition, he was able to show that his improved product overcame the variations in levels of the natural salts in milk that had been produced in different areas of the country or at different seasons of the year.

Roland[569] claimed a process by which he could control viscosity and at the same time stabilize suspended particles in such aqueous dispersions as milk, without having the food product gel or have a slimy mouthfeel. This was accomplished by first adding a polyphosphate glass, such as Graham's salt or SHMP, to an aqueous dispersion containing a di- or trivalent metal ion known to form gels with alginates. The water-soluble alginate salt was then added to the product. Roland claimed that all polyphosphates, from TSPP through long-chain polyphosphates, could perform the necessary functions in his invention. Gibsen[570] prepared a combination of sodium alginate, sodium carbonate, and a calcium salt that could be used in preparation of instant, cold-milk puddings upon the addition of a buffering agent, such as TSPP, and additional calcium salts; when blended with cold milk, the mixture formed a smooth pudding in about 15 min.

Gibsen[571] later prepared cold-water-soluble dessert gels of an improved, smooth, firm texture by implementing a four-component gelling system. He combined (1) slightly soluble, dibasic acids, such as adipic or fumaric acids, (2) sodium alginate, (3) a slowly soluble calcium salt, such as TCP or DCP, and (4) a gel-retarding or sequestering agent, such as SHMP.

Rocks[572] patented a carboxymethyl alginate with which he could prepare cold-water-soluble dessert gels that required only the slowly soluble calcium salt, such as DCP or TCP, and a weak acid, preferably fumaric acid; this mixture did not require the additional calcium sequestering agent. Rocks further reported that this composition was relatively unaffected by the varying mineral content of the water used to hydrate it. Hunter and Rocks[573] prepared a sodium phosphoalginate by making a wet paste of sodium alginate and TSP. This paste could then be dried and utilized to form milk puddings by the addition of a slowly soluble acid, such as glucono-delta-lactone, milk, flavorings, and sweetening agents.

Merton and McDowell[574] also used a four-component system composed of (1) sodium alginate, (2) TCP, (3) an alkali-metal carbonate, and (4) a free acid, such as citric acid, to prepare dry mixes for instant, cold-water-soluble dessert gels. Hunter[575] prepared chiffon pie fillings by utilizing sodium alginate as the gelling agent, skim-milk powder to provide calcium ions, and SHMP or a similar calcium-sequestering polyphosphate salt to retard the release of calcium in order to form calcium alginate gels. Miller and Rocks[576] prepared algin gels having high gelometer readings (high bloom strength) by combinations of sodium alginate, a calcium source such as calcium carbonate, STP, and adipic acid. In cases where additional gel strength was required, other gelling agents, such as gelatin, were added.

Andrew and MacLeod[577] reported that calcium was thought to form cross-links between the polymeric chains of the algin in a type of complex common to many polyvalent metals. They described a number of applications for alginates in food processes; for example, they reported that a delay in thickening for a canned product could be obtained by combining a low-viscosity sodium alginate with a slowly soluble calcium salt, such as DSP$\cdot 2H_2O$. If a starch-thickening agent were used in place of this system, the food product would thicken and gel during processing, thus becoming difficult to fill and reducing heat transfer during the sterilization process. They also prepared fruit- or vegetable-flavored particles by combining sweetening agents, coloring, flavoring, and sodium alginate in a syrup. The syrup was then deposited drop-wise into a calcium chloride bath, and small particles resembling fruit or vegetable pieces were formed. This technique has also been reported to be used to produce an artificial caviar. Andrew and MacLeod have reported several other unique food-processing applications for alginates.

Applications in Carrageenan Gel Systems

Carrageenan is extracted from a seaweed known as Irish moss. The gelling component is a strongly charged, anionic polyelectrolyte of large size and high molecular weight. Like algin, it can interact with other polyelectrolytes, such as proteins and/or the longer-chain polyphosphates. It is capable of forming complexes with proteins and other long-chain molecules that change their physical properties and form gels and sols of their solutions.

Glicksman[564] reviewed the chemical and physical properties of this gelling agent and its many applications in food products. Only some examples of combinations of carrageenan with phosphates to obtain specific physical characteristics will be reviewed here.

Stoloff[578] proposed the use of soluble phosphate salts to precipitate the calcium from milk; the calcium might have prevented or inhibited proper gelling of the carrageenan so that it would suspend chocolate or cocoa particles in chocolate milk drinks. He found DSP especially effective for this purpose. Stoloff[579] later prepared water gels with carrageenan if he first inactivated or sequestered sodium, calcium, and magnesium ions in the tap water used for hydrating the gum. The preferred sequestering agent was a potassium metaphosphate polymer containing 15 to 80 phosphate units per molecule. He reported that other potassium polyphosphate salts from potassium pyrophosphates through the potassium polymetaphosphate were also effective. Zabik and Aldrich[580] investigated the effects of the anions of various potassium salts on viscosities of carrageenan dispersions in deionized water. They reported that no significant differences in viscosity occurred with 0.05 *N* solutions of any of the potassium salts tested, including TKP. Differences were, however, observed at higher concentrations, i.e., 0.10 *N* and 0.20 *N*. The rate of shear appeared to have a more significant influence on viscosity of the carrageenan-potassium salt solutions than the presence of differing anions.

Applications in Other Gum Gel Systems

Two other vegetable gums have been mentioned as being improved for use in foods by the addition of phosphates. One of these is guar gum, which is obtained from the seed of the guar plant. Like the marine gums, guar gum is a long-chain polysaccharide. Weinstein[581] reported that ice cream stabilizers that would produce smooth, desirable characteristics in the ice cream, minimize separation of constituents, and inhibit the growth of ice crystals could be prepared by combinations of guar gum, carrageenan, sodium citrate, and SHMP. He gave several examples of varying formulas for ice cream stabilizers and recommended between 0.02 and 1% stabilizer mix be used, based on the total weight of the ice cream mix. Weinstein further revealed that locust bean gum could also be used in place of the guar gum with equal or possibly even improved results. He reported that the function of the SHMP was to stabilize the proteins of the milk so that they would not be denatured by the presence of the gums, which could act as polyelectrolytes in denaturing the protein.

Green and co-workers[582] reported that they were able to prepare multiphased food products by using various mixtures of gums if the food systems were stabilized by complexing their calcium ions with TSP, TSPP, STP, or SHMP. Calcium salts had to be added to some food systems to assure proper gelling, especially where the food did not contain sufficient calcium ion to form gels. The multiphased food contained an inner "core phase" and an "outer shell phase" and could contain more than two phases. Combinations of the gums, spices, flavoring agents, sugar, salt, and either vegetables or fruits could be used to prepare unique types of food products combining different flavored inner and outer phases. The preferred gelling agents were algin, low methoxyl pectins, carrageenans, and similar vegetable gums. The authors also reported that they could use starches, gelatin, and cellulosic materials, such as sodium carboxymethyl cellulose.

Calcium phosphate gels that contain no vegetable gum gelling agent can also be formed. Marcus[583] reported that he was able to prepare hydrated gels of calcium orthophosphate by combining calcium chloride, DSP, and TSP along with a stabilizing agent, or a compound capable of removing hydrogen ions. Examples of the stabilizing agents were reported to be calcium hydroxide, magnesium hydroxide, calcium carbonate, sodium bicarbonate, and sodium citrate.

Phosphate Applications in Meat Processing

Meat products are perhaps the most universally accepted foods in the human diet. They have formed a major portion of man's diet throughout history. The production, processing to edible forms, and preservation of meat have been major concerns of man since prehistoric times.

Meat contributes an important portion of essential nutrition in the diet. However, the degree to which the meat is accepted and utilized by the consumer is dependent on its palatability. The length of time during which the consumer considers the meat palatable also depends on the degree of preservation of the factors contributing to palatability. Thus, obtaining maximum palatability and preserving palatability are two major concerns in the technology of meat processing.

The palatability of meat is judged largely by three characteristics, namely, its color in the raw and prepared state, its tenderness, and its flavor. Each of these characteristics is directly affected by certain chemical and physical factors that can be measured and controlled to varying degrees.

Although the color of the cooked meat is often used to judge whether it is properly cooked and is directly related to other factors of acceptability for consumption, most consumers use the color of the raw meat as a measure of its freshness and palatability. Therefore, the biochemical reactions occurring between the heme pigments and oxygen in the presence of light and either natural or added constituents of the meat are of extreme importance in determining whether the meat retains an acceptable red or pink color that indicates freshness.

The consumer judges tenderness of the meat only after it has been cooked. However, the tenderness of the meat is directly affected by such chemical and physical factors as water binding, the texture of the meat fibers, the degree of binding if the meat has been ground, the degree to which the muscle fibers have been hydrolyzed through some physical or biochemical tenderization process, and the degree of fat emulsification, which contributes greatly to the eating quality of ground as well as whole meat.

The flavor of the meat is also judged after it has been prepared for consumption. Flavor is affected by such factors as the degree to which soluble proteins are retained within the meat, reduction of cook-out losses that can carry away numerous water- and fat-soluble flavoring compounds, the degree of oxidation of fat and other easily oxidized meat constituents, and the degree of microbiological deterioration.

The preservation of palatability is equally as important as assuring the initial palatability of the meat. The prevention of microbiological deterioration is often considered the most important factor in meat preservation. Another important factor involves the retention of the soluble constituents within the meat. In addition, preventing oxidation is very important in preserving palatability.

Each of these factors overlaps in its effects on the characteristics considered important in meat palatability. Each makes major contributions toward tenderness, flavor, color retention, microbiological preservation, and prevention of oxidation, as will be discussed in the following paragraphs.

The entire history of a piece of meat affects its palatability. Genetic factors that originate prior to the birth of the animal, the rate and amount of growth of the animal, its environment and diet, its treatment prior to slaughter, all processing treatments after slaughter, and the cooking just prior to consumption all contribute to meat palatability. Usually, by the time the piece of meat has been obtained from the animal, the influence of environmental factors is no longer capable of being controlled. However, the effects of the environmental factors can, to some extent, be compensated for through use of the various phosphates. Other factors that are controlled by the processing methods can also be improved through the use of the phosphates.

Studies of the influence of phosphates on the various palatability characteristics of meat have been numerous. A large amount of literature is available, and only a portion could be reviewed in this chapter. The importance of the effects of the phosphates on meat characteristics can be appreciated by reading the numerous reviews found in the literature. Each contains sections, often a major portion of the review, on the influence of phosphates on meat characteristics. Some of the more important reviews used in preparation of this section are those of Grau,[584] Hamm,[585] Saffle,[586] and Karmas.[587] A number of others will be mentioned in subsequent paragraphs.

The Biochemistry of Phosphate Interactions with Meat Proteins

Meat is the muscular tissue of an animal. It is composed of fibers of a protein complex called actomyosin. In living tissue actomyosin is a complex of two proteins, namely, actin and myosin. The ratio of actin to myosin is approximately 1:3. Actin represents approximately 13% of the total muscle protein; myosin represents approximately 38%. These proteins can be separated from extracted actomyosin complex by use of salt solutions. Investigations of their chemical and physical properties demonstrate that actin has an isoelectric point at pH 4.7, while myosin has an isoelectric point at pH 5.4. Actin can exist as a monomer with a molecular weight of approximately 70,000; it can also exist as a polymer, which varies in its number of units, and therefore its molecular weight, with the ionic conditions present in its environment. Myosin is a very large protein molecule, having a molecular weight in the area of 850,000. It is a negatively charged protein at physiological pH having a strong affinity for the calcium and magnesium ions in its environment. Myosin also either has the ability to hydrolyze ATP to ADP, or the enzyme responsible for this hydrolysis is indistinguishably associated with myosin.[588]

The contraction of muscles involves biochemical reactions that require energy obtained through the release of an energy-rich phosphate by ATP, its combination with actin to form actin-phosphate, and the simultaneous release of calcium by myosin. The calcium inactivates a protein fraction also existing in muscle known as the relaxation factor. Myosin also catalyzes the release of the energy-rich phosphate from ATP, thus forming ADP.[588,589] The physical result of these simultaneous biochemical reactions is that the contractile proteins in the muscle fiber slide over each other in a manner quite similar to the closing of a telescope, causing the muscle to contract.

Relaxation of the muscle is caused by the reverse of these reactions. ATP is resynthesized from ADP through the transfer of an energy-rich phosphate from phosphocreatine, or by the release of phosphate through aerobic or anaerobic breakdown of glycogen during normal metabolism. At the same time that ATP is resynthesized, calcium is again bound by myosin, releasing the relaxation factor, and the muscle relaxes.[589] The studies of the reactions of ATP with meat proteins have, according to Kotter,[590] led to the investigations of the reactions of the phosphates with meat proteins.

Once an animal has been slaughtered, however, a different series of biochemical reactions is initiated. The muscles in the freshly slaughtered animal are soft and pliable. As soon as death occurs, ATP in the muscle tissue begins to release a phosphate group forming ADP. The phosphocreatine that exists in the muscle resynthesizes ATP as long as any phosphocreatine exists. However, since metabolism has stopped, the amount of energy-rich phosphates available soon is depleted, and ADP begins to accumulate. At the same time lactic acid begins to accumulate due to anaerobic metabolism of glycogen; this results in a lowering of the pH of the muscles from their physiological pH to a pH ranging somewhere between 5.4 and 5.8. As the pH is lowered, anaerobic metabolism is inhibited, since the enzymes involved are inhibited in their activity as the pH decreases. The muscles now begin to contract through the sliding of the contractile proteins over each other in the same manner in which the muscles contract during life. The animal carcass finally becomes stiff and all muscle tissues become hardened, a state known as rigor mortis.

Meat in the state of rigor mortis is tough. It is, therefore, the practice to age some meats at refrigerated temperatures. During this time, the tissues again begin to soften due to numerous subtle and poorly understood changes in the biochemical and physical environment surrounding the muscle fibrils. Studies with x-ray and with the electron microscope indicate that the intact actin protein chain slides along the intact myosin protein chain to form contracted actomyosin molecules. These molecules have a different energy state than in living tissue due to formation of more numerous hydrogen bonds between the two protein molecules in the complex.[588]

The desirable characteristics of meat palatability are directly affected by the biochemical reactions undergone by the actomyosin complex after slaughter of the animal. Tenderness is influenced by the retention or dissociation of the actomyosin complex into its component proteins, actin and myosin. The dissociation of the actomyosin complex is influenced by the type and amount of salts present,[591-598] the pH of the system,[586,591,592] and, although this is disputed, possibly the amount of calcium, magnesium, and other polyvalent metal ions present in the system, as these ions would maintain the proteins in an insoluble state.[599-605]

Evidence that the chemical and physical state of the meat proteins affected some factors important to meat palatability can be found in the studies of water- and salt-soluble proteins in meat products. Turner and Olson,[606] Saffle and Galbreath[607] and Acton and Saffle[608] reported that pre-rigor meat contained 50% more salt-soluble proteins than post-rigor meat held at 4°C for 24 to 48 hr after slaughter. Acton and Saffle[608] also found that the salt-soluble proteins from pre-rigor meat would emulsify over 60% more fat than the salt-soluble proteins from post-rigor meat, indicating a significant difference in the chemical nature of these proteins.

Saffle[586] defined the salt-soluble proteins as those that are soluble in solutions of sodium chloride or similar ionic-strength salts and are insoluble in pure water. He cited actomyosin as an example of a salt-soluble protein. Water-soluble proteins are those proteins that are soluble in pure water and insoluble in water containing salt. The numerous sarcoplasmic proteins are water-soluble; they are contained within the sarcoplasm, a syrupy fluid containing enzymes, mitochondria, and the oxygen-binding protein, myoglobin, which gives muscle its characteristic color. Actin is also a water-soluble protein.

Szent-Gyorgyi, cited in Saffle,[586] extracted 10 to 15% actin with water from the total muscle protein prior to the onset of rigor mortis (termed pre-rigor muscle), while post-rigor muscle contained very little water-soluble actin. The presence of the free actin is believed responsible for the improved binding of pre-rigor meat over post-rigor meat when used in making sausage and other comminuted meat products.[584,586,593,595-597]

The water-soluble proteins and salt-soluble proteins differ in their abilities to emulsify fat, an extremely important factor in preparation of comminuted meat products, such as bologna and sausage; this is also a factor in fat retention in cured meats, especially when they are being smoked or cooked. The salt-soluble proteins are generally regarded as superior to water-soluble proteins in their ability to emulsify fat.[586,609-615] Kotter[609] and Hansen[610] found that salt-soluble proteins formed a stabilizing film at fat-globule surfaces. Carpenter and Saffle[613] reported that salt-soluble proteins would emulsify 70% more fat than water-soluble proteins. They explained that water-soluble proteins were globular in shape with their length-to-diameter ratio approximately 4:1, while salt-soluble proteins were more fibrous with a length-to-diameter ratio of 200:1. Swift[615] reported that the salt-soluble proteins would emulsify 30 to 400% more fat than water-soluble proteins. Most of these investigations have referred to the influence of water- and salt-soluble proteins on comminuted meat emulsions.

Similar advantageous properties of these proteins must also exist in meats that are processed and consumed without being comminuted, although the previously discussed investigations employed ground beef, pork, or rabbit muscle. Treatment of hams with curing-salt mixtures increased the retention of fat during the smoking process. Less fat was lost during cooking broiler carcasses and parts when they had been treated with salt, salt and phosphate mixtures, or phosphate overnight prior to cooking.[616]

A number of investigators have demonstrated that the addition of polyphosphates to sodium-chloride solutions increased the ability to extract salt-soluble protein. Bendall[593] proposed that the ability of phosphates to increase the extraction of salt-soluble proteins was due to their effect in dissociating the actomyosin complex. The most effective compounds were reported by Bendall to be pyrophosphate with tripolyphosphate next in effectiveness. The pyrophosphate was found to have twice the effect of a solution of sodium chloride of ionic strength equivalent to that of 1% sodium chloride and 0.5% pyrophosphate. Fukazawa and co-workers[595] reported that combinations of pyrophosphate, tripolyphosphate, and hexametaphosphate with salt increased the extraction of salt-soluble proteins; pyrophosphate was most effective, tripolyphosphate less effective but more effective than hexametaphosphate, which also was more effective than the salt control. Significantly greater quantities of protein were extracted by the salt-phosphate combinations in both native and denatured fibril protein. The work was extended by Yasui and co-workers,[596,597] who demonstrated that the pyrophosphate anion increased the solubility of myosin in suspensions of meat proteins. They also demonstrated that tripolyphosphate and hexametaphosphate did not cause actomyosin to dissociate to actin and myosin, except after these polyphosphates have been hydrolyzed to the pyrophosphate anion. In addition, these authors demonstrated that polyphosphates with chain lengths longer than tripolyphosphate were bound directly to the myosin protein except in the presence of high concentrations of sodium chloride or of calcium and magnesium ions. Their work also showed that the myosin molecule acted as an enzyme to hydrolyze tripolyphosphate to pyrophosphate but only in the presence of divalent cations and at high ionic strengths.

Several investigators have demonstrated that the action of polyphosphate on post-rigor muscle proteins caused the dissociation of actomyosin into its component proteins, actin and myosin.[593-598] In this respect, the phosphates, particularly tripolyphosphate and pyrophosphate anions, were similar to ATP in their action on actomyosin. The work of Fukazawa and co-workers[594,595] and Yasui and co-workers[596,597] demonstrated that the pyrophosphate anion was the effective anion in causing the dissociation of actomyosin and that the longer-chain polyphosphates could only exert similar activity if they had been hydrolyzed to pyrophosphate, either through enzymatic activity or by the pH of the solution. These investigators reported that pyrophosphate had an immediate effect in promoting the dissociation of actomyosin as shown by sharp increases in the soluble protein in the aqueous phase when muscle proteins were treated with a solution of pyrophosphate. The activity of tripolyphosphate extended over a longer period of time, indicating to Yasui et al.[597] that the tripolyphosphate first had to be hydrolyzed to pyrophosphate before the effects were noticed. Hexametaphosphate was effective over even longer periods of time, again indicating to the authors that it was first necessary for the hexametaphosphate to be hydrolyzed so that pyrophosphate anions would be present before the effect was noticed.

The effects of pH on the amounts of soluble proteins extracted have been thoroughly studied and reported.[586,591-597] Grau and co-workers[592] studied the effects of pH values below and above the normal pH of the muscle tissue. When acid was added to lower the pH from approximately 5.5 to 4.5, no change was apparent. However, as the pH was reduced below 4.5, the muscle tissue absorbed increasing quantities of water and finally took on the appearance of a gel with a greatly increased ability to bind water. When the pH was raised from 5.5 to 7.5, the meat retained a fibrous consistency up to pH 7.0 and then became paste-like as the pH was raised to 7.5. Fukazawa and co-workers[595] demonstrated, as shown in Figure 4, that pyrophosphate, tripolyphosphate, and hexametaphosphate extracted increasing amounts of protein as the pH was changed from 5.6 to 7.0 with the greatest increase occurring at pH 7.0. In all cases pyrophosphate was more effective than tripolyphosphate and hexametaphosphate. Controls containing 0.6 *M* sodium chloride solution also extracted an increased quantity of protein as the pH was increased, although to a lesser extent than in the presence of phosphate; this confirmed work of numerous other investigators. Yasui and co-workers[596] demonstrated that the extractability of proteins from myofibrils was a function of pH and salt concentration. In an extension of this work, they demonstrated that the ability of myosin to catalyze the hydrolysis of tripolyphosphate to pyrophosphate was dependent on the ionic strength of the solution, the presence of

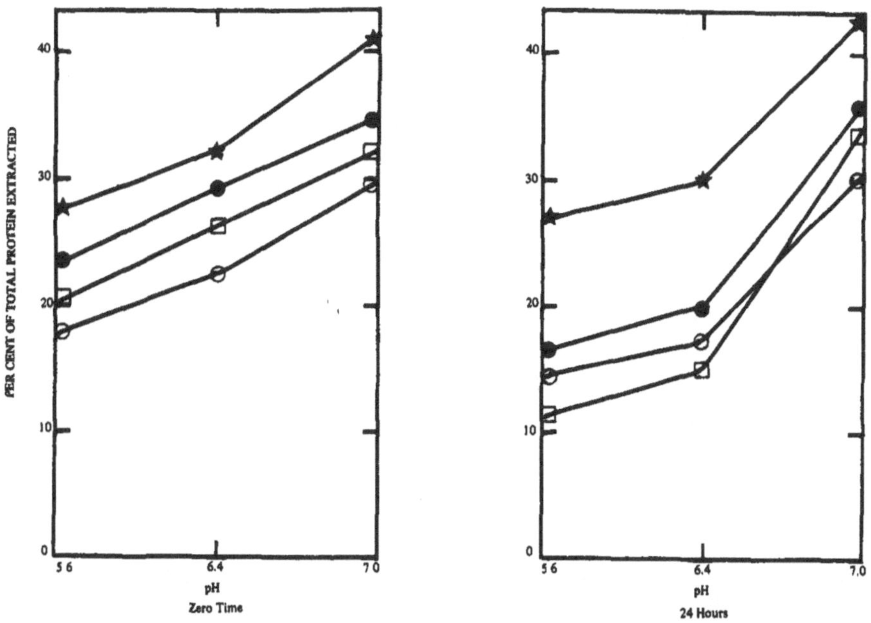

FIGURE 4. Effect of polyphosphates on the extraction of protein from intact muscle fibrils. ○ = Control; ★ = Pyrophosphate; ● = Tripolyphosphate; △ = Hexametaphosphate; Concentration of all polyphosphate = 10^{-2} M to muscle fibrils. (Adapted from Reference 595, Figure 7, with permission.)

divalent metal ions, and the pH.[597] This enzymatic activity was greatest at the more acidic pH values, which was the direct opposite of the effect of myosin as an enzyme for hydrolysis of ATP. The enzymatic hydrolysis of ATP was found to have its optimum at approximately pH 6.8. No enzymatic activity for the hydrolysis of tripolyphosphate was demonstrated in a myosin solution containing 0.6 M potassium chloride at any pH in the absence of a divalent cation.

There is a controversy over whether the effect of phosphates on meat proteins is the result of sequestering divalent cations, particularly calcium, magnesium, and zinc.[600,617] Conflicting data have been reviewed by Mahon[617] and the results of investigations supporting opposing opinions have been reviewed by Hamm,[585] Inklaar,[604] and Sherman.[618]

A number of investigators believe that one of the major functions of the polyphosphates is to sequester calcium from its complexes with the meat proteins, though many admit this is probably not the only function.[584,585,599,601,619-630] Hamm[585,620-626] published considerable evidence for calcium binding by the phosphates. He proposed that calcium was bound to meat proteins in such a way that it formed bridges between the protein fibers and, therefore, was directly involved in the folding of the protein molecule.[621] Thus, calcium may have inhibited meat hydration through its ability to maintain the folded protein structure that prevented the hydration of those hydrogen-bonding sites within the folds. When the bound calcium was replaced by sodium in an ion-exchange reaction with the subsequent formation of a calcium-phosphate complex, the protein molecule was able to unfold, exposing sites that could then form hydrogen bonds with water. Hamm et al.[620-628] developed this explanation for the water binding and swelling of meat proteins by the action of polyphosphates; perhaps magnesium was the principal cation involved in water binding and hydration of meat protein.

Zinc also was considered to be as important as calcium.[622,623] Swift and Berman[629] found a highly significant correlation between the content of bound zinc ion in fresh beef muscle and its ability to bind water; water binding increased in direct proportion to the decrease in zinc content. Swift and Berman also reported that water binding increased as the levels of calcium and magnesium ion decreased, supporting evidence presented by Hamm and others.

Epstein[601] reported that after slaughter the calcium in the animal tissue causes a hardening of the cell walls that prevents release of the meat proteins. He patented a composition of TSPP, TKPP, DSP, and one or more of the longer-chain polyphosphates to soften the cell membranes and, therefore, allow moisture to enter through osmosis. He believed that the phosphates replaced the calcium with sodium and potassium ions through ion-exchange reactions that solubilized the proteins, softened the cell walls, and allowed moisture to enter the cell. Baldwin and deMan[605] related mineral equilibria to some of the physical properties of meat. The untreated meat used in their investigation contained 42% calcium, 71% magnesium, 82% phosphorus, and 80% citric acid in a "free" or soluble form. The meat was tenderized when treated with DSP, TSPP, STP, sodium tetrapolyphosphate, or SHMP. This was not believed to be due to changes in pH since no correlation between the tenderness and pH of the meat was found. The addition of citrate and some of the phosphates caused significant changes in pH which could possibly have influenced the solubility of the muscle protein. When the meat was treated with phosphates, the soluble-protein-calcium complexes and colloidal-calcium increased with increasing phosphate chain length. Also, as the chain length of the added polyphosphate increased, the levels of calcium bound by the insoluble proteins decreased. The authors believed that phosphate removed calcium from its bound state in the insoluble protein. The longer-chain polyphosphates, including STP, sodium tetrapolyphosphate, and SHMP were the most effective calcium-complexing agents. Sodium citrate and TSPP were approximately equal in complexing ability, while MSP and DSP showed little or no calcium-complexing ability. Magnesium levels were not affected by these treatments.

Other investigations indicated that binding alkaline-earth metal ions does not explain the effect of phosphate on meat. Bendall[593] reported that the addition of calcium and magnesium ions at the concentrations found in tap water had no significant action on the water-binding properties of meat treated with pyrophosphate. This result might be explained by the fact that the levels (0.0025 *M*) of calcium and magnesium added were low. Swift and Ellis[631] reported that the water retention of beef was slightly improved when treated with a dilute solution of magnesium chloride. This confirmed a number of investigators' reports that magnesium ion caused beef and other meat to bind greater quantities of water. Wierbicki and co-workers[632] reported that calcium chloride and magnesium chloride increased the fluid retention of beef when heated to 70°C. The latter two investigations indicated that the presence of calcium and magnesium could have effects similar to those that occurred upon addition of the polyphosphates. Therefore, if the ability of the phosphates to complex alkaline-earth metal ions explained their effects upon meat protein, it seemed that phosphate addition should decrease, rather than increase, the water-binding properties.

Kotter[591,602,609] did not believe that the calcium-phosphate complexing mechanism caused these effects of phosphates on meats. He reported that several uncontrollable factors were measured at the same time which had nothing to do with the calcium-binding ability of the phosphates. Hamm[585] reported that the polyphosphates were capable of lowering the calcium and zinc content of meat to a lesser extent than the water-holding capacity. He reported that only the free anions, and not those bound in complexes, were capable of influencing the hydration of muscle proteins; e.g., calcium tripolyphosphate had no effect on the water-holding capacity of the meat.

Sherman[618] reported that the ability of the phosphates to complex calcium and magnesium ions could not be attributed to their significant effectiveness in improving fluid retention of meats. He compared the effects of treating meat with di- and tetrasodium EDTA with the effects of treating the meat with a commercial polyphosphate mixture. The data obtained are shown in Table 25. He found that the meat treated with the disodium EDTA had pH values ranging from 5.2 to 5.4, while the pH of the meat treated with tetrasodium EDTA was very similar to that of the meat treated with the commercial polyphosphate, which was significantly more effective in increasing the fluid retention of the meat. Tetrasodium EDTA increased water binding and disodium EDTA reduced water binding, indicating that there may have been some influence of pH involved. The addition of sodium chloride to the EDTA derivatives suppressed their ability to increase the water retention of the meat, while the addition of sodium chloride to the polyphosphate mixture either improved its water retention ability or did

TABLE 25

Comparison of the Effects of EDTA Sodium Salts and Polyphosphates on the Water Retention of Minced Pork[1]

Additive	pH of aged meat-additive mixture	% concentration of additive solution	% water retention at 0°C
Na_2 EDTA	5.4	1.0	-4.0
Na_4 EDTA	6.6	1.0	14.0
Polyphosphate[2]	6.5	1.0	23.0
Na_2 EDTA	5.3	2.0	-2.0
Na_4 EDTA	7.0	2.0	38.0
Polyphosphate[2]	6.9	2.0	57.0
Na_2 EDTA	5.2	4.0	-3.0
Na_4 EDTA	8.5	4.0	43.0
Polyphosphate[2]	7.5	4.0	58.0
Na_2 EDTA +	5.0	2.0	-7.0
NaCl		1.0	
Na_4 EDTA +	6.7	2.0	26.1
NaCl		1.0	
Polyphosphate +	6.4	2.0	58.2
NaCl		1.0	

[1] From Reference 618, Table 7, with permission.
[2] A commercial polyphosphate.

not change it. In addition, Sherman reported that the meat treated with EDTA derivatives was very crumbly, lacking the firm texture that was obtained with the meat treated with polyphosphate mixture.

Hellendoorn[603] also compared the effects of disodium EDTA and ammonium oxalate on meat to determine whether the binding of calcium and magnesium ions influenced the properties of the meat. He confirmed that disodium EDTA depressed the water-binding effect of sodium chloride on meat, while the addition of phosphates, particularly pyrophosphate and tripolyphosphate, improved water binding under identical conditions of pH and ionic strength. Although he obtained no depressing effect, as occurred with the EDTA salt, orthophosphates, hexametaphosphate, and longer-chain polyphosphates did not improve the water retention of the meat. Hellendoorn investigated the influence of the complete removal of calcium by precipitating it with oxalate. He found that the water-binding ability of meat treated with 2% sodium chloride plus 0.6% ammonium oxalate was identical to meat treated with only 2% sodium chloride, as shown in Figure 5. Thus, eliminating calcium from the meat had no effect on water-binding capacity. He also tested a combination of pyrophosphate with oxalate and found that the effect on water binding was almost identical with the effect of pyrophosphate alone. He was unable to perform similar experiments with magnesium and zinc ions because of the lack of suitable precipitating agents. However, when amounts of magnesium and zinc equivalent to those present in the meat were added along with salt and pyrophosphate, the treatment resulted in a slight depressing effect in water or fluid retention in the meat. A similar result occurred when calcium ion in the meat was doubled.

Wismer-Pedersen[633,634] reported that the addition of EDTA to fresh meat prior to freeze drying resulted in better penetration of water into the dried meat structure when it was being rehydrated, but did not increase the amount of water bound. Pyrophosphate, added to the fresh meat prior to freeze drying, increased the water

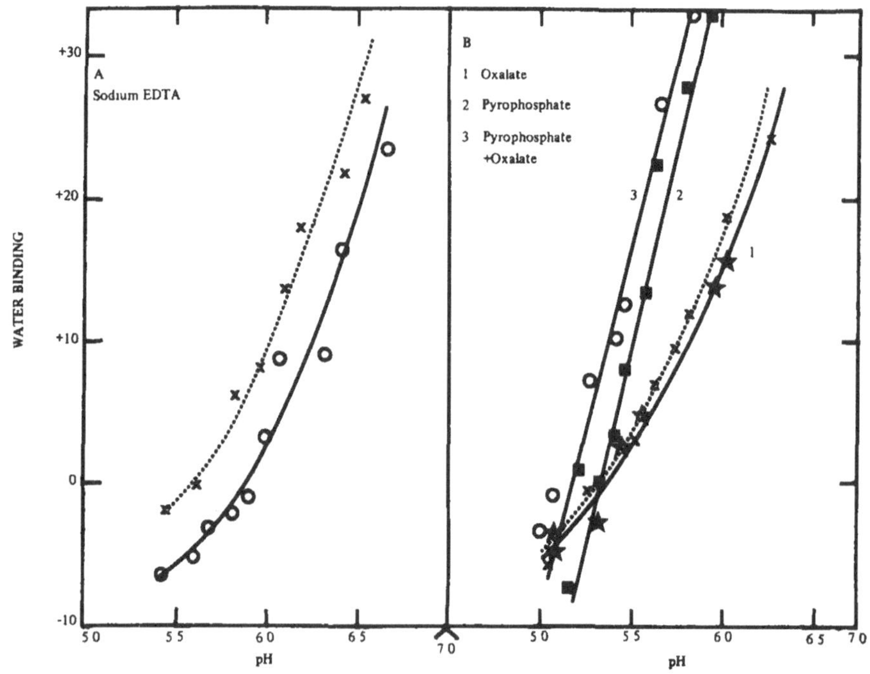

FIGURE 5. Effects of EDTA (A) and oxalate (B) on cooked meat in comparison to sodium chloride.

A. 2.5% NaCl + 0.6% $Na_2H_2EDTA \cdot 2H_2O$ ($\mu = 0.52$)
Dotted Line 3% NaCl alone ($\mu = 0.51$)
B. Curve 1. 2% NaCl + 0.6% $(NH_4)_2C_2O_4$ ($\mu = 0.49$)
Curve 2. 2% NaCl + 0.5% SAPP ($\mu = 0.48$)
Curve 3. 1.5% NaCl + 0.5% SAPP + 0.6% $(NH_4)_2C_2O_4$ ($\mu = 0.51$)
Dotted line 3% NaCl alone ($\mu = 0.51$)
From Reference 603, Figure 5, with permission.

binding of the meat proteins when they were already saturated. He reported that the EDTA apparently removed calcium and magnesium ions from the meat proteins by chelating these ions which seem to prevent the penetration of water into the dried meat fibers when the untreated meat is rehydrated. The effect of pyrophosphate seemed to be completely different from that of the EDTA. It increased the rehydration capacity to approximately the same degree as EDTA did as far as allowing penetration of water. However, the pyrophosphate-treated dried meat swelled to a greater extent in all of the saturated areas due to water binding by the meat proteins in the areas penetrated by the water. This effect of increasing water binding also would decrease the efficiency of freeze drying since it would cause the meat proteins to retain the water to a greater extent. The texture of the rehydrated, untreated, control samples was significantly tougher than the fresh meat after identical cooking processes, but the meats treated with EDTA and pyrophosphate were significantly less tough and many had tenderness readings falling within the range of fresh meat.

The biochemical effects of the polyphosphates on meat proteins which, in many cases, cause them to retain or absorb and bind greater quantities of water have caused considerable dispute and concern regarding whether they should be used in meat processing. Many governmental regulatory agencies, particularly in some European countries, fear that the consumer will be defrauded by purchasing greater amounts of water in phosphate-treated meats. However, Kotter[590] and Popp and Muhlbrecht[635] were firmly convinced that the polyphosphates in no way increased the water absorption capacity of meat beyond that which it normally possessed before the animal was slaughtered. The polyphosphates merely restored the water-holding ability of the meat which began to decrease with the onset of rigor mortis.[585] Water binding reached a minimum when rigor mortis was complete. Aging the meat restored water-binding ability, but never to original levels of the living tissue. Treating the meat with phosphates, in the presence of the proper types and levels of alkaline-earth metal ions and optimum ionic strength and pH, restored its original water-binding capacity.[590,635] Morse,[600] Ruf,[25] and Grau,[584] emphasized that the use of phosphates would neither improve the quality of poor meat nor overcome faulty production practices. To quote Morse,[600] they could only be used to "make a good product better, but cannot make a poor product good."

Numerous commercial blends and mixtures of polyphosphates with and without other salts are available, as will be discussed in following sections. Federal regulations allow polyphosphates in some products. It is highly advisable that any processor contemplating the use of polyphosphates contact the regulatory agencies in his country to determine whether polyphosphates are approved for use in his intended application.

General Meat Applications

Color Preservation

As discussed previously, the acceptability of meat is measured by the consumer by its pre-cooked color, its cooked tenderness, and its flavor. The phosphates can influence these characteristics and some general applications for fresh and cured meat will be discussed in this section.

Excellent reviews discussed the chemistry of the pigments which gave meat its characteristic red or pink color. Watts[636] discussed the factors involved in discoloration of both fresh and cured meats. These reactions involved oxidation and were influenced by numerous chemical and physical characteristics of the meat, such as its pH, the presence of reducing substances, and its exposure to oxygen. The presence of curing salts and metal ions also played important roles in protecting or discoloring the meat. The color of the meat was dependent upon the presence of and the chemical reactions of two pigments, myoglobin and hemoglobin, which are iron-containing proteins. Although a piece of meat has been well-bled, the blood pigment, hemoglobin, may amount to as much as 30% of the pigment in the piece of meat. Myoglobin is a constituent of the muscle tissues and can undergo numerous oxidation and reduction reactions which have dramatic effects on its color. Normally, a purple pigment in the reduced state, myoglobin is capable of forming a complex with oxygen at the freshly cut surface of meat which forms a bright red pigment, oxymyoglobin.[587,637] As discussed in the excellent review of the chemistry of meat pigments by Fox,[637] oxymyoglobin and myoglobin itself contained the ferrous form of iron and were easily converted to the ferric form through an oxidative mechanism to form the brown pigment, metmyoglobin. After freshly cut surfaces of meat were exposed to air for a period of time, the oxymyoglobin slowly began to form the brown pigment, metmyoglobin. In the eyes of the consumer, such brownish-colored meat is not as fresh as the bright red-colored meat, although this has no bearing on the freshness or edibility of the meat. Consequently, meat manufacturers are continuously searching for ways to stabilize the natural, bright red color of freshly cut meat.[638]

Fresh, aged meat normally ranges in pH between 5.2 and 6.6. Several investigators found that low pH accelerated the oxidation of the fresh meat pigments and formed brown-colored metmyoglobin pigments.[636] Optimum stabilization of the red color of the meat was obtained at the higher ranges of pH of fresh meat, i. e., in the area of pH 6 to 6.6. The patents by Hall[638] and Brissey[639] revealed that the color of meat could be stabilized if the pH was maintained in the area of 6.5[638] to 6.8[639] by the use of polyphosphates or dialkali phosphate, respectively. Chang and Watts[640] reported that the presence of polyphosphates at pH 5.6 had an adverse effect on color because they accelerated the oxidation of oxyhemoglobin to methemoglobin. However, the patents of Hall and Brissey specifically mentioned that the phosphates had no effect at these low pH's.[638,639]

Increasing Tenderness

The tenderness of meat is also an important characteristic of its palatability, involving a number of physical and biochemical changes in the meat tissue. The phosphates' ability to cause the dissociation of actomyosin into actin and myosin has an important effect on the tenderness of all types of meat. As mentioned previously, the onset of rigor mortis causes a shortening of muscle fibers as the contractile muscle proteins slide over each other and form highly stable hydrogen bond complexes as they stiffen. Pyrophosphate combined with magnesium can cause a relaxation of the muscle fibers.[641] Thus, the tenderizing effect of the polyphosphates might be explained by the action of pyrophosphate anions, formed by the addition of pyrophosphates, or through hydrolysis of longer-chain polyphosphates. Specific applications of polyphosphates in tenderizing fresh and cured meat will be developed in subsequent sections.

Increasing Binding

Rebinding of pieces of meat, whether small chunks or ground meat particles, is important in obtaining the desired texture in numerous meat products. Proper binding of meat chunks allows one to form large pieces which can be sliced in a manner similar to a whole piece of meat by causing natural exudates to come to the surface to bind the piece together.[587] Since these natural exudates contain the necessary soluble proteins, phosphates will improve binding.

Binding of ground meat is very important in preparation of such meat products as hamburger, meatloaf, and various types of fresh and cured sausage. Numerous investigators demonstrated that the salt-soluble proteins were the materials responsible for excellent binding in ground meat products. Fukazawa et al.[594,642] reported that water-soluble meat protein had no appreciable binding properties in sausage. Also, myosin-poor meat fibrils and the "ghost" fibrils remaining after the extraction of myosin, probably the "insoluble proteins" of Baldwin and deMan,[605] did not provide proper binding of sausage. True myosin provided the proper binding qualities to sausage products.[642]

Fukazawa[595] and Yasui et al.[596,597] clearly demonstrated that the polyphosphates were capable of increasing the solubility of myosin, particularly when pyrophosphate was combined with salt in the treatment of the ground meat. As reviewed previously, the pyrophosphate anion is the important anion in solubilizing myosin, and the longer-chain polyphosphates must hydrolyze to the pyrophosphate anion before they become effective. In addition, these investigators demonstrated that hexametaphosphate and long-chain polyphosphates are bound in complexes by the proteins; this binding is inhibited by both high salt concentrations and the presence of divalent cations.

Increasing Moisture Retention

The moisture retention, water binding, or water-holding capacity of muscle tissue is of considerable concern to the meat industry, as the loss of moisture can lead to serious economic losses. However, of greater concern is that the meats which lose moisture are less acceptable to the consumer from the standpoint of appearance, reduced tenderness, and reduced juiciness. Also, greater cooking losses occur when meat tissue has been exposed to the proper environment for excessive loss of moisture. Frozen meats can show increased "freezer burn" due to moisture loss in the frozen state. As a result, the meat industry is constantly searching for methods for preventing loss of moisture.[638]

The loss of moisture by meat increased as its pH approached the isoelectric point of its proteins, near pH 5. Freshly slaughtered meat gradually decreased in pH with the onset of rigor mortis.[585,592,636] The water-binding or water-holding capacity of meat increased as the pH was either decreased or increased away from the isoelectric point.[592] The water-holding capacity of meat dropped rapidly after slaughter.[585] According to Hamm, this can be prevented by salting ground or cut meat immediately after butchering, thus preserving the water-holding capacity for as much as one week. The loss in water-holding capacity does not appear to be due to the association or complexing of myosin and actin to form actomyosin since the presalting does not influence the formation of the actomyosin. Rather, the retention of high water-holding capacity appears to be due to the binding of chloride ions by the muscle proteins. This is especially true when ATP is still present in the muscle tissue as is the case immediately after the slaughter of the animal. Hamm[585] speculated that the combined action of the salt and ATP caused the peptide chains to unfold, leaving such large distances between the chains that the bivalent cations released by breakdown of ATP were unable to cross-link the chains. Thus, water was able to reach the numerous hydrogen bonding sites necessary for complete hydration of the proteins. Through their ability to cause dissociation of the actomyosin complex, and possibly, though this is disputed, through their ability to complex divalent cations, the polyphosphates were capable of increasing the water-holding capacity of the muscle proteins. Hamm and Grau[626] reported that the polyphosphates were active in the following order of *decreasing* activity: STP, sodium tetrapolyphosphate, SHMP, TSPP, cyclic sodium metaphosphate, and DSP. Sodium citrate was less effective than sodium pyrophosphate, but more effective than the cyclic sodium trimetaphosphate. Sodium oxalate was found to be approximately as effective as SHMP.

Flavor Improvement

The flavor of meat develops as the meat is being cooked, but its origin is in the proteins in the muscle tissue. Much of the flavor of cooked meat develops on the surface of the meat; however, the treatment of the meat prior to cooking also can significantly affect the flavor. For example, beef must be aged for at least a week before the most acceptable beef flavor is produced even though it is not tasted until after it has been cooked. Also, when meat is broiled or roasted, the surface crust of the meat contains the greatest amount of the typical flavor. Considerable variations in the flavor of meat can occur and it is, therefore, desirable to do everything possible to obtain maximum acceptable flavor. Attempts have been made to extract the flavoring components from meat, but these have been relatively unsuccessful. It has been hoped that meat flavoring components could be prepared either from low-grade cuts of meat or by synthesis so that these could be added to higher quality meat in order to standardize the flavor at the most acceptable levels.[587]

Phosphates can contribute toward the flavor of meat since they have such significant effects on the characteristics of the proteins and since the proteins in the meat contribute so significantly to the development of proper flavor. Phosphates have also been used in the preparation of meat-like synthetic flavors from proteinaceous materials by Giacino.[643] Combinations of vegetable protein hydrolysates, various fatty acids, a sulfur-containing amino acid, and orthophosphoric acid and its ammonium and potassium salts have been patented for preparations of beef, pork, and chicken flavors.

Preventing Off-Flavors

In addition to contributing directly to the formation of desirable flavors through their interaction with proteins and their use in synthetic meat flavor compositions, the phosphates are also useful in preventing the development of off-flavors in meat products. These will be discussed in considerable detail in following sections. Through their ability to act as antioxidant synergists, the phosphates prevent oxidation of fats and thus prevent development of rancid flavors in the meat fats. As a result, they inhibit development of off-flavors and consequently have an important role in flavoring meat products.[636]

Preventing Microbiological Spoilage

Prevention of the development of rancid flavors in meat could be considered a part of the preservation of meat. However, most people consider microbiological spoilage a major factor in spoilage of meat and, therefore, when one speaks of preservation of meat, it is usually in reference to preventing its microbiological spoilage. Here again, the phosphates play an important role. Healthy animals are capable of preventing invasion of microorganisms and, therefore, freshly-slaughtered meat is considered to be sterile by some investigators. However, immediately after slaughter, invasion by billions of microorganisms has begun and freshly cut meat cannot help but be heavily contaminated with these forerunners of spoilage.[587] As a result, one of the major objectives of the meat industry is to prevent its spoilage to the point of becoming inedible.

Keil and co-workers[644] developed a process which could be used on numerous foods, including any meat product capable of being dipped into and coated by a solution, e. g., sausage, pieces of fresh meat, and cured meat products of all types. The process involved dipping the meat into a heated solution (42 to 60°C) of a relatively high bloom-strength gelatin containing sodium polyphosphate and an edible acid capable of converting the sodium compound into polyphosphoric acid. The solution adhering to the surface then gelled and formed an antibiotic coating about the meat. One example in the patent described the use of 1,600 g gelatin, 400 g SHMP, and 120 g hydrochloric acid in 4,800 ml water. The thickness of the coating could be regulated by the temperature, since the viscosity of the solution decreased as the temperature increased. Thus, the lower the temperature, the thicker the coating. Other more specific microbiological preservation methods involving phosphates will be discussed in subsequent sections.

Applications in Fresh Whole Meat

Color Preservation

Kamstra and Saffle[645] injected or infused hams from freshly slaughtered hogs with solutions of SHMP and refrigerated them for 72 hr producing pork meat with high pH values, very dark-colored lean meat, higher levels of glycogen, and a sticky surface. The results indicated that the treatment produced "dark cutter" meat. This undesirable condition was corrected by infusing the hams with a combination of a 23.3% SHMP solution plus a solution of lactic acid. Paired hams from the same carcass were used in this study with one of the pair as a water-injected control, while the other was given the desired treatment. The pH and the color of the hams treated with a combination of SHMP and lactic acid solution were significantly improved.

Carpenter and co-workers[646] studied the effect of injecting solutions of various levels of SHMP into the left round cuts of beef carcasses and compared their color in a paired comparison experimental design against that of the right rounds from the same carcass which received an equal quantity of injected tap water. The results showed a significantly darker red color in the phosphate-treated rounds when compared to the water-treated rounds. Although a number of other investigators have reported the results of phosphate treatment of fresh whole meats, they apparently failed to observe or to report observations of the effects of the phosphates on the color of the fresh meat.

Voegeli and Gorsica[647] found that they could retain an attractive, uniform, bright red color in fresh meats if they treated the meat with a synergistic mixture which would contribute 0.02 to 2% of a phosphate buffering agent, 0.01 to 0.2% of an oxidation-reduction controlling poising agent, and 0.01 to 0.4% of a sequestering agent capable of inactivating any polyvalent metal ions present in the system. The authors preferred to use a polyphosphate which would act both as a buffer and a sequestering agent. Further work indicated that additional stabilization could be obtained in fresh meats if alkali metal salts of parahydroxybenzoic acid were added to the above mixture.[648]

Rust[649] investigated the effect of injected SHMP on the color of the crust and interior of cooked beef steaks. No effect on color could be correlated with the phosphate treatment at any of the levels investigated.

Increasing Tenderness

There is considerable evidence that the polyphosphates will tenderize fresh whole meat. In 1938 Allen[650] reported that the use of aqueous solutions of enzyme, MSG, sugar, TCP, and salt would tenderize meat. Mullins and co-workers[651] reported that hams injected with SHMP solutions were scored as more tender by panels than control hams although there was no statistically significant difference in the scores. Investigators found that meat was most tender immediately after slaughter, but became tougher with the onset of rigor mortis over the next 24 to 72 hr.[645] As a result, the tenderness of meat depended a great deal upon the reactions which occurred during the development of rigor mortis. Kamstra and Saffle injected fresh pork loins with 2.5 or 5% of a 23.3% SHMP solution and other hams with the same strength solution of SHMP but with 3, 6, 10, or 15.7% solutions of lactic acid until the weight of the hams increased 2.5 or 5%. The hams treated with the phosphate or phosphate and lactic acid solution were taken from one side of the carcass and the hams from the opposite side of each carcass were injected with water to increase their weights to the same amount as those of the treated hams for use as controls. All of the hams injected with phosphate solution, irrespective of whether lactic acid was present or not, were significantly more tender, as shown by taste panel scores and by Kramer Shear Press values. The authors made no attempt to determine the lowest level of phosphate which produced significant tenderness in the hams. The 23.3% solution was used because it was the highest level which could be injected without introducing a noticeable flavor due to the phosphate treatment. Carpenter and co-workers[646] extended this work to study the effects of phosphates on beef. Beef rounds were taken from the left sides of carcasses and injected with 10, 15, and 20% solutions of SHMP until the weight had increased 5% above the original weight. The rounds taken from the right side of the same carcass were injected to similar weight increases with tap water. Again, evaluations by taste panels demonstrated that there was a significant increase in tenderness in the phosphate-treated rounds. The differences were significant at the 1% level for rounds treated with the 10% solution and at the 5% level for those treated with the 15% solution. Warner-Bratzler Shear values again demonstrated that the phosphate-treated rounds were more tender than the paired control rounds, but the differences were not significant on a statistical basis.

Hopkins and Zimont[652] demonstrated a significant increase in tenderness when pieces of beef were treated with a solution containing approximately 1% by weight of sodium pyrophosphate and between 10 and 22% by weight of sodium chloride. The meat was either dipped in the solution for approximately 2 hr or injected with the solution. The optimum temperature for treatment with the dip was 34 to 42°F. Dipping the meat in the solution allowed it to take up approximately 3% of its weight in the form of a coating on its surface. Up to 10% by weight of the solution was injected into the meat, as the authors preferred not to inject any greater quantities. After treating the meat with the phosphate solution, it was necessary to store it for a period of time at refrigerated temperatures which allowed tenderization to occur. Slices of meat dipped in the solution were held from 18 to 24 hr. Larger pieces of meat, such as roasts, were held from 1 to 8 days, depending on the method of phosphate treatment. Both taste panel results and shear values with a Warner-Bratzler machine demonstrated the tenderization when compared against untreated controls. Williams[653] reported that the reason for the tenderizing action of hypotonic phosphate-sodium chloride treatments is that the internal salts in the meat cells draw water through the cell walls and retain it.

Rust[649] studied the effects of the injection of four levels of a 0.3 *M* solution of SHMP on the palatability of defrosted 2 in.-thick beef steaks cut from U. S. Standard Grade beef. After injecting the steaks with 5, 10, and 15% of their weight of the SHMP solution and controls with equivalent quantities of water, the steaks were aged for 18 hr at 3°C prior to cooking and evaluation. The phosphate-treated steaks were judged more tender than the untreated controls by a panel scoring them for initial tenderness impressions and for the number of chews required to completely masticate the steaks. Shear Press values also confirmed the increased tenderness of the steaks. The results indicated that maximum tenderization occurred in the steaks injected with 10% of the SHMP solution which was approximately equivalent to 0.18% SHMP based on weight of the meat.

Komarik[654] described a method for the tenderization of fresh meats, including beef, pork, lamb, and mutton which involved the application of between 2 and 12 oz of a basic pyrophosphate to the meat. TSPP or TKPP was preferred since it could be added either as a dust of the dry powder applied over the surface of the pieces of meat, or the meat could be injected with solutions of these tenderizing compositions. In addition to using the pyrophosphates alone, Komarik reported that combinations of the basic pyrophosphates and small quantities of a meat tenderizer, preferably papain, increased the effectiveness of the tenderizing composition. Since the tenderizing action of the basic pyrophosphates was considerably slower than that of proteolytic enzymes, the treated meats were allowed to stand at refrigerator temperatures for several days, and as long as a week without overtenderization. After dipping or dusting the meat, it was necessary to allow it to stand at least 6 hr for thin cuts to as long as 16 or 24 hr at refrigerated temperatures for thicker cuts.

Brown and co-workers[655] also reported a method for tenderization of meats by means of phosphate solutions. The phosphate solutions were injected into the meat, particularly beef, which was pumped with 15 to 40% of an aqueous solution, containing between 2 and 10%, preferably 2 to 7% by weight of salt, and between 1 and 4% by weight of the phosphate. The quantity of solution injected was preferably sufficient to add approximately 0.5% by weight of the phosphate and between 0.7 and 1.8% by weight of the salt to the meat. The preferred phosphates for tenderization of beef were STP, TSP, and DSP; SHMP was also effective, but to a lesser extent. Delaney[656] obtained significant tenderization of meats such as beef, pork, lamb, and poultry using an aqueous solution containing an alkali metal pyrophosphate or polyphosphate, common salt, and small amounts of proteolytic enzymes. The pyrophosphates, such as TSPP and TKPP, and the alkali metal polyphosphates were preferred, but the pyrophosphates were most effective. The meat was treated by dipping it for from a few seconds to several minutes in a solution containing a mixture of the phosphate, salt, and a proteolytic enzyme, then holding it at 38 to 44°F. The optimum dipping time ranged between 15 and 60 sec.

Huffman and co-workers[657] tenderized sheep meat by injecting the animals prior to slaughter with a solution of SHMP or TSPP so that they received 3 mg of the phosphate per pound of live weight. The meat of control animals was judged to be tougher than that of the phosphate-treated animals.

Miller and Harrison[658] attempted to repeat previous work on tenderization of steaks by marinating them in solutions of SHMP. They were unable to demonstrate any tenderization effect of marinating the steaks in a 0.03 *M* solution of SHMP for 1, 2, or 6 hr prior to broiling. The steaks were evaluated for tenderness by panels and by means of the Warner-Bratzler Shear Press.

Baldwin and deMan[605] studied the effects of DSP, TSPP, STP, sodium tetrapolyphosphate, and SHMP on the tenderness of beef. The meat was treated with one of the phosphates, or with citrate, and then evaluated for tenderness by means of a Kramer Shear Press. In every case, the phosphate-treated steaks were more tender than the untreated controls. Most of the phosphates were more effective than citrate in tenderizing the steaks. The tenderizing effect was not correlated with pH, since pH values did not change to any signficant degree for most of the phosphate treatments. In those treatments in which pH did change, the tenderness was not correlated with the degree of change in pH. Baldwin and deMan were unable to correlate the effects of the phosphates on tenderness with their ability to complex calcium and magnesium.

Increasing Binding

The phosphates improve the binding properties of fresh whole meat. Karmas[587] reported that smaller chunks of meat could be bound to each other to form larger pieces if they were first mechanically treated so that the natural exudates were brought to the surface of the chunks of meat. The chunks of meat then would be bound so tightly that the larger piece of meat, thus formed, could be sliced. The texture was reported to be similar to that of higher-grade cuts. Considering the influence of the phosphates on increasing the solubility of myosin which is important in meat binding, the phosphates would be expected to improve the binding of meats in this application.

Increasing Moisture Retention

The water-holding capacity of meat can influence such factors as the drip loss when meat is refrigerated or when frozen meat is defrosted, the amount of water bound by the proteins (which can affect tenderness of the meat when it is cooked), its juiciness when eaten, and the amount of moisture lost during cooking the meat. All of these factors have signficant influences on the palatability of the meat and, therefore, its acceptance by the consumer. Hopkins and Zimont[659] patented the use of dip solutions containing 10 to 20% sodium chloride and 0.5 to 5% of one of the phosphates from the group, orthophosphate, pyrophosphate, tripolyphosphate, or the longer-chain polyphosphates with or without ascorbic acid or sodium nicotinate. Pieces of fresh meat weighing approximately ½ lb were immersed in such a solution for 5 to 10 sec and 10 to 15 min for large roast cuts. Such treated meats could be prepackaged for sale in self-service counters in grocery stores without exuding unsightly red or pink juices into the package. This not only makes the meat more acceptable from an appearance standpoint, but also improves its tenderness, nutritional value, and other palatability factors. Beef, mutton, pork, venison, and poultry could be treated in this manner, according to the authors.

Rust[649] reported that steaks injected with increasing quantities of an SHMP solution increased in juiciness in proportion to the quantity of the phosphate injected. The results were significant to the 0.1% level. Significantly lower cooking losses were experienced in the phosphate-treated steaks, and the total moisture of the treated steaks was also greater than for the controls. Rust reported that the drippings from both raw and cooked steaks injected with 15% of her phosphate solution contained significantly ($P = 0.001$) more soluble protein (as expressed by increased protein nitrogen in the drippings) than the drippings of the other steaks. This indicated that more of the protein was solubilized by the phosphate treatment and that the soluble protein levels in phosphate-treated steaks were higher even though the steak had been cooked.

Miller and Harrison,[658] however, found no significant increase in water-holding capacity in steaks treated with marinades containing SHMP solution. Although juiciness scores were significantly increased for steaks marinated 2 hr or more, the cooking losses of the marinated steaks were increased over those of steaks given no marination treatment. These authors concluded that marinating steaks with SHMP solution was of little value in improving the eating quality of the steaks.

Wismer-Pedersen[633,634] demonstrated that increases in pH, the presence of magnesium and calcium ions, and the addition of pyrophosphate increased the water-holding capacity of meat fibrils. Pyrophosphate was found to increase the rehydration capacity of freeze-dried meats through increasing the swelling of meat fibers after they had been moistened. This author also reported that the ability of pyrophosphate to increase the swelling of meat fibers caused them to hold the moisture more tenaciously when the samples were being dried, thus decreasing the effectiveness of the freeze-drying process. Beef and pork meat reacted in the same manner in these studies. Treating beef with a solution of 10% STP prior to freeze-drying accelerated the drying and resulted in improved texture, odor, and taste when compared to an untreated control.[660] SHMP was also reported to have a similar effect on increasing hydration capacity of freeze-dried meat when the meat was treated with the SHMP prior to drying.[661]

Flavor Improvement

The protein in the meat and the manner in which it is processed and finally cooked contribute to the flavor developed in that meat. Certain reactions of other meat constituents that can occur will cause flavor deterioration. The oxidative rancidity of the fat in the meat can lead to serious off-odors and off-flavors in both the fresh and cooked meat. Watts et al. have contributed numerous studies of the stabilization of meat products against oxidative rancidity. Tims and Watts[662] studied the development of oxidative rancidity in refrigerated cooked pork by means of the 2-thiobarbituric acid reagent (TBA method). They reported that only those meats containing numerous polyunsaturated fatty acids developed rancidity after cooking. Pyrophosphate, tripolyphosphate, and hexametaphosphate exhibited protective action against development of rancidity, while orthophosphate exhibited no activity as shown in Figure 6. The phosphates were also reported to exhibit synergistic activity with ascorbic acid and sodium ascorbate in inhibiting development of rancid odors and flavors in the cooked meat products. Tarladgis and coworkers[663] reported that combinations of polyphosphates with sodium ascorbate, and also, the use of extracts of certain vegetable products, gave partial protection against development of rancid odors and flavors in cooked irradiated pork.

Chang and co-workers[664] reported that cooking uncured meats caused development of oxidative rancid flavors and odors upon refrigeration but less upon freezing the cooked meat. These are often described as stale, "left-over," or "rancid" odors and flavors. The triglycerides are not responsible for development of these off-odors and flavors; they result from the oxidation of lipoproteins and phospholipids. The reaction is catalyzed by the ferric heme pigments which develop in cooked, uncured meats due to the breakdown of the derivatives of myoglobin as a result of the heat treatment. Combinations of sodium ascorbate and tripolyphosphate were found to be most effective in eliminating this oxidative rancidity and resulted in greatly improved odor and flavor of refrigerated and frozen beef. These results are shown in Table 26. However, similar treatment did not benefit irradiated beef to the same extent.

When Ramsey and Watts[665] covered slices of cooked beef stored in the refrigerator with solutions containing 0.02 to 0.1% STP, they obtained excellent protection against the development of rancid odors and flavors. When they used a 0.5% solution of STP, it was only necessary to apply it as a dip to obtain effective control of rancidity. They also studied use of a number of vegetable extracts which naturally contain phenolic-type compounds. These were effective against the development of rancidity when used as cover solutions in some cases, but not as dip solutions. Zipser et al.[666] reported that the development of rancid odors and flavors did not occur during frozen storage, but rather that heme-catalyzed oxidation occurred during the preparation of the cooked meat for freezing and during the thawing of the meat after frozen storage. Watts[667,668] has reviewed much of this work.

Applications in Fresh Comminuted Meat Products

Color Preservation

Savich and Jansen[669] were able to extend the life of the red color in ground meat products to a significant extent if they incorporated a combination of a monobasic alkaline earth phosphate, the edible salts of propionic acid, and ascorbic acid into such items as ground beef. For example, they were able to maintain satisfactory color in ground meat for nine days when stored at 35°F by

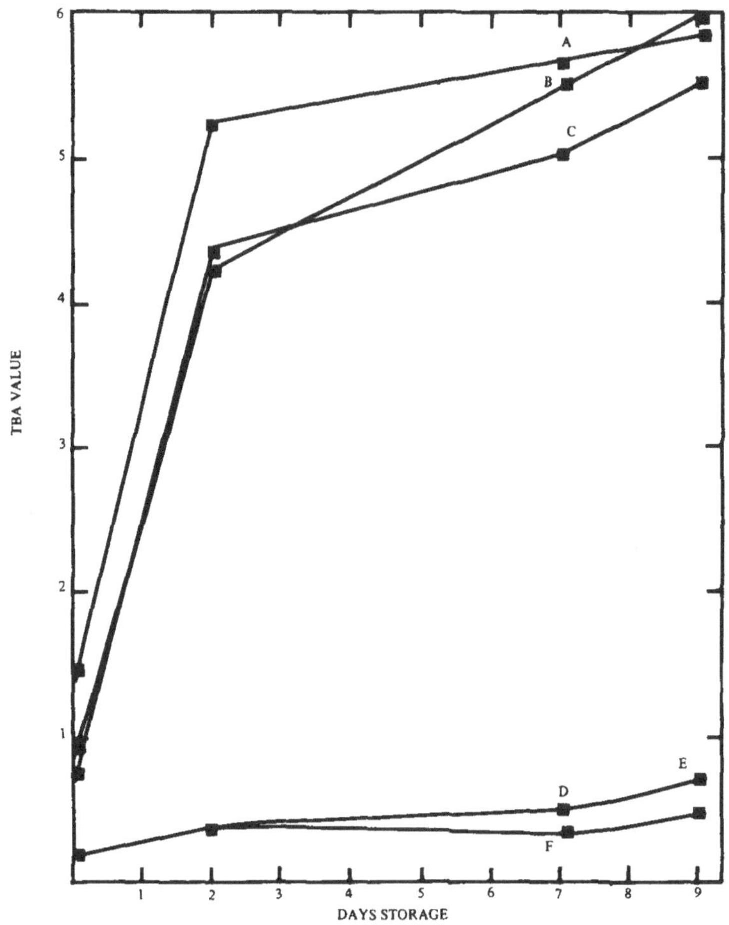

FIGURE 6. Effect of phosphate on TBA values of refrigerated cooked pork. A. Control; B. DSP; no pH adjustment; C. DSP + MSP to pH 6.4; D. STP; E. SHMP; F. SAPP + TSPP. (From Reference 662, Figure 1, with permission.)

incorporating 0.05% sodium propionate, 0.05% ascorbic acid, and 0.2% MSP. The meat was somewhat off color after 12 days and then became unacceptable in color. The untreated control had an unacceptable color at the end of the second day of storage.

Ziegler[670] prepared a red pigment from dried blood which could be used in coloring ground meat products to maintain the normal red color of the fresh, processed meat. The process involved collecting the blood under highly sanitary conditions and adding an anticoagulant such as an alkali metal salt of phosphoric or polyphosphoric acid; for example, sodium orthophosphates, sodium pyrophosphates, STP, or SHMP. The blood then was treated with a reducing agent, such as ascorbic acid, which was necessary to inhibit oxidation. If oxidation occurred, the ferrous pigments, myoglobin and hemoglobin, were converted to the brown-colored ferric pigments, metmyoglobin and methemoglobin, thus losing the bright red color. After treatment with the reducing agent, the blood composition was buffered to a pH between 7.3 and 8.0 by using suitable buffering salts. Among those found acceptable were the sodium phosphate salts which would provide such a pH range. Sodium nitrite was added to the composition to convert the hemoglobin to the desired compound, nitrosohemoglobin, which had a bright red color. The buffering of the blood to pH 7.3 to 8.0 was a critical factor since it was found that it must have this pH or else the red color would change to brown during spray-drying of the product.

Increasing Tenderness

The texture of ground meat products can sometimes be interpreted as tough or chewy. Jacoby and Berhold[671] obtained excellent tex-

TABLE 26

Effect of Antioxidant Dips on Roast Beef Slices Stored in the Refrigerator and the Freezer[1,2]

		No antioxidant (control)		Dipped in 1% $Na_5P_3O_{10}$ + .27% Na-ascorbate	
Storage temperature	Days in storage	TBA no.	Odor score	TBA no.	Odor score
1°C	1	11.5	–	0.3	–
	4	15.2	2.5	0.4	5.6
	7	19.5	2.6	0.3	5.3
	11	21.1	1.9	0.3	4.7
	18	21.2	–	0.3	–
–26°C	11	3.5	2.7	0.2	5.4
	31	2.3	2.4	0.2	5.5
	45	3.2	–	0.2	–
	81	4.9	2.2	0.2	5.9
	164	4.1	3.5	0.2	5.8

[1] At each storage period, the odor of the antioxidant-dipped meat was rated against its corresponding control. In all cases, the antioxidant-dipped rated higher than the control, and the differences were highly significant.
[2] Reference 664, Table 2, with permission.

tural properties and tenderness in ground or diced meat products, e.g., hamburger or stew meat, by adding a chelating agent, such as STP, to the meat. This was combined with an unmodified starch and a protease, preferably papain, in very small quantities to assure proper texture and tenderness.

Increasing Binding

Binding the particles of ground meat together in a homogeneous, uniform, and stable mass is important in the preparation of hamburger, meat loaf, fresh sausage, etc. The influence of solubilizing the salt-soluble proteins in improving the binding of meat products has been reviewed previously. Aged bulls were considered to provide the best type of binding meat for use in ground meat products. The polyphosphates improve the binding qualities of younger beef meat which normally would be considered poor grade for use in ground meat products because of their low binding properties.

Bickel[672] reported that it was possible to produce excellent textured, ground meat products that maintained their structure, and in which the meat particles swelled and bound to each other. He reported that the water-soluble salts of pyrophosphoric acid and longer-chain polyphosphoric acids exerted this activity on the ground meat products. SAPP, TSPP, STP, sodium tetrapolyphosphate, SHMP, and the very long-chain sodium and potassium polyphosphates, such as long-chain Graham's and Kurrol's salts, are applicable. Optimum levels of the polyphosphates appeared to be between 0.15% and 0.5%, with 0.3% the preferred level.

Fetty[673] patented the application of $DSP{\cdot}2H_2O$ and $MSP{\cdot}H_2O$ with combinations of wheat flour, nonfat dried milk solid, hydrolyzed vegetable protein, and sodium bicarbonate as binders that would prevent the shrinkage of ground meat products, such as hamburger, meat loaf, lamb patties, and cured sausage meats. The composition was added at levels between 3 and 6% based on the weight of the ground meat.

Increasing Moisture Retention

As with whole meats, the water-holding capacity of the meat proteins is extremely important in the comminuted meat products because it reduces the "drip loss" from fresh as well as thawing frozen meat, the drying and toughening of the meat, the reduction in cook-out loss, and the retention of highly nutritious soluble proteins in the meat during storage and cooking. The pH of the ground meat has been shown to be important in reducing the loss of fluid from fresh meats. Sair and Cook[674] reported that the minimum amount of drip loss occurred during thawing of ground meats when the pH was approximately 6.4, while a maximum amount of drip was obtained at a pH of 5.2.

The phosphates are even more effective in reducing the loss of moisture and soluble constituents of meat. Bendall[593] reported that pyrophosphate increased the swelling and the water-holding capacity of ground meat products, both cooked and uncooked, to the greatest extent. Orthophosphate, hexametaphosphate, and the very long-chain sodium and potasssium phosphate glasses also increased swelling and water-holding capacity of the muscle tissues to a lesser extent, but more than in the untreated meat. Bendall was one of the first to report that a combination of pyrophosphate and salt more than doubled the water-holding capacity of meat tissue over that of salt alone at equivalent ionic strengths. He suggested that the action of the phosphates could be explained by their ability to split actomyosin into its component proteins, myosin and actin and to convert these from a gel to a sol form.

Sherman[618] reported that the addition of sodium chloride or various phosphates to ground meat significantly improved retention of fluids, but that this could not be attributed to the ability of the phosphates to complex calcium or magnesium ions. There also was a statistically significant correlation between pH of the phosphate-meat mixtures and their ability to retain fluids. Sherman found a correlation between this effect and the absorption of anions by the proteins, particularly when sodium chloride was added to the meat. The rate of absorption increased with increasing ionic strength of the salt solutions added to the meats. When the meats were cooked, however, both anions and cations were released, although the anions were retained to a greater extent than the cations. Sherman correlated the retention of fluid in cooked meats with the amount of actomyosin that went into solution at 0°C during aging of the phosphate-meat mixtures. The actomyosin was, he believed, gelled during cooking, and this explained the lower release of moisture since this gel extended throughout the mass of meat.

In a further extention of this work, Sherman[675] reported that both cations and anions were absorbed to the same extent at 0 and 100°C from solutions of neutral salts. The concentration of the salts determined the degree to which ion absorption occured. The water-holding capacity of the meat tissues at either temperature was related directly to the amount of ions absorbed. The structure of the meat protein appeared to be opened by the absorption of the ions and thus water was able to penetrate the helical structure of the protein. Sherman believed that a different situation occurred when the pH values were in the alkaline range, such as when alkaline polyphosphates were used in treating the meat. Ion absorption was highly dependent upon pH. As the pH increased, the cations were absorbed at an increasingly greater rate. As a result, Sherman believed that the absorption of the phosphate ion was important at alkaline pH values. The water-holding capacity of the muscle tissue at 0°C was, therefore, influenced by the absorption of ions. However, the primary factor involved with water-holding capacity of cooked meat still depended on the degree of solubilization of actomyosin during aging of the meat at 0°C.

Hellendoorn[603] reported that pyrophosphate and tripolyphosphate depressed the water binding of raw meat at pH values below 5.5 and increased the water binding above this pH. Water binding of cooked meat was decreased when the pH ranged below 6 to 6.5 and increased at pH values above this range. He found that pyrophosphate was slightly superior to tripolyphosphate in improving the water-holding capcity of meat at pH 6 to 6.5. The superiority of the pyrophosphate increased over that of tripolyphosphate as the ionic strength was increased above 0.4. Orthophosphate and hexametaphosphate were less effective than these two salts and longer-chain polyphosphates, and the cyclic polyphosphates had no significant effect. Hellendoorn also reported that his results with SHMP, EDTA, and oxylate, as shown in Figure 5, did not indicate that calcium-binding was of any importance in the water-holding capacity of meat tissue.

The results reported by Inklaar[604] support the theory that calcium ions did not influence water binding in meat proteins and that the ability of phosphates to complex calcium ions had no effect on their ability to increase the water-holding capacity of meat proteins.

Applications in Cured Meat Products

Color Preservation

Several investigators reported that the phosphates will stabilize the color of cured meats, such as hams and cooked and smoked sausages stuffed in casings. A Dutch patent[676] reported that buffering a pickling brine for cured meat products which contained salt, sodium nitrate, and sodium nitrite with a combination of solutions of citric acid and DSP to maintain a pH between 3 and 5 produced a permanent bright-red color in the cured meat. Chang and Watts[640] reported that none of the phosphates were as effective as ascorbic acid in protecting meat against loss of color in the presence of hemoglobin at pH 5.6, the pH of the meat used in the experiments.

Hall[638,677] reported that polyphosphates, having a chain length of three or more, provided excellent protection against decolorization of frankfurters and other similar cured meats including whole unground meat such as hams, if the pH was maintained between 6.5 and 7.0. Decoloration occured whenever the pH was allowed to fall below 6.5. This patent has been one of the dominant patents in the applications of phosphates in cured meat products. Hall reported that the use of the polyphosphates in the curing salt mixtures increased the tolerance of the cured meat products to such variations as levels of curing ingredients, cooking temperatures, and the pH values of the meat.

Another dominating patent for the use of phosphates in meat products was obtained by Brissey.[639,677] This process involved the use of DSP to maintain the pH of the curing salt pickles at pH values between 6.6 and 6.8. Hams or ground cooked cured sausage products which had pH values above 6.8 had undesirable features, such as sliminess, while those maintained within the specified narrow pH range had excellent properties including resistance to decoloration.

Wilson[678] reported the results of studies on the effects of ascorbic acid and phosphates on color retention in bologna sausage products. DSP, SHMP, and combinations of DSP with ascorbic acid were used in this investigation. The author reported that meat treated with 0.5% anhydrous DSP had the same initial color and rate of fading as the untreated control. In one experiment, the application of 0.2% SHMP greatly enhanced the color of the bologna and, therefore, increased the period of time during which it had acceptable color. The combination of DSP and ascorbic acid increased the color shelf-life of the bologna up to 70% under fluorescent lights while stored at 34 to 40°F. Morse[600] cited a paper by Wilson, of the American Meat Institute, in which he reported that STP produced the best color of all the treatments tested in hams.

Mahon and co-workers[679] reported that a palatability panel was capable of detecting differences in the color of control hams vs. hams treated with a small quantity of DSP. The DSP-treated hams were considered to have the more acceptable color that was signficant at the 5% level. Fenton et al.[680] quoted Wilson[681] as reporting that a trained panel of judges significantly preferred hams cured with STP and SHMP over those which had no phosphate in the curing pickle. DSP had no effect on the color. In addition, Wilson reported that phosphates had similar effects on comminuted meat products such as bologna and frankfurters. Fenton and co-workers[680] found similar results when they added DSP to curing brines used on smoked hams. The DSP had no significant effect on the color of the finished hams.

Watts[682] reported that there was a direct relationship between fat oxidation and the loss of color in refrigerated hams, as shown in Table 27. The addition of either tripolyphosphate or hexametaphosphate to the brines used in pumping hams significantly increased the color retention of canned, baked hams and increased, but not to a similar extent, the color retention of canned and smoked hams. However, Watts also reported that the polyphosphates and similar inhibitors of color fading, had little or no effect on the fading due to light, such as fluorescent light.

Swift and Ellis[683] found that the conditions under which phosphate-treated meats were processed had a significant effect upon whether the phosphates decreased or increased the color stability. These authors reported that rapid processing of frankfurters, including brief tempering times and short smoking and cooking times, resulted in decreased color stability with addition of various phosphate mixtures. However, when longer periods of tempering and longer smoking and cooking times were used, the phosphates increased the color development in the same frankfurter formulas. Those frankfurters treated with SHMP alone and with a mixture of 70% TSPP and 30% SAPP had the most acceptable color. The authors believed that some of the discrepancies and confusion in the literature are due to significant differences in the processing variables used in testing the phosphates for their effects on cured meats.

Mullins and co-workers[651] reported that an evaluation committee could distinguish between the uniformity and stability of the color of phosphate-treated and untreated hams. Wasserman[684] patented the use of 0.05 to 1.0% STP and 0.005 to 0.05% sodium ascorbate in the curing salts used for ground and whole, cured meats. He reported that the cured meat took on an intensified red color after several days and that this high level of color development, therefore, increased the length of time that the meat retained an acceptable color before the color faded to a point making it unacceptable.

Gisske[685] reported that the use of a commercial phosphate mixture containing DSP, SHMP, and STP improved the color of hams to which it was applied in the pumping pickle. Schoch[686]

TABLE 27

Effect of Polyphosphates on Color Stability and Rancidity in Refrigerated Hams

Phosphate Used	Treatment	Days for 50% Color Loss	Days to Turn Rancid (TBA = 0.5)
None	Canned, baked	2	2
Hexameta	Canned, baked	>12	>12
None	Canned, baked	3	2
Tripoly	Canned, baked	>13	>13
None	Smoked, baked	3	3
Hexameta	Smoked, baked	>13	>13
None	Smoked, baked	6	6
Tripoly	Smoked, baked	>14	>14
None	Canned	10	9
Tripoly	Canned	>13	>13
None	Canned	10	13
Hexameta	Canned	>13	>13
None	Smoked	12	>14
Hexameta	Smoked	>14	>14
None	Smoked	12	>14
Tripoly	Smoked	>14	>14

Source: Watts, B. M., *Proc. 9th Research Conference*, American Meat Institute, Chicago, 1957, 61, Table 2, with permission.

patented the use of combinations of DSP and sodium polyphosphates, such as SHMP or longer-chain sodium polyphosphates to stabilize the color of cured meat products. The curing brines or pickles revealed in his patent contained one part of the DSP and from one to five parts of the polyphosphate. This mixture of phosphates was added to the brines in concentrations between 1 and 5% of the total weight of the brine as the phosphate salts.

Suri[687] reported that the addition of polyphosphates to meat curing brines stabilized the color. He described analytical methods which could be used to determine the levels of phosphates added to the meat based on analysis by paper chromatography. Bailey and co-workers[688] studied color retention ham slices that had been cured with ascorbic acid, combinations of ascorbic acid and nicotinamide, and ascorbic acid and niacin in the presence of orthophosphate buffers which maintained the hams at pH values between 6.2 and 6.8. The authors reported that those hams buffered at 6.8, and containing a combination of ascorbate and nicotinamide, were most stable in retaining the natural red color when exposed to fluorescent lights at refrigerated temperatures. All others were little or no better than the untreated controls. These authors also found that nitrosomyoglobin was more readily formed from metmyoglobin when nicotinamide was present than when it was absent and that it was more readily formed at pH 6.8 than at pH 6.2 even though the same additives were present.

Siedlecki[689] reported that the addition of 0.35% of polyphosphates stabilized the color of canned hams. Lauck and Tucker[690] reported that the addition of one part of an alkali metal trivalent metal phosphate such as SALP and approximately one to five parts of an alkali metal polyphosphate to curing salts used in comminuted meat products improved and stabilized the color as well as provided other improved properties in the meat product. Savic and co-workers[691] reported that the addition of polyphosphates to curing salt mixtures containing nitrites prevented nitrites from damaging the tin-plate coating and eliminated the resulting yellow-brown discoloration of the surface of canned pork products. Rahelic et al.[692] reported that the addition of 4% polyphosphate to muscle samples cured with brines containing salt, sodium nitrate, and sodium nitrite and stored at 4 to 8°C resulted in significant increases in the pH values and percent of nitroso pigment in the phosphate-treated meats compared to those treated with brine containing no phosphate. Sair and Komarik[693] reported that although low pH values (below the normal pH of meat) are generally considered to present adverse conditions for development of good emulsions in comminuted meat products, if an acid reacting phosphate is used to obtain the low pH, highly improved color stability is obtained with no adverse effects upon emulsion formation or stability. These authors reported that the phosphate radicals counteract the destabilizing effect of low pH on the formation of stable meat emulsions. Among the phosphates, which these authors reported can be used, are the acidic MCP, monoalkali metal salts of orthophosphoric acid, SAPP and SALP. Combinations of MSP, SAPP, or SALP with alkaline polyphosphates such as STP and TSPP can also be used as long as a sufficient quantity of the acidic phosphate is present to keep the pH low. In addition, combinations of other edible acids and alkaline polyphosphates can also be used.

Increasing Tenderness

The moisture and fat content of cured meat also has a significant effect on its juiciness and tenderness as shown by Swift and co-workers.[694] Therefore, since the phosphates allow increased moisture and fat to be incorporated in comminuted meats, their use is bound to increase tenderness, as has been shown by several investigators. The same effect of phosphates is found when they are used in curing whole meats such as hams. Mullins and co-workers[651] reported that panels scored hams treated with SHMP as moderately tender, while control hams cured without SHMP were scored slightly tender.

There may be a biochemical reason for the effect of phosphates on increasing tenderness of any meat product. Szent-Gyorgyi[695] speculated that any compound or process which would inhibit the formation of actomyosin by preventing the attachment of fibrillar actin to myosin would be likely to increase tenderization of meat. He later reported[696] that the formation of fibrillar actin from globular actin could be prevented with SHMP and that a solution of actomyosin would be dissociated into its component proteins if it were treated with inorganic alkaline phosphates.

Hams could also be tenderized if a small quantity of a bacterial protease was added to the pumping pickle along with one of the polyphosphates according to Schleich and Arnold.[697] The phosphate could be either SAPP or TSPP, STP, sodium tetrapolyphosphate, or SHMP. The polyphosphates prevented overtenderization of the hams by inactivating the enzyme over a period of approximately 3 hr.

Increasing Binding

The action of the phosphates on increasing the binding of comminuted cured meat products, such as bologna, frankfurters, and other typical sausages and sandwich meats which are stuffed in casings and then further processed by smoking and cooking, has been thoroughly reviewed in the literature.[25,30,584] As discussed previously, the action of the phosphates is believed to be due to their dissociation of actomyosin into its component proteins, thus releasing myosin to form its binding gel during processing of the meat. Swift and Ellis[683] demonstrated that various combinations of pyrophosphates, hexametaphosphate, and long-chain potassium polymetaphosphates increase the binding of meat in sausage products, as shown by the increased tensile strength or cohesion of the bologna products. Kotter[591] speculated that the pyrophosphates replaced ATP in promoting the dissociation of actomyosin that explained the binding action of the polyphosphates on sausage-type meat products. Bicket[672] reported that the use of combinations of polyphosphates and particularly of the long-chain potassium polyphosphates would cause the different meat constituents in sausage products to be bound together into a homogenous mass of uniform structure. Maas and Olson[698] used alkaline polyphosphates that caused whole pieces of meat or muscle tissue to become bonded together during heat processing. These authors reported that the phosphate ions were essential to obtain this binding that was undoubtedly due to the release of myosin. The released myosin can gel to bind the meat particles together to resemble larger whole pieces of meat. After cooking, the bound meat could be sliced just as larger whole pieces of meat.

Increasing Moisture Retention

The retention of water in cured meat products is very important to their palatability. This is true not only of whole meat products, such as hams, but also in the numerous types of sausages which are produced from comminuted meat. Numerous reviews discuss this ability of phosphates, particularly the polyphosphates, to increase the water-holding capacity of the meat proteins and a number of them have been mentioned previously.

Hall[638] specifically mentioned that more water was held in phosphate-treated frankfurters than in the untreated products, especially during smoking. This patent is one which dominated the applications of phosphates to meat products until it expired. In the second dominating patent for applications of phosphates to cured meat products, Brissey[639] also specifically mentioned the larger cooked yield and greater retention of soluble protein in the product after processing when he included dialkali phosphate in the curing salt mixture. The reduction of shrinkage in cooked and canned hams, pork shoulder picnic hams, and similar products[677] is reported to be one of the major advantages of including the various phosphates, DSP and the polyphosphates, in meat-curing salt mixtures. Moisture retention also is an important economic factor to the processor of such meat products.

Swift and Ellis[683] reported that when greater proportions of fat or moisture were retained in the cured meat product, the product had increased juiciness and tenderness. According to Morse,[600] the increase in moisture retained in the processed meat product did not give it a more moist appearance. The product actually appeared to be drier since the water was more completely bound by the meat protein.

Mahon and co-workers[679] and Leeking and co-workers[699] reported that the addition of 0.5 lb of DSP per gallon of curing pickle had no significant effect upon the water-holding capacity or cooking losses of cured, smoked ham. However, the amount of DSP used in their experiments was very low and this may explain the lack of effect on water-holding capacity.

Frank[700] found that the addition of phosphates to cooked meat had no effect upon its water-holding capacity. Therefore, it appeared that the phosphates must be added to the meat prior to any heat treatment. Swift and Ellis[701] reported that the most effective treatment temperatures for the addition of phosphate to ground meat ranged from 32 to 41°F and that the effectiveness of the phosphates in increasing the water-binding properties of the meat proteins decreased as the temperature increased above this range. They also reported that optimum effects were obtained when the meat was stored in contact with the phosphate solution for 16 hr, at the optimum temperature range. In later work, Swift and Ellis[683] reported that the addition of phosphates to bologna and frankfurter sausage mixtures had no effect on the water-binding properties or in reducing cook-out losses at the normal smoking conditions that raised the internal temperature of the meat to between 150 and 160°F. However, the phosphates markedly reduced the cooking-loss of the same sausage products when they were smoked to internal temperatures above 160°F. The authors thus demonstrated that the addition of phosphates to sausage products would allow use of higher temperature-shorter time smoking processes, a desirable factor for sausage processors. Swift and Ellis,[701] therefore, came to the conclusion that the phosphate additives had their effects on moisture retention of meat through their ability to solubilize muscle proteins. Such processing factors as temperature, time, ionic strength, and pH of the treatments were important considerations in obtaining maximum water retention by the meat proteins.

In his review on the use of phosphates in production of meat products, Grau[584] described experiments which demonstrated that the addition of small quantities of phosphates to sausage meats did not increase the water absorption of the meat above controls, but rather caused the meat proteins to bind the water more completely and, therefore, did not lose it during processing. This probably explains reports of work such as that of Wasserman,[684] · Grisske,[685] Volovinskaya et al.,[702,703] Epstein,[601] Rupp and co-workers,[704] and Turner and Olsen[606] that sausage products, and cured meat products in general, treated with phosphates, bind more water, are plumper products when encased in sausage casings, swell to a greater extent so they fill the casings and maintain a tight casing over the sausage product, and lose less water during smoking and heat processing as well as during cooking the product prior to eating it.

Although he acknowledged the ability of condensed phosphates to improve water binding and fat emulsification in sausage products, Zeller[705] felt that the health problems with addition of phosphates to meat were too great to warrant their use. Grau,[584,706] Kotter,[602,609] Hamm,[585] and other German researchers disagreed with Zeller. The small quantities of phosphates necessary to increase the water-holding capacity of the meat, and thus, to retain soluble proteins within the meat are too low to be a danger to human health and the benefits are considered too great. Grau[706] believed that the food analyst could eventually find ways to determine without question any abuse of the use of polyphosphates. This fear apparently was the major reason for Zeller's opposition to their use.

Mahon[617] studied the increase in volume of cured meat products because the volume of cured meat products is directly proportional to the amount of water which is bound by the meat proteins. There are an equal number of positive and negative charges on the protein of red meat at its isoelectric point and thus it has a neutral charge at this point. However, if the protein is acidified, it absorbs hydrogen ions and becomes more positive in its net charge. Upon the addition of alkali to the protein, the protein becomes more negative. Mahon reported that the volume of the meat increased as the pH of the meat either decreased or increased away from its isoelectric point, as shown in Figure 7. Mahon demonstrated that additions of small quantities of salt actually depressed the meat volume and the volume reached a minimum at approximately 0.5% salt, as shown in Figure 8. Beyond 0.5%, the volume of the meat increased to reach a maximum at about 5% salt concentration, after which it again decreased to another minimum in the area of 10% salt concentration. The addition of 0.5% STP to the salt, however, increased the meat volume, although the addition of 0.5% salt and 0.5% STP still produced the minimum meat volume for STP-salt mixtures. This minimum volume was, nevertheless, higher than that produced by the salt alone. Continued addition of salt increased the meat volume to a level substantially higher than that produced by similar levels of salt alone, as shown in Figures 7 and 8. Mahon suggested that his studies had shown that treatment with 4% salt with no added STP gave an

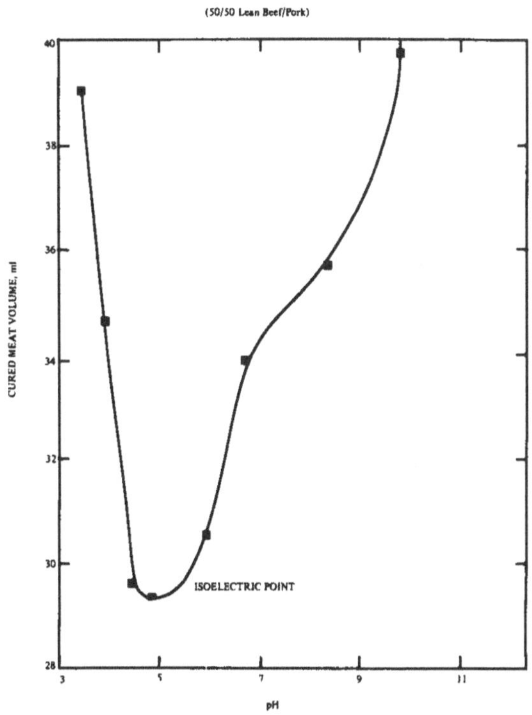

FIGURE 7. Effect of pH on cured meat volume - no salt. (From Reference 617, Figure 1, with permission.)

equivalent meat volume to that from treatment with 2% STP with no added salt. Since only 0.5% STP is legally allowed to be added to meat, Mahon reported that between 2 and 5% salt must accompany the STP to obtain maximum volume.[617] A study of Figure 8 confirms this.

Siedlecki[689] reported that the addition of 0.35% polyphosphates reduced the free juices in canned hams by 5% and improved their firmness. In addition, the polyphosphates reduced shrinkage of the hams and increased the yields during the curing of the hams without affecting losses during smoking. Lauck and Tucker[690] reported that a mixture of between one and fifteen parts by weight of an alkali metal polyphosphate and one part by weight of an acidic alkali metal-trivalent metal phosphate, such as SALP, improved the water-holding capacity of comminuted meat products, e.g., bologna and similar sausage products.

Mahon and Schneider[707] patented a blend of an alkaline SALP and condensed phosphates, preferably, 20 to 30% by weight of alkaline SALP and between 70 and 80% by weight of STP or TSPP to be added at between 0.3 and 0.5% by weight, based on the weight of the meat, to sausage mixtures such as Vienna Sausages. The resulting sausages, after processing, cooking,

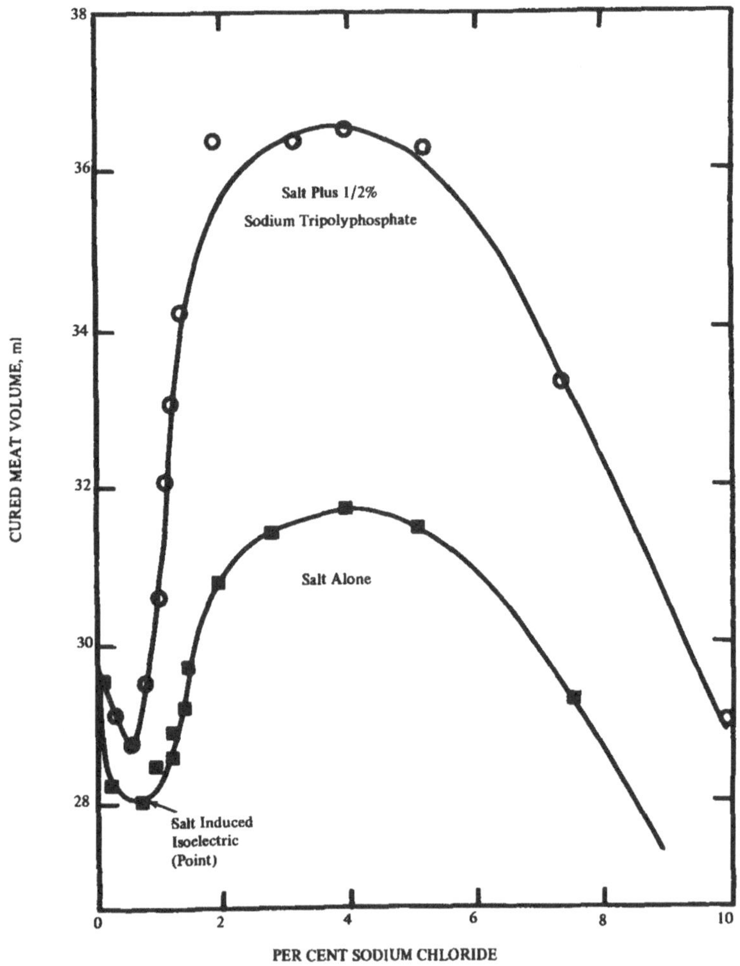

FIGURE 8. Effect of salt and salt + sodium tripolyphosphate on volume of cured meat (155°F). (From Reference 617, Figure 4, with permission.)

cooling, and removal of casings, contained 5% more moisture than the controls containing no phosphate in the curing preparations. Rongey and Bratzler[708] reported that the addition of phosphates to bologna products produced the greatest firmness (tensile strength) and highest yields among the various additives and binders studied in these investigations. Rahelic and co-workers[692] also reported higher water-holding capacity in meats cured with brines containing polyphosphate than in those cured with brines containing no polyphosphate. Sair and Komarik[693] reported a lower pH could be maintained in the cured meat during the final heating process in the presence of the phosphate radical introduced to the curing salt mixtures as phosphate compounds. These lower pH values would normally result in high cooking losses but resulted in minimum cooking losses in the presence of phosphates.

Improving Fat Emulsification

The ability of polyphosphates to increase the fat emulsification of comminuted meat products has also been thoroughly studied. Swift and Ellis[683] demonstrated that increasing the amount of fat and/or moisture incorporated into the sausage products, such as bologna, increased its juiciness and tenderness and, therefore, its acceptability to the customer. Keller[709] obtained excellent fat emulsification in meats treated with TSP, but a soapy taste occurred that made the meat unacceptable. Sodium metaphosphates also produced excellent fat emulsification and significantly reduced the loss of fat during cooking. He also reported that the degree of fineness to which the meat was chopped and the degree of mixing improved fat emulsification and reduced weight losses of the sausages produced from it. Those which were finely chopped and well mixed had low fat and weight losses, while those which were coarsely chopped and inadequately mixed had noticeably higher losses. The addition of polyphosphates, therefore, would overcome some of the processing variables which normally occur in any meat processing plant.

Morse[600] reported that certain phosphate salts, especially TSP, were capable of increasing the emulsification of fats in sausage products, but that the benefit of using the phosphates was not necessarily entirely that of fat emulsification. Kotter[609] reported that the actin and myosin, which are solubilized by polyphosphates, form stable protein envelopes about the fat globules or particles in meat emulsions and, therefore, stabilize the particles and prevent their coalescence. Later, upon heating, these proteins are formed into a framework which produces the firm, easy-to-cut texture of the typical sausage product. Similar effects attributed to the solubilized myosin were reported by Turner and Olson.[606] Hansen[610] reported that proper stabilization of fat globules within a sausage emulsion depended upon proper dispersion and breaking up of the fat globules into fine uniform dispersions within the protein matrix so that the protein enclosed the dispersed fat globules. Furthermore, the salt soluble proteins, myosin and actomyosin, appeared to be most effective in forming stabilizing membranes about the fat globules. These latter observations were confirmed by Swift and co-workers[611] and by Sherman.[710]

Pfaff[711] reported that a combination of heat coagulable and noncoagulating proteins, stabilized by approximately 0.1% of one of the salts of the ortho-, pyro-, meta-, or polyphosphoric acids, or their mixtures, produced highly stable fat-containing sausage emulsions. Hegarty et al.[612] reported that the stability of a sausage emulsion appeared to be directly proportional to the amount of protein appearing at the interface between the fat globule and the aqueous phase. Myosin and actomyosin produced emulsions with superior stability. These authors reported that in the absence of salt, actin was most effective in emulsifying the fat in sausage, while in the presence of 0.3 *M* salt, actin was least effective.

Trautman[614] reported that the water soluble proteins and the salt insoluble residues have little or no emulsifying activity which probably confirms the information reported by Hegarty et al. since Trautman's work was done in the presence of salt. Trautman also confirmed that the salt soluble proteins were most effective in fat-emulsifying properties in hams. Carpenter and Saffle[613] compared the fat-emulsifying capacities of various meat protein extracts. They reported that the water soluble proteins were very active fat emulsifiers, in contrast to the reports of Trautman. However, they reported that the ability of the water soluble proteins to emulsify fat was directly affected by the shape of the molecule; globular shaped molecules had lower emulsifying capacity than fibrous molecules. Changes in net charges of the water soluble proteins seemed to have little or no effect upon their fat-emulsifying capacities. In contrast, both the net charges and the shapes of the molecules had a significant effect upon the emulsifying properties of salt soluble proteins. These authors reported that myosin is a long rod-like molecule which spreads a film over a greater surface of the fat globule. High negative charges, which are produced at pH 9, resulted in the long fibrous, rod-like molecules, and changing the pH to 6 caused the molecules to form a more spherical shape and reduce their emulsifying properties. Sair and Komarik[693] reported that lower pH values resulted in excellent fat-emulsifying properties and improved emulsification.

Sherman[710] reported that the fat-emulsification properties of highly alkaline phosphates were due to the formation of soaps in the meat emulsion. In the presence of sodium chloride, normally present in most meat emulsion-curing salts, no soaps or very small quantities of soaps formed. Sherman, therefore, thought that the homogeneous distribution of fat as small globules during efficient chopping operations was most important in preparation of fat stable sausage mixtures. Excellent reviews of the various characteristics leading to optimum emulsifying properties of the meat proteins have been published by Swift[615] and Saffle.[586]

Flavor Improvement

Watts and her co-workers[636,682] reported that the polyphosphates were excellent synergists in the prevention of oxidation and rancidity of fats which would lead to off-flavors in cured meat products. The results of a series of experiments demonstrating the antioxidant effects of polyphosphates are shown in Table 27 in which Watts and her co-workers demonstrated that the polyphosphates, STP and SHMP, had significant effects in extending the shelf-life of canned, baked, and of smoked baked hams against development of rancidity. Similar results were also obtained in stabilizing the same products against color loss. Lesser effects, and in some cases, no beneficial effects were found with addition of similar phosphates to canned and to smoked hams. The improved taste reported in cured meats by Rupp et al.[704] was most likely due to the ability of polyphosphates to reduce the development of fat rancidity. Suri[687] reported similar protection against rancidity by addition of polyphosphates to "pickled meats." Zipser et al.[666] reported differences in the degree of lipid oxidation in cured vs. uncured frozen cooked pork. Eight to ten times higher ratios of peroxides to TBA numbers were reported for frozen cured meats in comparison to those in uncured frozen meats. The ferric hemes of cooked meat were responsible for the oxidative rancidity which did develop in both products and this heme-catalyzed oxidation in uncured meat occurred during the period of preparation for freezing and during its later thawing for testing. No development of rancidity appeared to occur during frozen storage. In contrast, the frozen, cooked, cured pork samples continued to develop oxidative rancidity during frozen storage. At the time of this work, the authors reported that they were unable to find a satisfactory antioxidant for frozen cured meat products. Later, Zipser and Watts[712] reported that a combination of STP and sodium ascorbate at about twice the level legally allowed (0.108%) produced high stability against development of oxidative rancidity in the cured, cooked, frozen pork products tested. If the maximum legal limit of sodium ascorbate (0.054%) was used, the meat samples became rancid in about four months. The authors were able to trace the disappearance of sodium ascorbate in the samples treated with the 0.054% level. The development of peroxides increased rapidly once the ascorbate had disappeared.

Preventing Microbiological Spoilage

Microbiological deterioration can also create off-flavors in cooked and cured meats.[586,587] Therefore, investigators have sought ways in which to stabilize the cured meats against all types of microbiological attack. The effects of the addition of orthophosphates to curing pickles were reported by Evans.[713] He reported that if the curing pickle is contaminated with *Staphylococci*, these organisms will not reproduce if the salt content is high enough and the temperature is maintained low enough, although some may survive for a week of more. He also found that nitrate had little or no effect on the organisms and that the addition of DSP, which raised the pH value of the pickle from 7.0 to 7.9, had no influence on their survival. The addition of SHMP, which lowered the pH value of the pickle from 7.0 to 5.3, also had little or no effect on the bacteria. Furthermore, his studies demonstrated that if *Staphylococci* were pumped into the center of the ham where they might be protected from the heat of the smoking operation, the presence of 0.5% of SHMP had a slight inhibitory effect on their anaerobic growth in this location.

Bickel[714] reported that meat and intestines (intended for use as casings for sausage-type meats) could be preserved against spoilage by treatment with between 2 and 10% by weight of a solution containing 80 parts of tartaric acid and 20 parts SHMP. Bickel also mentioned that the pyro-, tripoly-, tetraphosphoric acid and potassium metaphosphate derivatives were also effective. Volovinskaya and co-workers[702] reported that the addition of either TSPP or SHMP containing additional NaOH caused some reduction of microbiological growth in stuffed sausages. Epstein[601] obtained microbiological stabilization of a variety of sausage products prepared from cured comminuted meats.

Wollmann[715] reported that sodium pyrophosphate inhibited *E. rhusiopathiae, S. typhimurium,* and *M. tuberculosis* when incorporated in sausage meats that were subsequently cured and smoked. The sodium pyrophosphate, probably TSPP, may have had its effect because it increased pH, although this is not discussed. The addition of a commercial curing phosphate mixture containing 40% of a mixture of SAPP and TSPP, 10% SHMP, and 50% NaCl had a negligible effect. Nitrite curing salts increased the resistance of these organisms against the effects of the phosphates and Wollmann was able to trace this to the high levels of NaCl in the curing salt mixtures.

Miscellaneous Uses

Some of the phsophates have been found useful in maintaining the flow properties of the spice seasoning and curing salt mixtures used in meats. Allen[650] reported that TCP could be used to maintain a free-flowing mixture of enzymes, sodium glutamate, and sucrose for use in curing meats. Peat[716] reported that fine powders such as TCP, calcium silicate, or magnesium carbonate could be used as carriers for spice extractives. Gorsica[717] reported that soluble seasonings, flavorings, and coloring agents could be combined with TSPP as a soluble carrying agent for these compounds for use in seasoning and flavoring meat products. The TSPP would dissolve when placed into curing salt mixtures leaving the dispersed spice, flavoring, and coloring agents in suspension or solution.

Additional Patents in the Literature

The patent literature for the use of phosphates in meat processing is voluminous and in many cases appears to duplicate itself. Anyone contemplating the use of a phosphate in any type of meat processing would be well advised to consult the literature and make a thorough search before using it in a commercial process.

A number of patents have already been discussed in previous paragraphs of this section of meat applications of the phosphates. A few more will be described to demonstrate the number and variety of patents for phosphate applications in meat processing.

Phosphates have been patented for use as buffers to maintain optimum pH values for the greatest stability of meat-curing salts containing nitrite,[718] as sequestering agents,[719,720] and to prevent the caking of curing salt mixtures, especially those containing STP.[720-722]

A number of combinations of polyphosphates as well as polyphosphates combined with curing salts have been patented for their beneficial effects upon meats. For example, STP and its hydrate, SHMP, Kurrol's salt, and other condensed alkali metal phosphates have been patented for their applications in swelling and softening muscle fibers and creating uniform fat emulsification in sausage products.[723] The addition of SAPP and TSPP with potassium metaphosphate (Kurrol's salt) with or without other curing agents is reported to increase the swelling, have a tenderizing effect upon meat proteins, improve their binding properties, and improve the plumpness of cured meat products.[724] The use of TSP in curing-salt mixtures has been patented for its ability to improve color stability, water retention, yields, and to lower the shrink during processing.[725] Other patents have been obtained for use of the phosphates for improving the flavor and keeping properties of meats, especially sausages with combinations of sodium glutamate or glutamic acid plus an alkali metal polyphosphate with or without citric acid;[726] improving the properties of animal or fish meat by combinations of magnesium ion with polyphosphate;[727,728] eliminating the need for nonmeat binding agents through the addition of TSP, TSPP, SHMP or KHMP, or STP or mixtures of these alone or blended with spices;[729] phosphate mixtures containing MSP, DSP, SAPP, and TSPP in combination with other curing salts for curing ham;[730] solutions of STP hexahydrate, SAPP, and TSPP decahydrate to create a plasticizing effect on meat which improves its handling properties and palatability;[731] and for the use of combinations of pyrophosphate or tripolyphosphate with nontoxic organic acids to improve the water-binding properties of sausages.[732]

Patents have also been obtained for various forms and mixtures of polyphosphates with other curing compounds for applications in meat processing. For example, patents have been obtained for spraying a solution of a phosphate on sodium chloride crystals in order to form a crystalline complex of the phosphate component with the crystalline structure of the sodium chloride;[733] treating canned, cooked, or scalded sausages with a liquid medium containing TSPP or other condensed phosphates;[734] spray-dried long-chain polyphosphates for treating blood and meat products;[735] free-flowing compositions of alkali metal tripolyphosphate and a strongly basic alkali metal compound prepared in such proportions that a 1% aqueous solution has a pH value between 10.5 and 12;[736] preparation of solutions of STP and NaOH in such proportions that when they are combined during pumping of cured meat products, TSPP is formed in the meat;[737] combining solutions of sodium erythorbate and an alkali hydroxide, carbonate or othophosphate during pumping cured meat products;[738] and crystallizing sodium nitrite on the surface of crystals of STP.[739]

Other examples of the applications of phosphates which have been patented are combinations of STP with cyclamate which provide artificial sweetening and replace sugar in the curing process;[740] combining phosphates with a coloring agent so that when they are blended with other curing salts, one can determine when a homogeneous blend has been obtained by the uniformity of the color;[741] and the stabilization of flavor intensifiers against the action of phosphates in the meat by first inactivating them by means of condensed phosphoric acids or their ammonium or alkali metal salts, such as TKPP, KTP, or SHMP.[742]

Phosphate Applications in Poultry Processing

Identical chemical and physical reactions occur in postmortem poultry muscle, in beef, and in similar animal muscles.[743,744] As the oxygen supply to the tissue is cut off by death of the animal, ATP and its phosphate-bond reservoir, creatine phosphate, are first maintained at their normal levels through anaerobic glycolysis. When no more glycogen is present, the ATP rapidly disappears and the muscle becomes rigid as rigor mortis sets in. At the same time, the production of lactic acid through the metabolism of the glycogen reduces the pH to levels of about 5.5 to 6.0. The major difference in the case of poultry is that rigor mortis sets in much faster and the aging process to obtain subsequent maximum tenderization is much shorter than for red-meat animals. For example, while beef requires 10 to 20 days to develop optimum tenderness, poultry requires only about 12 to 24 hr.[745] deFremery and Pool[743] and deFremery[744] published the results of their own investigations and excellent reviews of the biochemistry of chicken muscle and effects of rigor mortis upon it. The same palatability factors important for red meats are important to the consumer of poultry meats, including color, tenderness, and flavor.

Color Preservation

The color of fresh poultry meat is a major factor utilized when selecting meat. Several investigators have reported the effects of polyphosphates on the color of fresh poultry meat. When polyphosphates were fed to chickens prior to slaughter, an undesirable dark color or discoloration was observed in the meat.[746] Klose and co-workers[747] reported that soaking freshly-chilled chicken carcasses in polyphosphate solutions changed the yellowish cast of normal skin pigmentation in the control to a disagreeable bluish-white appearance. Mahon[748,749] reported that a favorable color of fresh poultry meat on the chilled, cut surfaces in refrigerated, cooked meat was obtained when poultry meat was treated with polyphosphate solutions during the aging of the carcass. Mahon[750] patented the treatment of poultry in the raw state with solutions or a dusted coating of fine, dry polyphosphates selected from the group of straight-chain polyphosphates having a ratio of alkali metal oxide to P_2O_5 from 0.9:1 to 2:1. The fresh appearance and color of fresh, frozen, and refrigerated poultry meat are reported to be retained for a longer time if the freshly killed poultry is soaked in solutions of phosphate or dusted with powdered phosphate during the chilling and aging of the poultry carcass.

Froning[751] reported the treated raw, uncooked meat had an unappetizing bluish-white glistening appearance similar to that reported earlier by Klose et al.[747] Froning noted, however, that the cooked meat was affected much less than the uncooked meat. In a later investigation, Froning[752] evaluated the color of cooked chicken rolls with a Gardiner Color Difference Meter standardized with a white plate. He reported that the color of the meat was significantly darkened by the addition of polyphosphates and that the meat darkened in proportion to the level of phosphate added. The color of the polyphosphate-treated meat was acceptable when 0.5 to 1.0% polyphosphate was added to the meat, but was unacceptably dark upon the addition of 2%.

Increasing Tenderness

The tenderness of poultry meat is significantly influenced by the reactions that accompany the development of rigor mortis, just as with red meats. However, the reactions occur more rapidly under the conditions used in processing poultry. Dodge and Stadelman[753] reported that cooking poultry meat prior to the start of rigor mortis resulted in significantly more tender meats, again, just as has been reported for red meats.[645] It, therefore, is important to arrest those reactions occurring during the onset of rigor mortis that are responsible for toughening or to counteract their effects to obtain tender meat.

deFremery and Pool[743] investigated the relation of rigor mortis to tenderness of poultry muscle. Rigor mortis began to appear when the level of ATP was reduced by approximately one half and continued to increase as the ATP level dropped to zero. Glycogen began to disappear after death of the animal. The onset of rigor mortis could be followed by the disappearance of glycogen and the accompanying lowering of the pH due to formation of lactic acid during anaerobic metabolism of the glycogen. deFremery and Pool demonstrated that any conditions that caused a more rapid development of rigor mortis increased the toughness of the cooked meat. The muscle of 10- to 16-week-old chickens became rigid with rigor mortis at room temperature within 2 to 4½ hr after death and attained a pH between 5.8 and 5.9. They compared various postmortem processing variables upon toughening, such as immediate excision of the muscles prior to onset of rigor mortis, mechanical beating, freezing the muscle prior to onset of rigor mortis with subsequent thawing, exhaustive electrical stimulation, irradiation, and injections of sodium monobromoacetate. Every treatment that resulted in more rapid loss of ATP than occurred during the normal development of rigor mortis caused increased toughening of the cooked muscles. However, when the enzyme systems required for the breakdown of glycogen were poisoned by injections of sodium monobromoacetate, the resulting muscle failed to become tougher and was as tender as the prerigor muscle even though ATP was lost as rapidly as in untreated meat. deFremery[744] demonstrated that rapid depletion of glycogen in the live bird, prior to slaughter, caused little or no loss in the initial tenderness of the cooked meat. When live birds were injected with 1 mg/kg epinephrine 15 hr prior to slaughter, the carcasses were found to undergo more rapid loss of ATP and onset of rigor mortis, but to have little glycogen left in the carcass. The pH of the meat fell very little from its original value immediately after slaughter, indicating that little or no glycolysis occurred and little or no lactic acid was thus formed. The treated carcasses produced meat which was as tender immediately after onset of rigor mortis as the meat of the control which had been fully aged. deFremery also reported that chicken muscle was more tender if it was cooked immediately after death of the bird than if it was aged for 1 hr. This indicated that post rigor mortis muscle was tender, but became tougher as rigor mortis occurred and finally became tender upon subsequent aging for 12 to 24 hr.

As a result of these investigations, deFremery[744] concluded that the disappearance of ATP and rapid onset of rigor mortis in poultry meat do not result in toughness, but rather that the rapid disappearance of glycogen and the subsequent increase of lactic acid in the muscle were related to toughening. In contrast, the poultry meat in which some treatment occurred that inhibited postmortem glycolysis was as tender and remained tender during aging, as the controls given normal aging. Thus, toughness of poultry meat seemed to be involved with glycogen breakdown after death of the bird. deFremery reported that similar biochemical changes were undoubtedly responsible for toughness developing in red meat with onset of rigor mortis.

May and co-workers[754] found that chilling the carcasses in a 3% solution of polyphosphates significantly improved the tenderness of light and dark poultry meat. Klose et al.[755] reported that soaking chicken, fowl, or fryers in a chilling solution containing 5% polyphosphate had no effect on tenderness of the resulting meat. The chicken meat was evaluated either 3 or 22 hr after the killing periods. Similar tests were also run on turkeys and again no beneficial effect of the polyphosphates on tenderness was reported. Wells and co-workers[746] reported that feeding SHMP and other salts to chickens prior to slaughter did not affect the tenderness of the resulting cooked meat. Klose and co-workers[747] later demonstrated that there was a small, statistically significant increase in tenderness when poultry, both fowl and fryers, was chilled for 22 hr in a 5% solution of a mixture of STP and TSPP.

According to Spencer and Smith,[756] the chill treatment of fryers with approximately 7.5% polyphosphate solution for 6 hr significantly increased the tenderness as determined by taste panels and shear measurements. Baker and Darfler[757] reported that treating leghorn fowl and fryers with phosphates improved the tenderness of both the fresh and frozen-thawed meat. Schwall and co-workers[758] reported that suspending a water-soluble alkali metal polyphosphate salt in a nonaqueous fat or similar system and injecting it into the muscular portions of poultry resulted in significant tenderization of the meat.

Increasing Binding

The binding properties of poultry meat have also been demonstrated to be significantly improved by treatment with polyphosphates. The effect appears to be due to the increased solubilization of the muscle protein, as in red meat.

Froning[751] reported significantly increased binding when polyphosphate was added to a soak solution in which the aged carcasses were held for 15 hr prior to cooking. After evisceration, the birds were chilled for 6 hr in ice water, then packaged in plastic bags and frozen. Later, the birds were thawed at refrigerated temperatures and soaked for 15 hr in a 6% solution of polyphosphate maintained at 1 to 3.5°C, whereas the control carcasses were soaked in plain water. The carcasses were boned, formed into rolls, and cooked to an internal temperature of 85.5°C. The resulting chicken rolls exhibited evidence of increased binding because they were less crumbly and the slices were more even when polyphosphate-treated meat was used than those of untreated controls. In a later experiment, Froning[752] reported that the polyphosphate could be added directly to ground chicken meat and formed into chicken sausages or rolls. The treated poultry meat exhibited increased binding properties when polyphosphates were added. In these experiments, the birds were eviscerated, chilled in an ice slush bath, packaged in plastic bags, and frozen. One month later, the carcasses were thawed at refrigerated temperatures, boned, and the meat was ground. Various binders were mixed into the meat, and the mixture was placed in casings and cooked to an internal temperature of 82°C. Judged by tear strength measurements, 1% polyphosphate was found to be superior to 10% milk solids, 3% gelatin, or 6% gluten flour as binders.

When Schwall and Rogers[759] treated pieces of poultry meat with a surface coating of sodium chloride, potassium chloride, or salts of the various sodium phosphates and mixed them thoroughly for a period of time, a surface coating of the soluble proteins from the poultry meat was formed which then caused the pieces of meat to be bound firmly together into solid pieces when placed in forms or casings and cooked. In a second patent, the same authors[760] reported that combinations of sodium chloride and edible phosphate salts were superior to any of the individual salts for this same application.

Schnell and co-workers[761] and Vadehra and Baker[762] provided excellent studies of the chemical and physical factors involved in the binding of chunks of meat. The results of these studies are applicable to all types of meats, not just to poultry. These authors reported that in all their experiments binding was improved by heating the pieces of meat. Phosphate treatments increased the amounts of proteins which would be found in the fluids on the surface of the chunks of meat, but sodium chloride did not increase this protein. Both sodium chloride and food grade phosphates reduced the amount of fluid which could be expressed from the chunks of meat and improved the binding properties of these chunks of meat. Vadehra and Baker[762] concluded that binding is a complex phenomenon involving the water-holding capacity of the meat, the cell disruption and breakage, the release of intracellular material, the type and qualities of myofibrillar and connective tissue protein, and extraneous sources of proteins. They also concluded that sodium chloride and the polyphosphates appeared to have similar effects on the binding properties of meat, but that their mechanism of action differs because the chemical composition of the cookout from the two treatments is also different.

Schlamb[763] patented a process for improving binding properties of pieces of poultry meat by subjecting them to mechanical action to break the cellular structure on the surfaces of the pieces, coating them with a water-soluble polyphosphate and sodium chloride, and then pressing them together to form larger pieces and cooking them.

Increasing Moisture Retention

Just as with red meat, the phosphates also had significant effects on the moisture retention or water-holding capacity of poultry meats. Numerous investigations have confirmed these effects. However, the treatment of poultry meat during this chilling stage with solutions of 5 to 15% polyphosphate tended to lower the chilled yield of the raw poultry meat indicating that the meat picked up somewhat less moisture than the controls soaked in ice water.[747,750,751,754-756,764-767] May and co-workers[754] reported that low quantities, i.e., 3% of polyphosphate in the chilling solution, caused the poultry meat to absorb greater amounts of moisture than the controls, while quantities normally used, ranging from 5 to almost 10% solutions, caused a reduced uptake of moisture when compared to the controls. Schermerhorn and Stadelman[765] reported that levels of polyphosphates above normal levels, e.g., a 12%

solution, caused significant lowering of moisture uptake.

Klose and co-workers[747] reported that there was much less difference in water uptake between water-chilled controls and poultry meat first chilled for 6 hr in ice and water and subsequently treated 7 hr with a polyphosphate solution. A significant reduction in water uptake occurred by chilling the meat in the polyphosphate solution for the full 23 hr. This indicated to the authors that the most critical rate of water absorption occurs during the early stages of chilling.

The polyphosphates also were found to reduce the drip loss upon thawing treated, frozen poultry meat.[747,749] Table 28, taken from Mahon's patent,[750] demonstrates the effects of SHMP, STP, and pyrophosphate soaks or dips on thawing drip-loss, cooked yields, and TBA values. The freshly killed and eviscerated chickens were cut in half. One half, labeled A, was treated with water as a control, while the second half, labeled B, was given the indicated phosphate treatment. As shown, all phosphate treatments except SHMP significantly reduced the thawing drip-loss, increased the cooked yield, and reduced the TBA value. These and other data in the patent indicate that STP was most effective. The data in this table also demonstrate that the quantity of polyphosphate solution taken up by the poultry meat was either less or not significantly greater than the amount of water absorbed by the controls. This is an important point as it has been argued that the treatment of meat with polyphosphates allows "loading" the meat with water. Obviously, this is not the case.

Table 29, also taken from Mahon's patent,[750] demonstrates that increasing the level of STP in the chill water from 1% to 5% significantly improves each of the benefits studied. Mahon, therefore, stated that STP was the preferred phosphate for treating poultry meat, including chicken, turkey, geese, ducks, capons, Cornish hens, squab, guinea fowl, and pheasants.

Between 2 and 8% polyphosphates were shown to significantly reduce cooking losses or cooking shrink.[747,749,750,755,764-770] Thus, although the polyphosphates did not cause raw poultry meat to pick up a significant amount of moisture, they did reduce the amount of moisture loss during the cooking, therefore increasing the yields. This was found to occur whether the meat was cooked in boiling water, steam, cooked under pressure, or cooked by microwave ovens as shown in the graphs in Figures 9 and 10. Treatment of the poultry meat with polyphosphates was also reported to increase the juiciness of the cooked meat when eaten, as evaluated by taste panels.[754,756,758] Feeding polyphosphates to the bird prior to slaughter had no effect upon the water-holding capacity or rehydration of freeze-dried poultry meat.[746] The effects of the polyphosphates on moisture retention were noticeably different for different types of poultry. Broilers were affected to a greater extent than heavy fowl,[766] as were lighter fowl compared to heavy fowl.[767] Figures 9 and 10 demonstrate the effects of phosphate treatment on loss of fluid and, therefore, yield when broilers and hens were cooked by four different methods. As shown in Figure 10, hen meat was much less affected by phosphate treatment than broiler meat shown in Figure 9.

Mahon[768] reduced cooking time and thus increased the deboned yield of fowl cooked in an aqueous solution of up to 4% of a noncyclic polyphosphate, preferably STP, although KTP, TKPP, TSPP, and SHMP were also effective. Montemayor and co-workers[769] demonstrated that salt and salt plus a commercial mixture of phosphates greatly reduced the cookout of frozen poultry meat as compared to untreated controls.

Increasing Fat Emulsification

The fat emulsifying properties of the salt-soluble protein of light and dark meat of broilers, hens, turkeys, and ducks were studied by Hudspeth and May.[770] These authors reported that dark meat from each of these fowl was capable of emulsifying greater quantities of oil than light meat. This was a surprising discovery since greater quantities of salt-soluble proteins were extracted from the light meat than from the dark meat of the same type of fowl. It, thus, appeared to the authors that the greater emulsifying capacity of the salt-soluble proteins from the dark meat counterbalanced the greater amounts of extractable salt-soluble proteins from the light meat.

Improving Flavor

A number of investigators have commented upon the flavor characteristics of phosphate-treated chicken meat. No significant flavor changes were found between phosphate-treated poultry fed

TABLE 28

Comparison of Effects of Different Polyphosphates on Drip-Loss, Cooked Yield and TBA Values of Poultry Meat

Chicken No.	No. 1 dip solution 22 hr	Dipped weight, percent	No. 2 dip solution	Dipped weight, percent	Thawing drip, percent	Thawed weight, percent	Cooked weight, percent	TBA value Cooked ground meat 7 days at 40°F	TBA value intact cooked breast 14 days at 0°F
1A	Water	105.9	–	–	3.1	102.8	76.8	15.0	9.0
1B	6% hexameta[1]	105.3	–	–	1.0	101.3	76.1	8.4	0.4
2A	Water	105.9	–	–	4.1	102.3	77.0	14.0	8.1
2B	6% tripoly[2]	104.8	–	–	0.9	103.9	81.8	1.0	0.2
3A	Water	109.6	–	–	4.9	104.7	77.5	13.0	7.4
3B	3% N pyro[3]	106.4	–	–	2.0	104.4	81.1	2.6	0.3
4A	Water	106.0	–	–	3.9	102.1	68.8	15.0	7.5
4B	6% 50/50 N/A pyro[4]	105.5	–	–	1.5	104.0	75.0	1.5	0.2
5A	Water	107.5	Water (¼ hr)	107.5	5.9	101.6	71.8	13.0	6.6
5B	do	106.0	15% tripoly (¼ hr)	105.4	1.5	103.9	78.0	4.2	0.3
6A	do	109.0	Water (1 hr)	109.9	6.3	103.6	73.5	13.0	6.3
6B	do	107.0	15% tripoly (1 hr)	106.3	1.6	104.7	77.7	1.5	0.2

[1] $(NaPO_3)_6$.
[2] $Na_5P_3O_{10}$
[3] $Na_4P_2O_7$ practical saturation at 40°F.
[4] 3% $Na_4P_2O_7$, 3% $Na_2H_2P_2O_7$.
From Reference 750, Table 4, with permission.

TABLE 29

Comparison of Effects of Different Levels of STP on Drip-Loss, Cooked Yield and TBA Values of Poultry Meat

Chicken No.	Soak Solution 17 Hours	Soaked weight, percent	Thawed weight 7 days, percent	Thawing drip, percent	Cooked flesh, percent	Increased flesh yield,[1] percent	Added tripoly, percent	TBA value 7 days, 40°F.
1A	Water	106.2	–	–	53.2	–	–	12.0
1B	1% tripoly	109.3	–	–	56.6	+ 6	0.04	3.9
2A	Water	107.1	–	–	52.3	–	–	13.0
2B	3% tripoly	107.9	–	–	57.1	+ 9	0.34	2.1
3A	Water	106.7	–	–	52.3	–	–	10.0
3B	5% tripoly	106.6	–	–	60.9	+16	0.48	0.8
Soaked Chicken Frozen 6 Days at 0°F; Thawed Overnight to 40°F Before Cooking								
7A	Water	106.6	102.0	4.6	53.4	–	–	10.5
7B	1% tripoly	106.6	103.1	3.5	57.2	+ 7	0.05	9.6
8A	Water	106.7	102.1	4.6	54.0	–	–	10.5
8B	3% tripoly	106.2	104.8	1.4	60.6	+12	0.34–	2.9
9A	Water	106.8	102.0	4.8	46.9	–	–	8.2
9B	5% tripoly	105.2	104.9	1.3	58.6	+25(?)	0.58	0.6

[1] Water-treated sample taken as 100% cooked meat yield.
From Reference 750, Table 5, with permission.

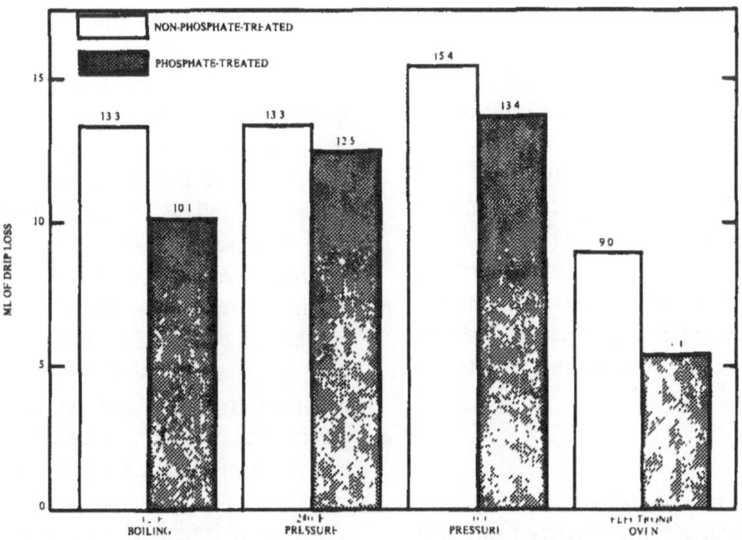

FIGURE 9. Average volume of drip in minced broiler muscle cooked by four different methods, phosphate-treated compared with non-phosphate-treated. (From Reference 766, Figure 3, with permission.)

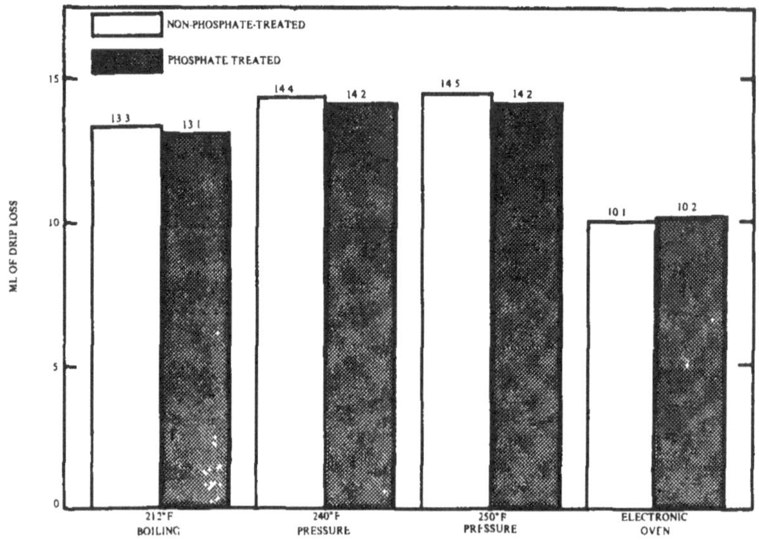

FIGURE 10. Average volume of drip in minced hen muscle cooked by four different methods, phosphate-treated compared with non-phosphate-treated. (From Reference 766, Figure 4, with permission.)

the phosphates prior to slaughter and untreated controls.[746] Klose and co-workers[755] and Spencer and Smith[756] found no significant flavor changes when phosphates were used to treat poultry meat prior to cooking. Flavor differences were reported by Froning[752] and later by Klose and co-workers,[747] but these were thought to be due to a difference in saltiness which enabled tasters to distinguish between the phosphate-treated meat and the controls. The flavor differences were not considered to be advantageous or otherwise. Significant improvements in the flavor of the phosphate-treated meat over untreated controls were reported by a number of investigators.[749,750,757,758,768] Improvements in the odor of the cooked refrigerated poultry meat were also reported by a number of investigators.[749,750,755,768,771,772] The reasons for the improved flavor and odor characteristics have been attributed to the ability of the polyphosphates to inhibit the oxidation of lipids in aqueous suspensions as measured by significantly reduced TBA values in the treated meat when compared to untreated controls.[749,750,768,772]

Preventing Microbiological Spoilage

The microbiological state of poultry meat also has an important effect on its odor and flavor acceptability. Phosphates, and particularly polyphosphates, have been found to aid in controlling the microbiological population on the surfaces of poultry meat. Jimenez and co-workers[773] reported that combinations of hexametaphosphate and chlortetracycline-HCl were effective in preventing microbiological spoilage of poultry meat, especially when used as dips on the eviscerated poultry. Melnick and co-workers[774] reported a synergistic action of H_3PO_4 and MSP with sorbic acid on extending the microbiological stability of refrigerated poultry meat. Treatment of poultry meat with combinations of MSP with ascorbic acid was reported to triple the shelf-life over that of poultry meat treated with either compound alone. Spencer and Smith[756] reported that microbiological spoilage was reduced on chicken fryer carcasses chilled for 6 hr in ice water containing 10 oz of polyphosphate per gallon. The reduction in microbiological spoilage was confirmed by plate counts, UV fluorescence, and reduction in off-odors. Shelf-life was increased by one to two days by use of the phosphates alone. A possible explanation for the reduction in fluorescent microorganisms on the surfaces of the poultry meat has been proposed by Kraft and Ayres.[775] These investigators found that *Pseudomonads* required a minimum of 1 ppm iron present in their environment to produce spoilage on poultry. This quantity of iron is usually provided by the water used in processing and would be enough to allow rampant growth of these organisms. The polyphosphates are excellent sequestering agents and may prevent iron from being utilized by the microorganisms.

Tomiyama and co-workers[776] reported that combinations of hexametaphosphate or pyrophosphate as well as other chelators of iron, copper, and cobalt were effective when combined with antibiotics and antioxidants in preventing deterioration by microbiological spoilage of chicken meat as well as fish flesh. Tylosin was found to be inactivated by iron. The effect of the iron could be overcome by treating the flesh with pyrophosphates.

Elliott and co-workers[777] further reported on the microbiological inhibition provided by polyphosphates on chicken flesh. They found that nonfluorescent *Pseudomonads* were completely inhibited by the presence of polyphosphates in their medium. Fluorescent strains grew, but only after a lag period. The pH was not the controlling factor in inhibition of the growth of these spoilage organisms. The authors reported that addition of magnesium ion would reverse the inhibiting effect of the polyphosphate. When polyphosphates were added to the ice used in chilling chicken carcasses overnight at 3 and 8% levels, the shelf-lives of the chicken carcasses were increased 17 and 25%, respectively. Holding the carcasses in continuous contact with the same polyphosphate solution at refrigerated temperatures, increased shelf-life by 17 and 67%, respectively. Using antiseptic ice for the 8% polyphosphate soak increased shelf-life by 60% over the use of ordinary ice water. The authors reported that the 3 and 8% polyphosphate soaks prevented nonfluorescent strains of *Pseudomonads* from growing, but did not completely inhibit the growth of fluorescent strains.

Phosphate Applications in Seafood Products

The Biochemistry of Phosphate Interactions with Seafood Flesh

The proteins of fish and other seafood muscles apparently undergo similar biochemical reactions to those experienced by the proteins of the postmortem muscles of mammals and poultry. It is, therefore, not surprising that the polyphosphates have similar effects when they are used to treat fish filets and mammal or poultry flesh. A number of investigations have sought to explain the effects of the polyphosphates upon fish muscle. For example, Linko and Nikkila[778] reviewed the findings of numerous investigators that demonstrated that even low levels of sodium chloride reduce the solubility of the myosin fraction of fish muscle and thus denature it. However, the addition of polyphosphates, such as the pyrophosphates, tripolyphosphate, and hexametaphosphate inhibits the denaturation of the muscle protein by salt. These authors also reported that citrates have a similar effect. The authors speculated that there are two mechanisms that may explain the inhibition of denaturation by phosphates and citrates. One possibility may be that the salt is directly bound by the protein, thus increasing the polar groups in the protein. Condensed phosphates are known to be bound by actomyosin, leading to an increase in the solubility of the protein. A second possible mechanism may be that the phosphates and citrates remove calcium and zinc ions that had been bound to the muscle proteins, again releasing polar groups. The authors ruled out the second possibility since they reported that EDTA, which strongly binds calcium and zinc, had no effect in preventing the denaturation of myosin by sodium chloride.

Kuusi and co-workers[779] investigated the rate of incorporation of radioactive pyrophosphates into Baltic herring filets dipped in the pyrophosphate solution just prior to freezing them. The filets incorporated pyrophosphate slowly at the skin side and much more rapidly at the cut surfaces. However, even after 34 days' storage, the inside of the filet was free of the radioactive salts. Most of the radioactivity remained in the soluble fraction. In contrast, Nikkila and co-workers[780] reported that radioactive pyrophosphate containing P^{32} completely penetrated into all parts of Baltic herring filets in 24 hr when they were allowed to stand in the phosphate solution at +4°C. In fact, quantitative measurements of the radioactivity demonstrated that it continued to increase over a five-day period. Again, the skin retarded penetration, while the cut surface allowed rapid penetration. An exchange of a small amount of P^{32} also occurred with lipid and nucleic acid components of the filet. The authors found that there was no change in the rate of penetration whether sodium chloride was present or not.

Scheurer[781] investigated the rate of penetration of STP containing P^{32} and Na^{24} and sodium nitrite containing Na^{24} into haddock filets. The filets were dipped into the solutions for either 10 sec or 10 min, then held for 24 hr at 33°F. The concentration of STP and nitrite decreased with distance from the surface and the lowest levels were found in the centers of the filets. The quantities of each chemical at any point in the filets were directly proportional to the concentrations of the salt in the dip solution and also increased as the length of the dipping time increased.

Love and Abel[782] reported that there were similarities in the denaturation of the protein and the development of toughness of the fish flesh when it was cooked. As the fish proteins became more and more insoluble or could not be extracted from the flesh, the flesh became tougher and fluid exuded more easily from it. The authors reported that the main structural proteins of the fish muscle became increasingly denatured or insoluble when the fish muscle was thawed and the denaturation increased in direct proportion to the length of time the fish muscle was stored in the frozen state at any given temperature. Treatment of fish filets with three polyphosphates, a combination of TSPP and SAPP, STP, or SHMP failed to inhibit denaturation with or without the presence of salt. According to these authors, the effect of the polyphosphates in reducing drip loss and preventing the dehydration of the fish muscle which leads to toughening was due to the solubilization of certain proteins in the fish surface tissues. These proteins then came to the surface of the fish flesh, gelatinized, and sealed it to prevent further loss of fluid. The authors reported that a solution of 5% sodium chloride was more effective in extracting the fish muscle proteins than the polyphosphates or combinations of polyphosphates and sodium chloride, but that the disadvantage of using sodium chloride was that it diffused into the tissues and lowered their freezing point. This caused a greater amount of liquid phase to remain in the fish muscle. As a result, oxygen diffused more rapidly through the liquid phase and gained access to the muscle fats, causing them to undergo oxidative rancidity. This, in turn, accelerated the denaturation of the muscle proteins.

Akiba and co-workers[783] reported that mixtures of approximately 5% by weight of sugars with alkaline polyphosphates at approximately 0.5% of the weight of the fish meat were effective in maintaining high-quality frozen, raw, ground fish muscle. The combination prevented the denaturation of the fish muscle proteins, a desirable factor in the preparation of fish sausage and fish ham.

Spinelli and co-workers[784] studied the quantity of dip solution absorbed by fish filets or steaks when dipped for 15 to 30 sec in solutions containing 2% salt and either 7.5% or 10% STP. They reported that the fish flesh absorbed between 2.5 and 3.0% of the dipping solution. Comparison of treated and untreated filets indicated that the STP uptake was fairly uniform as measured by concentration of STP in the filets. The authors also determined the amount of phosphorus absorbed by the fish flesh when Pacific Ocean perch filets were dipped in a solution of STP-salt ranging in concentration from 0 to 10%. They found that there was a linear uptake of STP by the filets that was directly proportional to the level of STP in the dipping solution. They reported that the filets absorbed 10 mg of phosphorus for each 2.5% increase in STP concentration.

A review of the results of numerous replicate experiments in which Pacific Ocean perch filets had been dipped in a solution containing 2% salt and 7.5% STP demonstrated that the filets absorbed an average of 32 mg STP phosphorus per 100 g of fish which is equivalent to absorbing 0.13% STP based upon the weight of the fish filet. The authors reported that the surface of the filets became very slippery to the touch and assumed a gelatinous characteristic over the entire surface. This made it difficult to handle the fish because of the slipperiness of the filets. The slippery characteristics decreased during storage but were still noticeable after three weeks.

Treated and untreated filets of Pacific Ocean perch and English sole were compared for amounts of extractable myofibrillar protein over a period of four weeks of storage. The authors reported no significant difference in quantities of proteins extractable from either treated or untreated filets of both species of fish. This does not concur with results reported by Nikkila and co-workers[778-780] and others, but does agree with the work reported by Love and Abel.[782]

Ravesi and Anderson[785] studied the ability of various salt solutions to extract proteins in an attempt to correlate protein solubility with the textural characteristics of the fish. The salt solutions tested were combinations of sodium chloride and sodium bicarbonate and a mixture of KCl, DSP, and MKP. The solutions containing phosphates usually extracted less protein than those containing the sodium chloride and bicarbonate. The authors were not able to correlate the amount of protein extracted with textural characteristics.

Kuusi and Kytokangas[786] studied the changes in the mineral constituents of Baltic herring filets stored one to five days at 4°C in solutions of 3% sodium chloride, 1% TSPP, or a combination of 3% sodium chloride plus 1% TSPP. TSPP was found to reverse the effects of sodium chloride on cation excess, which the authors interpreted to explain the protective action of pyrophosphate on myofibrillar proteins.

Color Preservation

A number of investigators reported that the polyphosphates improved the color of both the raw and cooked fish flesh treated with them. Meyer[787] reported a significant improvement in the color of salted herrings intended for marinating and other purposes where the color may be important to acceptability of the final product. He also reported that adding a polymeric phosphate to the brine in which the herring were preserved after harvesting and prior to further processing greatly improved the color of the fish when ready for further processing. The phosphates reported to be effective ranged from TSPP to the long-chain potassium polyphosphates known as Kurrol's salts.

Yamaga and co-workers[788] reported that injection of solutions containing 0.25% Curafos (85% STP, 12% SHMP, and 3% TSPP), 0.15% sodium citrate, and 0.25% sodium bicarbonate into the vertebral foramen and the hemal canal of the caudal peduncle of tuna and yellow-tails with storage at -20° for 55 to 57 days gave the most favorable retention of flesh color of the fish. A solution containing STP and sodium bicarbonate was also effective. Immersing the living fish in sea water containing dry ice for 5 to 10 min also preserved the color, but less effectively than when solutions were injected into them.

Sen and Lahiry[789] reduced the brown or yellow discoloration of sun-dried mackeral that had been treated with a dressing largely containing sodium chloride with 0.5% sodium benzoate, 3.0% MSP, 0.25% sodium nitrite, 0.2% BHA, 0.2% BHT, 0.5% propyl gallate, and 1.0% ascorbic acid prior to drying. The treatment was used to inhibit mold growth on the surface of the fish because they were only dried to about 45% moisture which would still allow growth of mold.

Miyauchi and co-workers[790] reported that dipping filets of Petrale sole, English sole, Pacific Ocean perch, halibut steaks and treating crab meats and oyster meats with STP-salt solution dips prior to irradiation preserved the original appearance of the products. In an extension of this work, Spinelli and co-workers[784] delayed the fading of the bright flesh hues of freshly cut fish filets if they were treated with phosphate dips. The filets were treated with a phosphate and then irradiated and stored either in cans or in polyethylene bags. The authors reported that the filets maintained their original hue during the entire storage period. In addition, the phosphate-treated filets retained a superior color to the untreated filets after they were cooked. The untreated filets turned a darker brown after approximately two weeks of the four-week storage test.

Increasing Tenderness

The addition of polyphosphates to seafoods has been reported to increase tenderness of the stored products. Sen and co-workers[791] reported that the addition of organic acids to curing salts used in preparation of dried-salt cured fish increased toughness of the fish after cooking. However, use of sodium biocarbonate, sodium carbonate or SHMP in the curing salts softened the texture and made the fish less tough. Garnatz et al.[792] increased the tenderness of cooked and frozen shrimp to a significant degree by treating them with sodium and potassium phosphates and other salts.

Sen and Lahiry[789] obtained significantly more tender salt-cured dried mackeral when they added sufficient sodium bicarbonate to the curing salt mixture to raise the pH approximately 0.5 points, but could not raise it any further without increasing microbiological spoilage. They, therefore, added SHMP to the curing salt mixture and increased the tenderness further without changing the pH. MacCallum and co-workers[793] reported that the addition of STP to dips for cod filets significantly improved the texture and increased tenderness of the treated filets over controls.

Miyauchi et al.[790] preserved the original texture of halibut and Pacific Ocean perch filets by dipping them in an STP-salt solution, irradiating them, and storing them at refrigerated temperatures for up to four weeks. Controls were significantly less desirable. In a continuation of this work, Spinelli and co-workers[784] reported that treating the halibut and Pacific Ocean perch filets with a dip in a 10% STP and 2% salt solution prior to irradiation resulted in significantly higher scores for tenderness of the filets when evaluated by a trained panel. The difference in tenderness over untreated controls became apparent after two weeks of the four-week storage period and the treated filets continued to score higher for the balance of the test. Similar results were obtained when a 7.5% STP and 2% salt dip was used prior to irradiation.

Increasing Binding

The Japanese are very fond of fish paste, or "kamaboko," also used to prepare fish sausage and fish hams. An excellent review of how these products are manufactured and some of the problems in manufacturing them has been published by Tanikawa.[794] It is necessary for the fish flesh to bind properly to prepare the most desirable types of fish paste, fish sausage, and fish hams. The polyphosphates have been found to increase the binding properties of these products as shown by numerous investigations.

Okamura and co-workers[795] reported that TSPP and STP at concentrations of 0.1 to 0.3% were highly effective in increasing the binding (jelly strength) of kamaboko. SHMP and DSP were somewhat less effective, and TSP, sodium tartrate, and sodium succinate had no effect on the binding of the fish meat. The effect of the phosphates on the pH had no relation to their effects on binding of the meat and none of the phosphates had any influence on its flavor. Although the pH values of some of the cakes were slightly elevated by the addition of these salts, the change had no influence on microbiological spoilage.

A later report by the same authors[796] indicated that the binding properties of fish cakes were lowered by the addition of excess starch, but that the polyphosphates overcame this defect. The most significant binding effects were obtained with the addition of 0.3% of a combination of 6 parts TSPP and 4 parts STP. Comparative studies[797] demonstrated that the addition of 0.3% of mixtures of 8:2, 7:3, and 6:4 parts of TSPP to STP increased the kamaboko jelly strength to a greater extent than 0.3% of TSPP or STP alone. A 1:9 mixture of TSPP and STP had less effect on jelly strength than the individual polyphosphates. The addition of 0.2% SHMP, but not less, was also shown to affect jelly strength to the same degree as 0.3% TSPP or STP alone, while 0.1% STP plus 0.2% SHMP produced the greatest increase in jelly strength although the plastic flow of the jelly was also increased. The addition of 0.1% to 0.3% TSP increased jelly strength without affecting plastic flow to a degree dependent on the pH of the jelly. DSP also increased jelly strength, but was more useful in mixtures with other phosphates and could be used to decrease plastic flow. Thus, it appeared that the jelly strength and plastic flow of kamaboko could be adjusted to any desired degree by proper selection of mixtures of ortho- and polyphosphates. Okamura[798-801] extended these studies by reporting[798] that minimum values in jelly strength of kamaboko were obtained by the addition of 0.0083 to 0.028 *M* potassium orthophosphate, 0.0025 to 0.005 *M* TSPP, or 0.0013 to 0.0066 *M* STP during manufacture. Comparison of levels of 0.049%, 0.11%, and 0.22% sodium tetrapolyphosphate[799] indicated that jelly strength was increased at concentrations above about 0.05%. The ability of 0.2% or more SHMP to increase jelly strength was confirmed.[799] Addition of 0.1% SHMP produced a sharp decrease in jelly strength, while a second, less distinct minimum occurred at 0.15%. The use of 0.2% SHMP appeared to be a practical maximum as higher levels produced an unpleasant taste.

Further work by Okamura[800] demonstrated that the jelly strength could be increased or decreased by varying the mixture of types and ratios of ortho- and polyphosphates added to kamaboko. He also reported[801] that ortho- and pyrophosphates would not improve the gel strength of high pH kamaboko containing 0 to 20% starch, while the use of STP, SHMP, potassium iodide, or magnesium chloride improved its strength. All these salts improved the gel strength of acidic kamaboko containing starch. STP was reported to provide the greatest improvement.

Miyake and co-workers[802] reported that 0.5% of a commercial polyphosphate mixture gave the greatest jelly strength to kamaboko but 0.2 to 0.3% was preferable for organoleptic reasons. When the polyphosphates were added along with magnesium ion, greater jelly strength was obtained than when either was added alone. A decrease in jelly strength was produced when calcium ions were added with the polyphosphates. These authors found the effectiveness of the polyphosphates on jelly strength to be as follows in decreasing order of effectiveness at 0.3% addition: TSPP, STP, SHMP, TKPP. The authors also found a significant degree of dissociation of actomyosin with a corresponding decrease in protein viscosity upon addition of polyphosphate plus magnesium ion and no such dissociation upon addition of polyphosphate plus calcium ion. The presence or absence of alkaline earth ions may well explain differences in the effects reported for the same polyphosphate by different investigators. Akiyama[803] reported that the addition of polyphosphates increased the binding of cooked whale meat.

Polyphosphate dips of individual fish fillets prior to forming and freezing into blocks have been reported[804] to lubricate the fillets while forming the block and to act as a bonding agent in forming a solid block during freezing. Both the lubrication and the bonding are believed to be due to the ability of the polyphosphates to dissolve surface proteins and form a continuous protein gel over the complete surface of the fillet.

Increasing Moisture Retention

The water-holding capacity of fish muscle is important to the organoleptic acceptance of the product. Cut surfaces of fish muscle, such as filets and steaks, commonly exude a slimy liquid which is unsightly and detracts from the acceptability of the product. The exudation of liquid is increased if the fish has been preserved by freezing, as the liquid drainage increases in content when frozen fish products are thawed. The loss of this liquid is believed to be directly related to the toughness of the fish product.

Seagran[805] reported the results of his analyses of the protein constituents of the drip obtained from thawed fish muscle. A comparison of paper electrophoresis patterns obtained from the drip of thawed rock fish filets and extracts of low ionic strength from fresh rock fish filets demonstrated that the same components were present in the drip in the fresh fish flesh. Although there had been speculation that some of the drip was due to denaturation of the contractile protein actomyosin, no actomyosin was present in the drip liquid.

The British firm of Albright and Wilson, Ltd.[806] patented the treatment of fish with an aqueous solution of sodium or potassium salts of polyphosphoric acids having molar ratios of H_2O: P_2O_5 of 1:1 to 2:1. The loss of moisture upon thawing the frozen fish and during cooking, which causes loss of high proportions of soluble protein, minerals, and other nutrients, is prevented by treating the fish prior to freezing with a 10 to 15% solution of the polyphosphates. The patent also mentions that it is advantageous to use a mixture of polyphosphates with an alkali metal chloride.

Mahon[807] patented the use of dips containing polyphosphates and salt, and especially containing STP and salt to prevent drip-loss upon thawing and to reduce loss of yield upon cooking frozen fish filets. This patent is one of those dominating the use of polyphosphates in the U.S. for treating fish and other seafood products. Mahon included a considerable amount of data demonstrating the applications of various phosphates by themselves, as well as in numerous mixtures, the amount of dip taken up by the filets, the changes in pH of the fish meat, the actual amount of phosphate absorbed, and the amount of drip-loss upon thawing the frozen filets. The effects of water, salt, STP, water plus STP, and salt plus STP dips on the percent uptake of dip, fish pH, percent thawing drip, percent added phosphate, and TBA values are shown in Table 30. Figure 11 shows the differences in thawing weight loss for filets of ocean perch and flounder dipped in a solution of 12.5% STP and 4% salt. Figure 12 demonstrates that these dips significantly reduce the protein loss from the treated filets which Mahon states leads to more nutritious filets. This process is now in commercial use[808] and reported to have increased the sale of fish filets.

Tanikawa and co-workers[809] studied the effects of treating cod filets with polyphosphate dips prior to freezing. They reported that a mixture of TSPP and STP reduced the drip from thawing cod tissue in a significant manner as shown in Figure 13. In addition, treatment of the filets with TSPP or with a TSPP-STP mixture reduced the amount of deoxyribonucleic acid phosphorus in the drip-liquid as shown in Figure 14, indicating that the cod flesh protein became less soluble. They also reported that the flesh proteins increased in their ability to reabsorb the liquids formed during thawing.

Mahon and Schneider[810] further reported the results of dipping haddock filets in various solutions of polyphosphates and other salts. The authors reported data demonstrating that alkaline salts of citric acid, carbonic acid, and similar salts do not reduce thawing drip until the pH of the fish flesh is considerably above 7.

The effects of the various phosphate solutions

TABLE 30

Effect of Salt and Sodium Tripolyphosphate on Thawing Drip of Haddock

Fillets Frozen at 0°F For 43 Days

No.	Dip Solution No. 1	Dip Solution No. 2	Dip Time, minutes	Dip Uptake, percent	Fish, pH	Thawing Drip, percent	Added Phosphate, percent	TBA
1	water	–	4	2.4	6.6	5.3	–	0.27
2	water	–	2	2.5	6.5	6.3	–	0.21
3	water	–	1	1.6	6.6	5.1	–	0.26
4	4% NaCl	–	4	5.5	6.5	10.3	–	0.18
5	4% NaCl	–	2	5.2	6.5	8.7	–	0.26
6	4% NaCl	–	1	3.5	6.6	7.3	–	0.32
7	12.5% $Na_5P_3O_{10}$	–	4	5.3	6.8	2.3	0.59	0.19
8	12.5% $Na_5P_3O_{10}$	–	2	1.9	6.7	2.7	0.30	0.14
9	12.5% $Na_5P_3O_{10}$	–	1	2.1	6.7	5.1	0.12	0.11
10	water	12.5% $Na_5P_3O_{10}$	2+2	4.8	6.8	1.8	0.44	0.24
11	water	12.5% $Na_5P_3O_{10}$	1+2	5.5	6.8	4.1	0.37	0.12
12	water	12.5% $Na_5P_3O_{10}$	½+2	6.2	6.9	2.6	0.33	0.18
13	4% NaCl	12.5% $Na_5P_3O_{10}$	2+2	3.2	6.7	2.2	0.25	0.35
14	4% NaCl	12.5% $Na_5P_3O_{10}$	1+2	6.0	6.7	2.9	0.45	0.19
15	4% NaCl	12.5% $Na_5P_3O_{10}$	½+2	6.1	6.7	2.6	0.41	0.11

Source: Mahon, J. H., U.S. Patent 3,036,923, 1962, with permission.

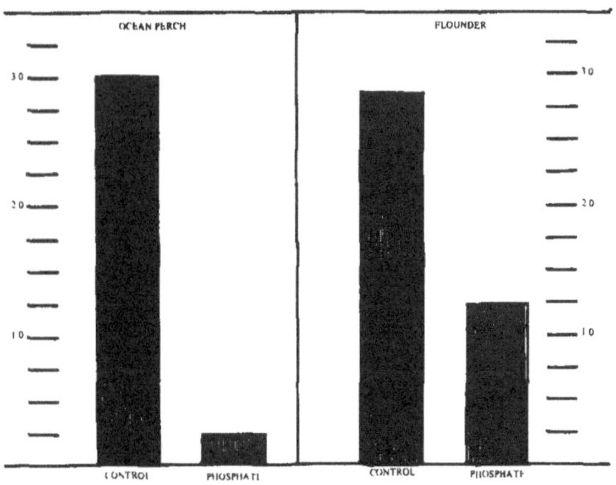

FIGURE 11. Thawing weight loss of frozen fish fillets after 6 months' storage. (From Reference 808.)

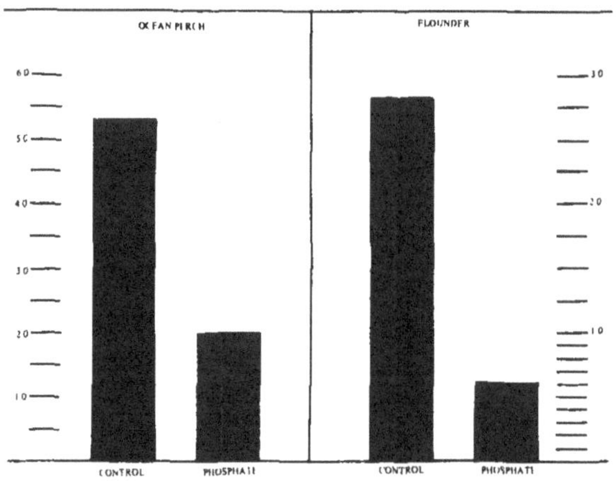

FIGURE 12. Protein loss from frozen fish fillets on thawing after 6 months' storage. (From Reference 808.)

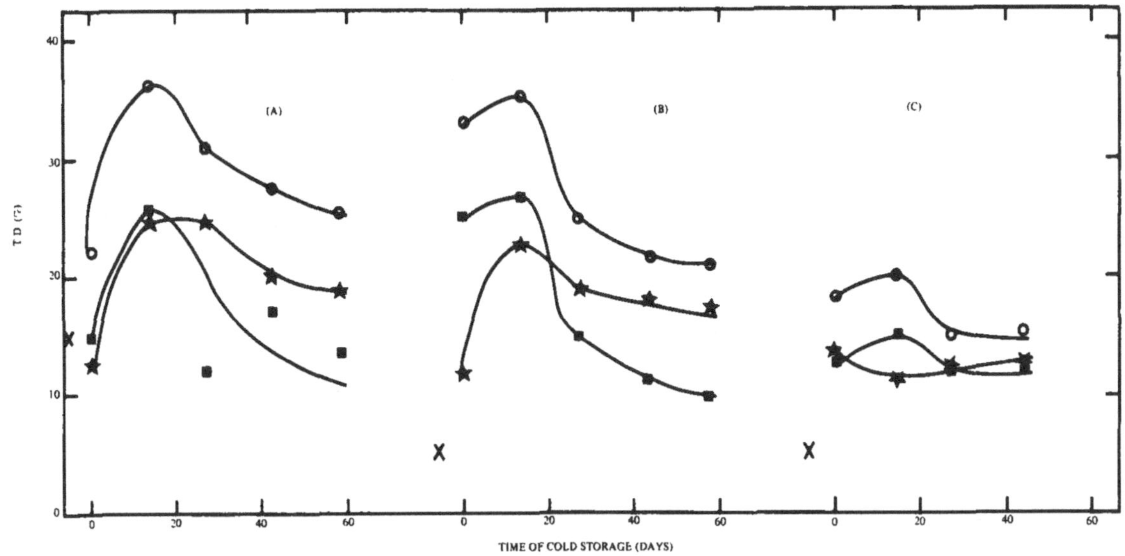

FIGURE 13. Variation in the amount of total drip (TD) during cold storage of frozen cod fillets (A) not treated; (B) pretreated with sodium pyrophosphate; (C) pretreated with a mixture of sodium pyrophosphate and sodium tripolyphosphate; ○ frozen at −5°C; □ frozen at −30°C; ★ frozen at about −90°C in acetone-dry ice; x TD in raw material before freezing. (From Reference 809, with permission.)

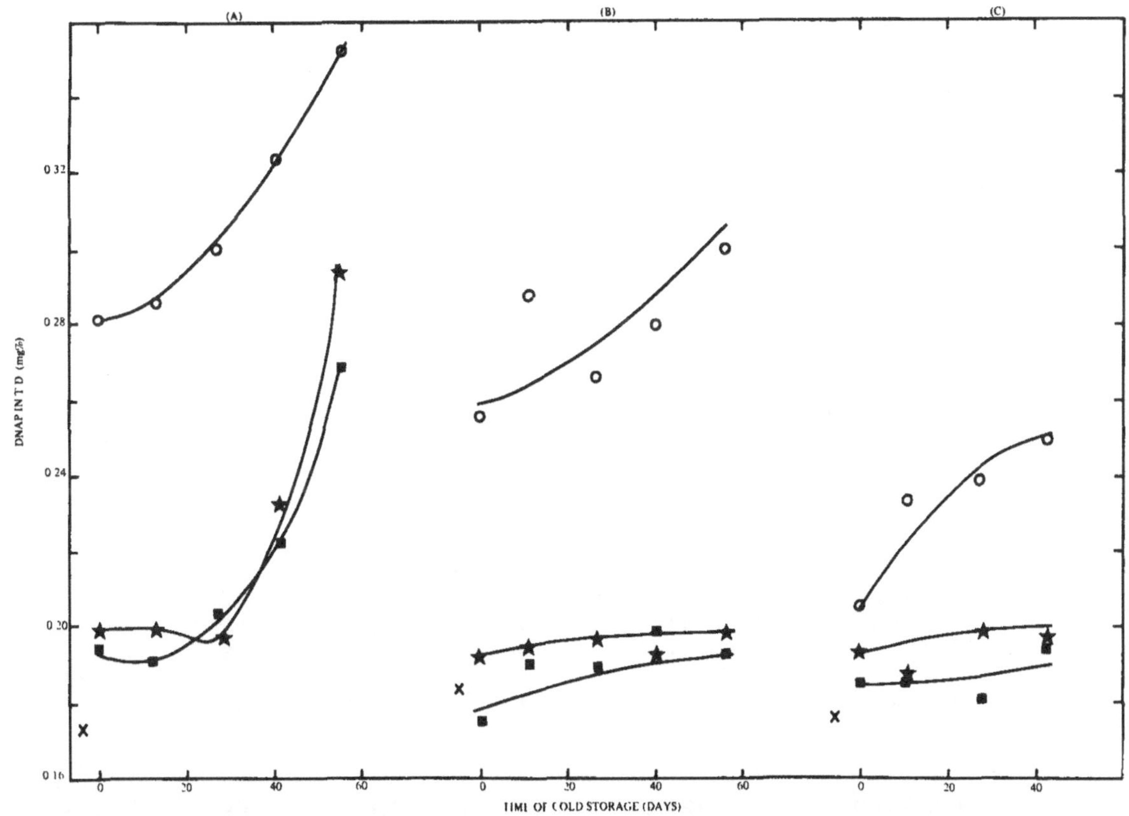

FIGURE 14. Variation in the amount of DNAP in TD. Same as in Figure 3, as to the marks (A), (B), and (C), and different symbols. (From Reference 809, with permission.)

and other alkaline salts used as 2-min dips for haddock filets on the thawing drip, uptake of solution, and the resulting pH of the fish are shown in Tables 31, 32, 33, and 34. The data demonstrate that STP was the most effective in reducing the drip-loss. The data also demonstrate that lower quantities of the salt solutions than of water were taken up by the filets. A calculation of the quantity of STP absorbed by the filets indicates that a maximum of 0.2% was absorbed. The data suggest that the effectiveness of the phosphates in reducing thawing drip is not just a pH effect since the phosphate dips caused insignificant changes in the fish pH. The nonphosphate

TABLE 31

Effect of Immersing Haddock Fillets for Two Minutes in Sodium Hexametaphosphate Solutions on Thawing-Drip Following 34 days of Storage at 0°F

Concentration of sodium hexametaphosphate	Solution pH	Solution uptake by fillet	Thawing drip	Treated fish pH
Controls (5) Range	–	3.0-4.0%	5.7-7.1%	6.6-6.8
Av.	–	3.2	6.5	6.7
5% hexametaphosphate	5.5	2.4	3.5	6.7
10%	5.4	3.2	3.6	6.7
20%	5.2	2.2	0.9	6.7
30%	5.1	2.0	0.9	6.6
40%	4.9	2.9	0.7	6.6

From Reference 810, with permission.

TABLE 32

Effect of Immersing Haddock Fillets for Two Minutes in Tetrasodium Pyrophosphate, Trisodium Orthophosphate, and Sodium Hydroxide Solutions on Thawing-Drip Following 34 days of Storage at 0°F

Treatment solution used	Solution pH	Solution uptake by fillet	Thawing drip	Treated fish pH
Controls (5) Range	–	3.0-4.0%	5.7-7.1%	6.6-6.8
Av.	–	3.2	6.6	6.7
1.5% $Na_4P_2O_7$	9.7	2.7	6.6	6.6
2.0%	9.6	2.1	3.1	6.7
4.0%	9.5	4.4	5.9	6.8
1.5% Na_3PO_4	11.7	3.4	6.7	6.7
2.5%	11.7	4.3	3.9	6.8
5.0%	11.5	–	1.4	7.5
0.75% NaOH	12.5	3.9	2.9	6.8
1.5%	12.3	5.2	4.3	7.3

From Reference 810 with permission.

TABLE 33

Effect of Immersing Haddock Fillets for Two Minutes in Trisodium Citrate, Sodium Bicarbonate and Sodium Carbonate Solutions on Thawing-Drip Following 34 days of Storage at 0°F

Treatment solution	Solution pH	Solution uptake by fillet	Thawing drip	Treated fish pH
Controls (5) Range	–	3.0-4.0%	5.7-7.1%	6.6-6.8
Av.	–	3.2	6.6	6.7
3% Na_3 Citrate	7.0	2.2	6.0	6.7
6%	7.0	1.9	5.0	6.7
9%	7.0	2.5	3.1	6.8
12%	7.0	2.2	3.4	6.8
1.75% $NaHCO_3$	8.0	4.0	4.5	7.0
3.5%	8.1	2.8	2.0	7.1
2% Na_2CO_3	11.2	2.7	4.7	6.9
4%	11.3	3.3	3.2	7.2

From Reference 810 with permission.

TABLE 34

Effect of Immersing Haddock Fillets for Two Minutes in Sodium Tripolyphosphate Solutions on Thawing-Drip Following 34 days of Storage at 0°

Concentration sodium tripolyphosphate	Solution pH	Solution uptake by fillet	Thawing drip	Treated fish pH
Controls (5) Range	–	3.0-4.0%	5.7-7.1%	6.6-6.8
Av.	–	3.2	6.6	6.7
2.5% tripolyphosphate	8.5	2.9	5.1	6.7
5.0%	8.4	2.0	3.4	6.8
7.5%	8.3	2.9	2.3	6.8
10.0%	8.3	1.7	1.7	6.9
12.5%	8.2	1.2	0.3	6.8

From Reference 810 with permission.

salts had little effect until the pH of the fish reached 7.0 or above. Table 35 demonstrates that the thawing drip of other types of fish can also be reduced by the treatment with STP.

Boyd and Southcott[811] reported that pretreating filets of dover sole, Pacific cod, halibut, red snapper, and chinook salmon with STP prior to freezing reduced drip-loss during thawing and increased the yield of the frozen and thawed fish. Swartz[812] reported increased yields of tuna treated with STP prior to cooking.

Sutton[813] reported the results of his experi-

TABLE 35

Effect of Treating Various Types of Fish Fillets with Sodium Tripolyphosphate on the Reduction of Thawing Drip [1]

Fish used	Controls	Tripoly-phosphate treated
Cod	3.2%	0.1%
Flounder	4.6	0.3
Haddock	4.0	0.1
Perch	7.0	0.7
Pollock	5.0	1.3
Scallops	5.0	0.8
Av.	- 4.8%	0.55%

[1] From Reference 810, Table 5, with permission.

ments on the effects of STP and TSPP on improving the quality of frozen fish. He found STP more effective than the pyrophosphate and reported that the polyphosphate had a twofold action, first by reducing the loss of fluid during thawing of frozen filets and second, by increasing the water-retention in the filets and thus improving textural properties of the cooked fish. Sutton reported that unprocessed prerigor mortis fish untreated by polyphosphate had drip-losses ranging between 2 and 10% after up to five days' storage in ice. Dipping the fish in polyphosphate solution significantly reduced the thaw-drip loss. Fluid retention increased with an increase in the time of dip and with increasing concentration of the TSPP up to a 6% solution. No further increase in fluid retention occurred with higher concentration of TSPP. He recommended dipping the fish for 30 sec in a 4% solution of STP for optimum results through improved water retention and improved texture, factors confirmed by large experiments. He also reported that the texture was improved only in the fish treated with STP which had low pH, while only improved water retention and no improvement in texture was obtained if the fish muscle had a high pH. Sutton further reported that the presence of sodium chloride had no improving effect over the effects of the phosphate alone.

Sutton and Ogilvie[814] reported an excellent investigation that quantifies the effects of polyphosphate dips on fish muscle and their retention of the muscle's natural fluids during storage. The authors reported significant reductions in drip loss of fish muscle treated by phosphate dips of various concentrations for a standard 2-min dip period over the drip-loss of untreated controls. The authors found that both sodium and phosphate ions were absorbed into cod muscle during dipping and their retention in the muscle during freezing, storage, and subsequent thawing is directly proportional to their ratio in the dip solution. Fluid-retention was also found to be directly proportional to the concentration of the ions in the solution. The authors reported that when dilute solutions of polyphosphate were used as dips, the polyphosphate anion and the muscle phosphate undergo exchange reactions with a resulting net loss of phosphorus from the dipped muscle. Furthermore, the authors reported that their calculations indicated 260 additional moles of water were retained in fish muscle treated by dipping for each mole of STP absorbed into the muscle, when compared with undipped controls.

Miyauchi and co-workers[815] reported that irradiated fish filets of Pacific Ocean perch, English and Petrale sole, as well as halibut steaks could be prevented from dripping and subsequently losing their internal juices if they were first treated with mixtures of STP and salt before irradiation. The authors reported that the products retained their preirradiation quality for four weeks with this treatment. In later work, Miyauchi et al.[790] reported that dipping filets of ocean perch, English and Petrale sole, and halibut steaks in polyphosphate (STP)-salt solutions effectively reduced the drip of irradiated filets during as much as four weeks of storage at refrigerated temperatures. They reported that the STP-salt dip treatment and subsequent irradiation would allow the filets to be kept a sufficient time so that they could be distributed throughout the U.S. for sale.

Spinelli and co-workers[784] reported that preliminary work with Petrale sole filets demonstrated that treating them in a 10% STP solution containing 4% sodium chloride prior to irradiation practically eliminated drip-loss of the filets stored at refrigerated temperatures for 28 days. Later, experiments with filets from Pacific Ocean perch, Petrale sole, and English sole demonstrated that the formation of drip was effectively retarded only after concentrations of 5% or more STP were used in the dip solutions. Increasing the STP concentration from 7.5 to 10% produced no appreciable further decrease in drip. Similar experiments with halibut steaks produced considerably more variation in results. The authors thus concluded that there are differences among various species in the effects of polyphosphate dips on drip-loss of fish muscle. The authors also confirmed the results of Mahon[807] that the addition of salt increases the effectiveness of the STP in reducing drip-loss. However, Mahon reported that the most effective treatment was using a 2-min dip of the fish muscle in a 12% solution of STP containing 4% salt. Spinelli and co-workers found the drip-loss could be effectively reduced by a 22 sec dip in a 7½% solution of STP containing 2% salt. Studies of the physical properties of the proteins by these authors indicated that the STP treatment had little influence on the rate of protein denaturation. The authors concluded that treatment with STP is an economic advantage and highly beneficial, in that, during storage, drip-loss is significantly reduced and the fish retains a significantly better appearance and texture. Spinelli and Weig[816,817] effectively reduced drip-loss of both fresh and frozen fish filets by use of a machine that they developed in their laboratories for the treatment of fish filets with STP. The machine consists of a chain-type of conveyor belt on which the filets pass a set of five adjustable spray nozzles, two above and three below the conveyor belt. The nozzles are such that the pressure of the spray could be varied between 0 and 100 psi and the speed of the belt could also be varied in order to adjust the quantity of STP solution applied to the filets. Although the machine described by the authors did not incorporate any type of recycling system for the spray solution which ran off the filets and equipment, they reported that such a system could be developed but may not be economical. During tests reported in these papers, fresh, refrigerated fish filets sprayed with solutions containing between 7.5 and 10% STP plus 2% salt had a drip-loss between 0.5 and 1.0% compared with 3 to 6% drip for untreated controls. The machine was also useful for applying a glaze to filets to be frozen if it was not being used for applying STP solution. A major advantage of the spray application of STP over dips is that the spray eliminates the contamination of the dips. Dissolved proteins, bacteria, and fish particles accumulate in the dips that require changing the solution at frequent intervals and contamination increases as more and more fish filets are dipped. None of these disadvantages occur with the spray application. The authors further point out that the dips are most effective if one uses a 12% solution of STP containing 4% salt which may cause a slight increase in cost over the spray since less STP and salt are required. A similar spray machine has been developed and is being marketed by the Calgon Corporation.[818,819] The fish filets are coated by a mist of the STP solution while being flexed in a drum that allows the solution to penetrate deeply into the filet, thus improving the frozen quality of the fish.

Barnett and co-workers[820] reported that treating halibut, silver salmon, and black cod filets with STP and salt solutions prior to kippering (hot smoking) significantly reduced the weight-loss during the kippering process. Combinations of 2% sodium chloride and 2% STP were more effective in reducing weight-loss than combinations of 2% sodium chloride with 7.5% STP. The 2% salt-2% STP solution reduced weight-losses of halibut, silver salmon, and black cod by 4.5%, 3.5%, and 1.6%, respectively. The 2% salt-7.5% STP solution reduced weight-losses of the same fish by 3.5%, 0, and 1.0%, respectively.

Not all investigators of the effects of phosphates on fish agree that they reduce drip-loss. For example, Dyer and co-workers[821] did not find any differences in extractable protein, lipid hydrolysis, or thaw drip when frozen cod filets had been treated with tripolyphosphate dips prior to freezing. They obtained a greater yield of the frozen and of the thawed filets due to water uptake during dipping the filets. They did not use combinations of polyphosphate and salt, and speculated that this might have made the effects quite different since Mahon[810] and others have reported that the polyphosphates are most effective in the presence of salt.

Love and Abel[782] studied the mechanism of phosphate activity in preventing drip-loss since there are considerable discrepancies in the reported results by other investigators. Using a cell fragility technique, the authors investigated the effects of three phosphates, a combination of SAPP and TSPP, STP, and SHMP on the denaturation of the fish muscle proteins during cold storage. The values for the cell fragility technique with the phosphate treated fish muscle were so low initially that no further changes occurred during storage of the fish, indicating a great deal of denaturation had resulted. The values for the

SHMP-treated filets were slightly lower than those for the other phosphates and the authors reported that this was not due to a difference in pH. Studies of the effects of polyphosphates upon protein extractability indicated that the protein extractability was by no means proportional to the effect on cell fragility. Variable results were obtained with the polyphosphates. The authors reported, however, that the polyphosphates all exerted a strong solvent action on the muscle proteins. This solvent action was not as great as that of a 5% solution of sodium chloride and was slightly less than the solvent action of a mixture of sodium chloride and the same polyphosphate. Further studies by means of electron microscopy showed that the phosphate solutions removed a portion of the protein from myofibrils and deposited it on the surface of the fish filets. The resulting film of gelled protein effectively sealed the fluids inside the filet and thus prevented them from leaving the filet as thaw-drip. The authors pointed out that although the polyphosphates had a beneficial effect on the storage of fish filets, they are no substitute for using the best possible processing conditions for freezing and cold storage. They will not improve fish proteins which have already been denatured.

The discrepancies in the results of numerous studies of the effects of phosphates on fish filets were reviewed by Love[822] who quoted some work indicating that polyphosphates did not reduce the thaw-drip of filets that had been dipped in them prior to freezing and that they also did not prevent denaturation or insolubilization of the fish protein as reported by some investigators. Love concluded that the work of MacCallum and co-workers[793] and others that demonstrated that the muscle proteins of cod were not split or hydrolyzed by the polyphosphates, indicated that phosphate dips did not increase the fish muscle protein solubility. Love quoted further work in his laboratories[823] that confirmed that the polyphosphates did not increase the salt solubility of the fish muscle protein, but rather that their effect was through swelling of the surface cells of the filets to eliminate extracellular spaces through which the fluids could exude from the interior of the filets. Thus, the surfaces of the filets were effectively sealed to prevent fluids from leaving them. The author concluded that the earlier observation by Linko and Nikkila[778] that phosphates inhibited denaturation of proteins caused by sodium chloride solution was explained by the fact that the phosphates blocked the entry of sodium chloride into the tissue through eliminating the intracellular spaces. Thus, the polyphosphates appeared to prevent exudation of fluid as well as entry into the filet.

The phosphates have been found highly effective in increasing the water-holding capacity of the Japanese kamaboko, a steamed fish meat jelly, by aiding in formation of protein gels which are important to its formation. Okamura and co-workers[797] demonstrated that various ratios of TSPP and STP markedly increased the water intake of kamaboko. The effects of the TSPP and STP alone were shown to be different and dependent on the effect they had on the pH of the fish jelly. Okamura[798] later determined the minimum effective levels of DKP, TSPP, and STP that influenced water absorption of fish meat. He also demonstrated[824] that strong electrolytes decreased the water intake of fresh fish meat and that this decrease was directly proportional to the ionic strength of added salts. However, he found that the polyphosphates were exceptions to the rule and that they, instead, increased the water intake.

In later work Okamura and co-workers[825] reduced the water intake of water-washed raw fish flesh for use in kamaboko by the addition of sodium chloride, magnesium chloride, magnesium sulfate, potassium phosphate, sodium pyrophosphate, and sodium citrate salts to the solution.

Tanikawa et al.[826] observed a decrease in thaw-drip and an improvement in elasticity of cooked pastes, such as kamaboko, prepared from fish and those used in making fish sausages and hams by the addition of polyphosphates, sugar, starches, egg whites, and shortening oils. The most effective combination was sugar (cane sugar, glucose, or sorbitol) and polyphosphate, although egg white was also effective, especially with low elasticity raw fish pastes. Starches and shortening oils improved these properties but required the presence of sugars and polyphosphates too.

Ueoka and co-workers[827] reported that pretreating freeze-dried fish with salt solutions maintained the salt-soluble proteins in a soluble state and increased the water-holding capacity of the fish meat through formation of protein gels. Soaking the fish in solutions containing 0.01 to 0.03 *M* each of sodium citrate, sodium pyrophosphate, EDTA, neutralized EDTA, and sodium pyroposphate containing 10% sucrose for 5 hr at 2°C prior to freeze-drying was effective for this purpose. The greatest effect was obtained with the 0.03 *M* salts. The 0.03 *M* sodium pyrophosphate containing 10% sucrose produced the best results of all treatments studied.

A further report from the Torry Research Station in Aberdeen, Scotland[804] suggested that although the polyphosphates are highly effective in reducing or preventing escape of fluid from fish filets during storage and do improve the appearances of retail fish filets dipped in the polyphosphate solution through less fluid release in the package, where optimum freezing and cold storage practices have been used in processing the fish, there will be very little drip from the filets. Also, where the processing conditions have not been optimum, fluid will be released whether the filets have been dipped in polyphosphate or not. However, the polyphosphates will prevent the fluid from escaping from the filets and retain it within the flesh by their effect on swelling the surface cells and forming a film of gelled protein about the surface of the fish. Thus, polyphosphates have little or no effect on good quality fish, but do reduce weight-loss and improve quality through fluid retention in poorly processed fish.

The water-retention of seafoods other than fish is also increased. Takei and Takahashi[828] reported that air-dried squid meat would recover its appearance and texture after contact with water if it was treated prior to drying with a 5% solution of a mixture of SHMP with either TSPP or STP in a 2:1 ratio. The ability to absorb water decreased somewhat after storage for 50 days at 5°C.

Mahon and Schneider[810] reported that the thawing-drip of frozen scallops was significantly reduced by a dip in a solution of STP as shown in Table 35. Miyauchi and co-workers[790] reported that STP dips prior to irradiation reduced the drip of crab meat and Pacific oysters.

Ichisugi and co-workers[829] reduced or inhibited the development of stiffness resulting from desiccation during storage of crab meat at -22°C for 120 days if the crabs had been heated in a 3° Be. saline solution for 30 min and then dipped in a solution containing either 0.3% sodium acetate or 0.8% TSP for 30 min. The same result was obtained when the crab meat was dipped in the acetate or phosphate solutions prior to heating. The water-holding capacity of the crab meat was effectively retained by both reagents.

Mathen[830] found the phosphates reduced the drip-loss of frozen prawns. He studied the effects of MSP, DSP, TSP, TSPP, STP, SHMP, and sodium metaphosphate. MSP and sodium metaphosphate did not reduce drip-loss. The others increased yields of thawed prawns by 8 to 10% by reduction of thawing-drip, but produced adverse effects on their organoleptic properties. The most satisfactory treatment that prevented dripping and improved organoleptic qualities was with a solution of STP neutralized with either MSP or citric acid.

Jones[831] studied the effects of the addition of varying concentrations of STP at two pH values to canned Alaska king crab. The addition of 0.25 and 0.35% SAPP increased moisture-retention and, therefore, yield of the crab meat by 2%. Levels under 0.25% failed to improve the product, while levels over 0.35% had an adverse effect on flavor and produced no significant increase in water retention or yield.

Improving Flavor

The flavor of seafood products has also been reported to be improved by treatment with various phosphates. Sjostrom[832] reported that treating fish products with an alkali such as DSP after cooking and preferably after cooling but before sealing them in packaging containers significantly retarded the development of undesirable flavors during sterilization.

Treating sun-dried mackeral with a curing mixture of salt, sodium benzoate, DSP, sodium bicarbonate, SHMP, and ascorbic acid improved the flavor and odor of the resulting dried fish to a significant extent during its storage period.[789] The palatability of frozen crab meat was reported to be improved if heated in a saline solution for 30 min or if the meat were dipped in the salt solutions first, then heated, frozen, and stored at -22°C.[829]

The effects of the polyphosphates on improving flavor of seafood products are, however, disputed.[804] Claims for improved flavor may be due to the dipping effects masking some of the symptoms of deterioration of poor quality seafood products and may only be an impression that the quality is better. Spinelli and co-workers[784] detected no differences in flavor or odor when they compared fish filets or steaks dipped in polyphosphate solutions with undipped controls.

However, Tarr and co-workers[833] reported that

an enzyme that hydrolyzes the 5′-nucleotides is strongly inhibited by EDTA, pyrophosphate, potassium fluoride, and zinc chloride at levels between 1 and 10 mM. Since the 5′-nucleotides are good flavor intensifiers, the inhibition of the enzyme that would inactivate them may offer some explanation for flavor improvement due to the presence of pyrophosphate.

Polyphosphates have been reported to be effective antioxidants that prevent the development of rancidity in foods containing fats in an aqueous environment. The results of numerous investigations demonstrating the effectiveness of the polyphosphates in preventing development of rancidity in aqueous fat food products have been reviewed in a previous section. A number of investigators have reported the effectiveness of polyphosphates in preventing the development of rancidity in seafood products. Watts[667] reported the results of numerous studies of the development of rancid odors and flavors in ground, cooked mullet meat. She states that the TBA values were an excellent guide to rancidity development and that the detectable limit of rancid odors for all cooked meat, i.e., fish, beef, pork, veal, lamb, and chicken , is a TBA value of 0.5 to 1.0. Watts reported that if the lipids were extracted from the cooked mullet tissue, the development of rancidity in the chloroform-extracted lipids was much less than in the cooked, unextracted, whole mullet tissue as shown in Figure 15. Even the total lipids, extracted from the raw fish, developed rancidity at a lower rate than in the whole tissue. Watts noted, too, that once lipid oxidation began, less lipid material could be extracted as shown in Table 36. The oxidized lipids apparently could no longer be extracted by the same solvent system.

Zipser and Watts[834] reported that TBA determinations demonstrated that the fats in cooked mullet muscle began to oxidize very rapidly after cooking. Tissues containing high levels of fats and heme pigments were especially affected. The authors retarded the development of rancidity by limiting the oxygen supply and by freezing the fish. However, the most effective treatment was the treatment of the ground fish muscle with a combination of STP and sodium ascorbate. The combination of the two salts was more effective than either of them alone or in combination with other curing salts.

In an extension of this work, Ramsey and Watts[665] reported that direct addition of the polyphosphate, STP, to the meat either prior to or after cooking had a slight advantage over storing the meat in an STP solution. Table 37 demonstrates the effects of the levels of STP on TBA values and odor scores. As shown, the TBA values, even at the 0.5% level, were above the threshold of odor detection, demonstrating that the fish lipids oxidize more rapidly than beef which was protected by levels as low as 0.1% for eight days. Akiyami[803] also used the TBA values to demonstrate that the addition of 0.5% polyphosphate to cooked, fatty whale meat was more effective in preventing rancidity than the addition of 0.02% BHA or BHT.

Mahon[807,808] claimed the treatment of fish with a combination of 12% STP and 4% salt effectively inhibited development of rancidity in the fish and, therefore, improved flavor and odor. This flavor and odor improvement may be due to the influence of the polyphosphates on retarding the rancidity of the lipids in the fish flesh. Mahon[807] has shown that the TBA values remain significantly lower, compared to controls, over the same period of storage by the polyphosphate treatments as shown in Table 30, indicating the rancidity of lipids was retarded in the treated fish. Thus, flavor and odor would be expected to be improved. Sen and Lahiry[789] reported that including DSP and SHMP in curing salts for sun-dried mackeral resulted in decreased development of rancidity.

Dyer and co-workers[821] demonstrated no effect on lipid hydrolysis due to dipping fish filets in an STP solution prior to freezing with subsequent thawing. Boyd and Southcott[811] also reported that treatment of fish filets prior to freezing with a 12.5% TSP and salt had no effect on preventing the development of rancidity. As discussed previously, the dipping treatment of seafood products is thought to mask some of the symptoms of deterioration and poor quality, rather than provide actual protection from that deterioration.[804] Spinelli and co-workers[784] detected no differences in the development of rancidity in irradiated fish filets whether they were treated with STP and salt dips or with salt alone.

Greig et al.[835] inhibited the development of rancidity in ground cisco (fresh water herring) flesh stored at 0°F more effectively with ascorbic acid than with other soluble antioxidants, such as STP, propyl gallate, and monosodium glutamate. All of the water-soluble antioxidants were some-

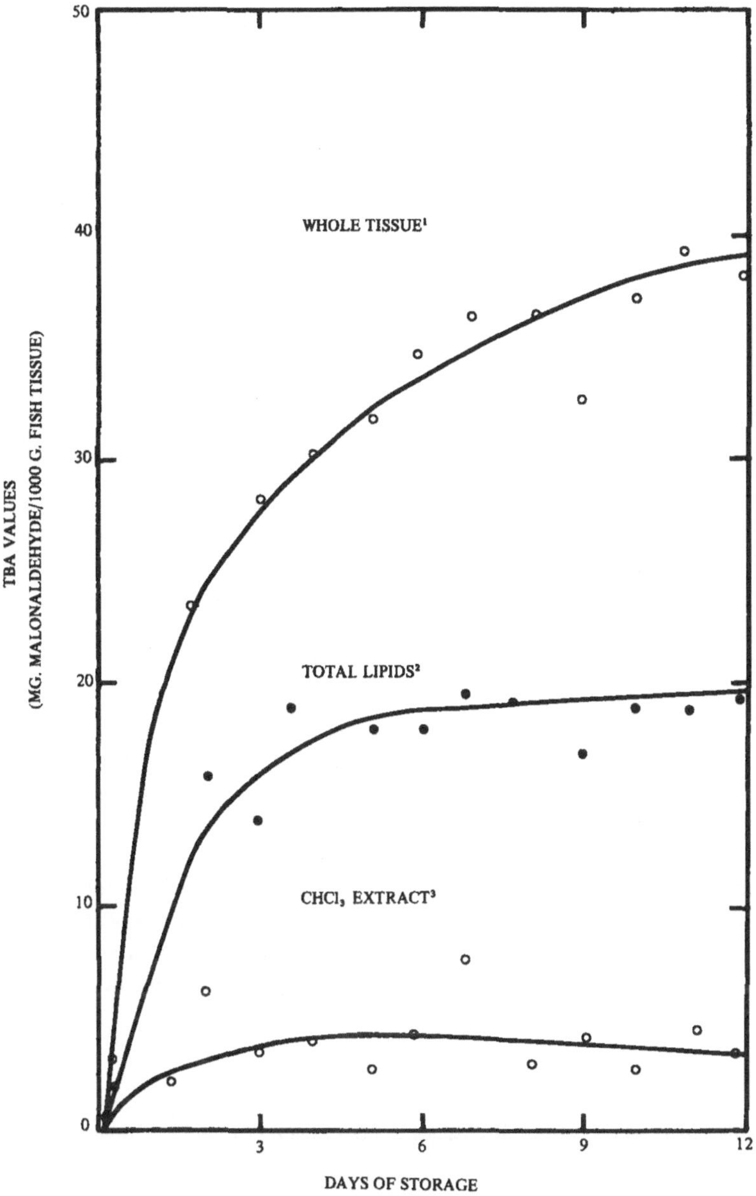

FIGURE 15. TBA values of cooked whole mullet tissue, total lipids, and chloroform extracted lipids. 1. Whole, cooked fish tissue, no lipids extracted. 2. Total lipids extracted from fresh fish tissue with chloroform-methanol-water solvent techniques. 3. Lipids extracted from cooked fish tissue with chloroform and a dehydrating agent. (From Reference 667, Figure 1, with permission.)

what more effective than untreated controls, but the ascorbic acid was significantly more effective than the others. Perhaps there are species differences as is the case in water retention.

Preventing Microbiological Spoilage

The inhibition of microbiological deterioration of seafoods has also been shown to occur by treating them with various phosphates. Hempel[836] reported that if dry sea salt, often used for preservation of fish, is not properly treated, a pink or red microorganism rapidly develops in the brine in the presence of the fish protein. If the salt crystals are coated with between 0.25 and 4% of a finely powdered alkali metal phosphate such as MSP or DSP or mixtures of these compounds, the red microorganisms fail to develop.

A Norwegian patent[837] claims the use of a

TABLE 36

Total Lipids Extracted from Cooked Red Lateral Line Tissue of Mullet

Treatment	Total lipids %	TBA (Whole tissue)
Raw	8.22	–
Cooked (0 days)	7.87	3
Cooked (3 days)	6.85	60
Cooked (5 days)	6.51	86

From Reference 667, Table 2, with permision.

combination of carrageenan and phosphates in sufficient proportions to form thickened solutions of the aqueous salts (500 to 3,000 cps) used to coat fish will inhibit microbiological deterioration of the fish. Uchiyama and Amano's[838] studies of the softening spoilage of fish sausage by *Bacillus circulans* demonstrated that the addition of sodium pyrophosphate to the sausage did not inhibit the germination of the spores, but rather inhibited the development of the vegetative phase of the organisms. They reported that 0.4% sodium pyrophosphate added to a medium in which the microorganism grew well restricted its consumption of glucose. A series of photomicrographs and electron micrographs demonstrated that sporulation could occur, but that vegetative growth was inhibited in the presence of the pyrophosphate as well as sorbic acid.

Ozawa and co-workers[839] reported that the polyphosphates were effective synergists with tetracycline as fish and other food preservatives, with SHMP being most effective, followed by STP, and the least effective polyphosphate was TSPP. The authors reported that the condensed phosphates inhibited the growth of *Staphylococcus aureus* and *Bacillus subtilis* in broth medium and also prevented their growth on fish. Again, SHMP was most effective, followed by STP, and TSPP.

Tomiyama[776] also reported the synergistic effect of the polyphosphates, particularly SHMP, with antibiotics such as chlortetracycline and Tylosin. The antibiotic effects of combinations of the antibiotics with SHMP and other chelating antioxidant compounds were tested in clam meat, cooked fish paste, and in poultry products. Iron and copper ions are potent inactivators of the antibiotics and can be effectively removed and inactivated by pyrophosphate and similar chelating agents.

Evidence that chelation may explain at least some of the microbiological effects of polyphosphates in preserving fish products is also presented by Levin[840] who reports the effectiveness of EDTA as a preservative of fish products. Odor and taste evaluations indicated that a 1% dip of tetrasodium EDTA increased the shelf-life of the fish from five to nine or ten days and significantly

TABLE 37

Effect of Level of STP on TBA Values and Odor Scores of Cooked Ground Mullet

	Rancidity After Storage			
	2 – 3 Days		8 Days	
Phosphate concentration (%)	TBA no.	Odor score	TBA no.	Odor score
0.5	1.8	5.0	10	4.2
0.1	2.2	5.4	14	2.4
.03	2.6	4.6	20	3.5
.01	5.8	4.0	25	2.5
0	16	3.5	41	3.0

From Reference 665, Table 5, with permission.

decreased the development of trimethylamine and volatile basic nitrogen compounds. Spinelli and co-workers[784] reported that polyphosphate-treated, irradiated fish filets had predominantly *Achromobacter* microorganisms on them after 17 days of storage, while the undipped irradiated filets were predominantly contaminated with *Lactobacillus*. However, when the filets were held under refrigerated storage until they spoiled, the *Lactobacillus* predominated in both the treated and untreated filets.

Preventing Struvite Crystals

Canned, cooked seafoods such as lobsters, shrimp, crab meat, haddock, cod, and salmon develop, over a period of time, a quantity of transparent crystals of magnesium ammonium phosphate, known as struvite. The crystals are actually harmless both physically and nutritionally, but because they appear to be sharp pieces of glass, the product is often rejected. The quantity of magnesium in many seafoods and especially in the water used in processing the seafoods can be sufficient to cause the formation of these crystals during the normal shelf-life of the product.

McFee and Peters[841] reported that as little as 0.015% by weight of the normal pack of magnesium ion can result in the formation of these crystals. Many of the canned seafoods containing 80% water and 20% solids have sufficient magnesium in them to form the crystals of struvite. These authors reported that the acidification of the water used in washing the cooked seafood product could leach away the magnesium and thus prevent the formation of the glass-like crystals. Thus, acidification with citric, hydrochloric, and acetic acids, or the use of sequestering agents in the water, such as alkali metal metaphosphates, particularly sodium acid pyrophosphate, and amino-polycarboxylic acids, such as EDTA will prevent the quantity of magnesium to be sufficient for the formation of this compound. Attempting to extract the magnesium compounds with water, either by rinsing or washing, or by boiling the seafood in ordinary water is not sufficient to remove the quantity of magnesium necessary to prevent formation of struvite crystals. McFee and Peters recommended the use of SHMP, ammonium metaphosphate, STP, or the potassium salts equivalent to the same products as well as EDTA for these purposes. In a later patent, Kreidl and McFee[842] recommended the use of between 0.5 and 1.5% of a sodium polyphosphate compound, such as SHMP, in the liquid used in preparing the canned seafood product as a means of preventing the formation of the glass-like crystals of struvite during its shelf-life. Quantities of the soluble alkali-metal phosphate as low as 0.25% were found to be effective based on the total moisture content of the canned seafood.

Two years later, McFee and Swaine[843] reported that the addition of SHMP to the liquid in the canned product proved to be the most practical method for preventing the formation of struvite crystals in such canned seafoods as lobster, shrimp, crab, and flaked fish such as pollock, cod, and haddock. They reported five advantages to the use of SHMP. These were:

1. It is a neutral salt and, therefore, does not create acidic or alkaline flavors in the canned product.
2. It is a GRAS compound and, therefore, is safe for use in all food products.
3. It can be applied either as a solution or in the dry form with equivalent results.
4. It can either be mixed with the food before canning or added to the cans during their filling.
5. A very small quantity is necessary to prevent the formation of struvite, thus making it very economical in use.

The authors reported that canned seafoods containing SHMP had a shelf-life of three to four years.

Yamada[844] reported that the minimum quantities of SHMP required to prevent the formation of struvite in canned king crab meat ranged between 0.13 and 0.22% of the total product. Jones[831] reported that adding between 0.25 and 0.35% of SAPP to king crab meat inhibited the formation of struvite during the observed storage period of one year in canned products containing salt.

Phosphate Applications as Microbiological Inhibitors

The applications of the phosphates as microbiological inhibitors in various types of food products have been reviewed in several sections of this chapter. There are, however, a number of references to food applications for polyphosphates as microbiological inhibitors and these will be reviewed in the following section.

Applications as Inhibitors of Bacterial Growth
General Inhibitory Effects

Kelch and Buhlmann[845] reported that commercial mixtures of phosphates normally used in the curing of meat products (Curafos® and Fibrisol®, manufactured by J. A. Benckiser) act as microbiological inhibitors of certain food spoilage microorganisms when tested in optimum nutrient media for these microorganisms. The Curafos used in these experiments contained 15% TSPP, 70% STP, and 15% SHMP. The Fibrisol used in the experiments contained 25% SAPP, 15% TSPP, 10% SHMP, and 50% NaCl. Each of the phosphate mixtures was tested against the various organisms at levels of 0.3%, 0.5%, and 1.0%. Controls containing no phosphates were included in the tests. In addition, the media containing the inoculations of microorganisms with and without the phosphates were heated to 50°C, 55°C, 60°C, 65°C, and in the case of sporulating bacteria, to 75 and 80°C. *Staph. aureus* and *Strep. faecalis* were inhibited or destroyed by all levels of phosphate and at all temperatures tested from 50 through 65°C in the presence of the phosphates, but grew rampantly at each temperature in the absence of the phosphates in the media. The *Strep. faecallis* grew in the unheated media containing all levels of phosphate, while only the 0.3% and 0.5% Fibrisol-containing media supported the growth of *Staph. aureus* without heating. All levels of Curafos prevented growth of *Staph. aureus* with or without heating. Growth of the spore forms of *Bacillus subtilis* was inhibited with and without heating in the presence of 0.5% and 1.0% Curafos. *Bacillus subtilis* spores were able to germinate and grow, although with limited success, in the presence of 0.3% Curafos and with increasing success with time in the presence of the Fibrisol at all levels. The vegetative forms of this organism were strongly inhibited by the presence of 0.5 and 1.0% Curafos with and without heating. They grew to a limited extent after five days' incubation or more in the presence of 0.3% Curafos. They were capable of growing after prolonged incubation in the presence of 0.3% and 0.5% Fibrisol, but not in the presence of 1.0% Fibrisol, with heating. They were capable of growing with prolonged incubation in the presence of all levels of Fibrisol when the medium was not heated. *Clostridium sporogenes* spores had difficulty germinating in the presence of 0.5 and 1.0% Curafos, but were able to grow, especially after prolonged incubation, in the presence of all levels of Fibrisol with and without heating. *Clostridium bifermentans* had considerable difficulty growing in the presence of all levels of Curafos with any degree of heating. It was capable of growing after prolonged incubation in the presence of 0.3%, but was completely inhibited from growing in the presence of 0.5 and 1.0% Curafos without heating. It was capable of growing, although requiring prolonged incubation, at 0.3 and 0.5% levels of Fibrisol with and without heating. It was completely inhibited from growth with and without heating in the presence of 1.0% Fibrisol.

The effects of orthophosphates as lytic agents for various bacteria have been studied and reviewed by Pacheco et al.[846,847] $TSP \cdot 12H_2O$ was found to be the most effective lytic agent in concentrations between 0.001 and 0.1 *M* solutions. $MSP \cdot H_2O$, $DSP \cdot 2H_2O$, and $DKP \cdot H_2O$ were also found to be effective against *S. typhosa, E. coli*, and *Staph. aureus*. Lytic action of the phosphates is puzzling since the orthophosphates have been demonstrated to support rampant growth of microorganisms,[102,103] and particularly are essential to spore germination of some of the spore-forming bacteria as determined by Heiligman and co-workers.[848]

Williams and Hennessee[849] reviewed the earlier work and studied the apparent heat resistance of spores of *Bacillus stearothermophilus* in the presence of orthophosphate buffer. The authors reported that the organisms had greater heat resistance at phosphate concentrations of *M*/120 or lower than in distilled water or in solutions with higher phosphate concentrations. Their work appears to have confirmed some of the inhibiting effects of higher concentrations, for example *M*/15, reported in the earlier literature which they review.

Wollmann[850] studied the effects of various salts including sodium pyrophosphate and SHMP as well as Fibrisol, a commercial phosphate-curing salt mixture, and other similar salts upon the heat resistance of *E. rhusiopathiae, S. typhimurium*, and *M. tuberculosis*. The salts were tested at 0.5% concentrations at various temperatures to determine their ability to inhibit the growth or to completely kill the bacteria. *E. rhusiopathiae* was completely inhibited by sodium pyrophosphate and potassium iodide, while Fibrisol and sodium citrate were less effective. SHMP and sodium citrate caused an increase in heat resistance of the organisms. Only sodium pyrophosphate and potassium iodide were effective upon the addition of 0.2% sodium chloride to the medium. *S. typhimurium* was significantly affected by all of the salts investigated but was most affected by sodium pyrophosphate and potassium iodide. However, the addition of 0.2% sodium chloride completely overcame the effects of all of the salts, including the sodium pyrophosphate and potassium iodide. The heat resistance of *M. tuberculosis* was lowered only by sodium pyrophosphate and potassium iodide. Initially, Fibrisol also lowered the heat resistance, but its effect was overcome by the addition of sodium chloride.

Post and co-workers[851] studied the effects of SHMP on pure cultures and wild cultures of bacteria. The authors reported that concentrations of 0.1% SHMP were effective in preventing the growth of most gram-positive bacteria when added to the media in which they were grown. Gram-negative bacteria were found to be able to grow at concentrations of SHMP as high as 10%. However, some gram-negative bacteria (e.g., *Pseudomonas fluorescens*) were lysed when SHMP contacted the cells. This lysis was inhibited or prevented by the addition of sodium chloride or magnesium sulfate. The authors also reported that the growth of the gram-positive *Sarcina lutea* could occur in the presence of levels of SHMP which normally prevented growth if magnesium sulfate was added to the medium. The authors speculated that SHMP interfered with the metabolism of divalent cations, especially with the magnesium ion and thus inhibited cell division and caused the loss of the cell-wall integrity.

Gould[852] reported that concentrations of 0.2 to 1.0% SHMP containing only small amounts of ortho-, pyro-, or tripolyphosphate prevented the normal germination and growth of spores of bacteria. The higher concentrations prevented growth before the rupture of the spore wall and thus prevented the development of the spores into vegetative cells. At the lower concentrations, the spore germination was not inhibited, but the vegetative cells were distorted, failed to develop in the normal manner, and did not multiply. The effect of SHMP at pH 6 was 1.5 to 2 times as great as it was at pH 7.

Kohl and Ellinger[102] obtained a number of foreign patents on the application of medium chain-length polyphosphates averaging in chain length between 16 and 100 and preferably between 16 and 34, as microbiological inhibitors in numerous food products. The medium chain-length polyphosphates have been found to be more effective than the SHMP previously reported in the literature. The activity of the polyphosphates increases as the chain-length increases from 16 through the high 20's and then begins to decrease in effectiveness as the chain length increases. Certain bacteria were effectively inhibited in growth or completely destroyed by levels as low as 0.1%, while others required levels considerably higher, ranging in the area of 1% to as high as 5%. *Staph. aureus* was very effectively destroyed by levels of 1% of the sodium polyphosphates having average chain-lengths between 16 and 34, as demonstrated in Table 38. The polyphosphates were found to be effective to varying degrees against different species of bacteria as demonstrated in Table 39 which shows the effect of polyphosphate chain-length upon the growth of *E. coli.*

Effects of Metal Chelation

Considerable speculation about the effects of polyphosphates as microbiological inhibitors centers upon their ability to sequester the essential nutritional mineral elements required for normal growth of microorganisms. Calcium, magnesium, and iron are three of the more important of the essential metal ions. The polyphosphates are known to be excellent sequestering agents for these ions.[853,854] The polyphosphates are capable of sequestering calcium in the presence of oxalate and this ability increases as temperature increases within the pH range of 5 to 12. However, the ability to sequester calcium decreases with chain-length as the pH changes. At pH values of 9.5 and higher, STP and TSPP were the most effective calcium and magnesium sequestrants. Magnesium, in particular, is important to the cell-wall integrity of some of the common food microorganisms.[855] Increasing or decreasing the level of magnesium in the media has been shown to change *Lactobacillus* species from the typical rod forms to filamentous forms and vice versa. The number of the metal cations as well as lipopolysaccharides is apparently essential to the structural integrity of the cell walls of some of the *Pseudomonas* and *Alcaligenes* species of organisms

TABLE 38

Effect of Polyphosphate Chain Length on Inhibition of Growth of *Staph. aureus*[1]

Phosphate additive	Chain length	Turbidity (measure of growth)	Bacterial count per ml (by plating) after			Percent bacteria surviving after		
			24 hr	5 days	10 days	24 hr	5 days	10 days
$Na_2HPO_4 + KH_2PO_4$		Heavy turbid	Est. 10^{8+}	10^{8+}	10^{8+}	$100^+\%$	$100^+\%$	$100^+\%$
Sodium acid pyrophosphate + tetra sodium pyrophosphate		Turbid	2×10^6	10^{8+}	10^{8+}	30%	$100^+\%$	$100^+\%$
Sodium tripolyphosphate		Clear	2×10^5	2×10^3	50	3%	0.03%	0.001%
Sodium polyphosphate	12	Clear	10^5	20	20	2%	0.0003%	0.0003%
	16	Clear	4×10^3	0	0	0.06%	0	0
	18	Clear	4×10^3	0	0	0.06%	0	0
	34	Clear	2×10^3	0	0	0.03%	0	0
	37	Clear	2×10^3	100	40	0.03%	0.002%	0.01%
NaCl (control)		Heavy turbid	10^8	10^8	10^8	$100^+\%$	$100^+\%$	$100^+\%$

[1] From Table 1, British patent 1,154,079, Reference 102, with permission.
[2] Initial bacterial count was 6×10^6.

TABLE 39

Effect of Polyphosphate Chain Length on Inhibition of Growth of *E. coli*

Phosphate additive		Amount of additive	% growth initial bact. population 80 per ml	% growth initial bact. population 800 per ml	% growth initial bact. population 8000 per ml
$Na_2HPO_4 + KH_2PO_4$		1%	100+	100+	100+
Sodium acid pyrophosphate + tetrasodium pyrophosphate		1%	40	40	39
Sodium tripolyphosphate		1%	1	1	3
Sodium polyphosphate	12	1%	21	29	33
Sodium polyphosphate	16	1%	19	28	32
Sodium polyphosphate	18	1%	1	1	4
Sodium polyphosphate	34	1%	25	35	32
Sodium polyphosphate	37	1%	37	37	37
Potassium polyphosphate in 2% vitrafos	1000+	0.5%	36	41	40
Potassium polyphosphate (1000+) in 1% vitrafos		0.5%	29	36	46
Potassium polyphosphate (1000+) in 0.5% vitrafos		0.5%	70	70	73
NaCl (control)		1%	85	85	85
No additive			100%	100%	100%

From Table 2, British patent 1,154,079, Reference 102, with permission.

as the addition of EDTA at alkaline pH values effectively solubilizes their cell walls.[856] Sequestration of metal ions also appears to increase the permeability of the bacterial cell walls to compounds which can interfere with the growth of the organism.[857] The fact that EDTA inhibits the growth of organisms, as reported by Brown and Richards,[858] may provide support for the metal chelation theory. However, there appear to be factors other than metal-chelating ability involved in the microbiological inhibition. Clues might be found in the reports of Burkard and co-workers[859] which indicate that polyphosphates are capable of interfering with or at least decreasing the acceptor activity of soluble RNA in yeast. These authors demonstrated that if one eliminates polyphosphates from yeast RNA preparations, the acceptor activity of the RNA is usually satisfactory. However, if one adds polyphosphates to the purified RNA preparations, the acceptor activity is decreased in proportion to the amount of polyphosphate added.

Tomiyama[860] and co-workers, Ozawa and co-workers,[839] and others have demonstrated the synergistic activity of the polyphosphates with various types of antibiotics in the preservation of foods through the ability of the polyphosphates to sequester metallic ions that interfere with the antibiotic activity of the organic compounds. A related effect is demonstrated in the patent obtained by Kooistra and Troller[861] in which the effect of organic preservatives, e.g., the edible acids and salts of propionic acid, sorbic acid, and benzoic acid with its methyl and ethyl esters, is potentiated by addition of the phosphate, carbonate, chloride, pyrophosphate, and other edible salts of iron, manganese, zinc, tin, and silver. In this case, the polyvalent metal ion is probably slowly released to the solution and produces some type of potentiating effect upon the microbiological activity of the organic salt.

Applications as Inhibitors of Yeast and Fungal Growth

The ability of STP to interfere with the fermentation of glucose by yeast has been known for some time.[862] The work of Vishniac indicated that STP inhibits the enzyme, hexokinase, and that the STP inhibition can be reversed if ATP or magnesium is added to the medium. Vishniac suggested that the STP chelates the magnesium that is essential to the activity of the hexokinase. The longer-chain polyphosphates, however, have been shown to be even more effective in inhibiting yeast fermentation of fruit juices, such as apple cider, by Kohl and Ellinger.[509] As in the case with general microbiological inhibition by medium chain-length polyphosphates, the most effective range of chain-lengths appears to be between 16 and 34. The degree of inhibition increases as the chain-length increases from 3 through approximately 30 and then decreases with increasing chain-length.

The polyphosphates have been found to be effective inhibitors of the growth of fungi in certain applications. Post and co-workers[508] reported that dipping fresh cherries into a 10% solution of SHMP, STP, TSPP, or sodium tetrapolyphosphate inhibited or delayed the spoilage of the cherries by such fungal spoilers as *Penicillium expansum*, *Rhizopus nigricans*, and *Botrytis* species. Kohl and Ellinger[103] obtained several foreign patents for the applications of medium chain-length polyphosphates, including those having average chain-lengths between 16 and 34 phosphate units as inhibitors of fungal spoilage in beer and wine, refrigerated doughs, malt and other grain products, processed cheese, fruits, meats, poultry, seafood, and vegetables.

Applications as Inhibitors of Viruses

The polyphosphates have been found to be effective inhibitors of viruses. Several applications of phosphates to inhibit viruses in dairy applications have been reviewed in the section on Phosphate Applications in Dairy Products. The use of TSP to inhibit potato virus X and tobacco mosaic virus has been reported by Brock.[863] The use of a 10 or 20% solution of TSP was found to be more effective than equivalent concentrations of formalin and the TSP solution was more effective if the treated inoculum containing the virus was held for 5 min prior to inoculating the plant. A 10% solution of TSP, in addition, was found to be an effective sterilizing agent for instruments and hands during experimental work with the viruses.

Phosphate Applications in Processing Food Protein

Proteins represent essential components of human diets. Their importance in proper nutrition is very well known. Phosphates have been shown to improve numerous characteristics of the proteins so that they can be more useful in their intended applications. Combinations of orthophosphates are often used to obtain optimum pH values for desired protein characteristics. Less known, but equally important in food applications, are the interactions between proteins and the polymerized phosphates. Numerous specific interactions between proteins of cereals, dairy products, meats, and other protein constituents of food systems have been reviewed in the sections dealing with those specific systems. However, some of the general interactions between proteins and the polyphosphates have also been reported in the literature and should be reviewed since they can have important applications in any food system containing proteins.

The shorter-chain polyphosphates, such as pyrophosphate, tripolyphosphate, and tetrapolyphosphate are usually quite specific and have been reviewed in the specific food systems. The interactions of the longer-chain polyphosphates from hexametaphosphate to the very high molecular weight, highly polymerized polymetaphosphates will be of major concern in this section.

That polyphosphates interact with proteins has been known since the work of Berzelius in 1916 and excellent reviews of this early work have been prepared by Horvath[237] and Leach.[239] Horvath reviewed work done through 1945, while Leach concentrated on work beginning in the late 1930's through 1962.

Some of the very early work, e.g., that of Fuld in 1902,[864] indicated that the proportion of polyphosphate bound to the protein is determined by the quantity of basic amino acids present in the protein, in other words, lysine, arginine, and histidine. The work of Briggs[234] and Perlmann[236] has confirmed that the polyphosphates form very strong, non-ionizing, salt-like bonds with the basic groups of the protein to such an extent that they mask the fundamental effects of the protein and shift the dissociation constant of the carboxyl groups to a lower pH region.

The applications of the interaction of the polyphosphates with proteins have resulted in reports of precipitation and coagulation of egg white and ovalbumin, precipitation of blood serum proteins from dilute solutions, precipitation of gelatin-polyphosphate complexes from dilute solutions that are resistant to microbiological decomposition, the precipitation of milk proteins and especially casein, and the precipitation of various albumins and peptones from aqueous solutions.[237] In each case, the reaction results in a flocculent precipitate, which is composed of the protein-polyphosphate complex. The quantity of polyphosphate bound to the protein increases in direct proportion to the amount of polyphosphate in the solution and as this quantity is increased, the bound polyphosphate reaches a maximum that is different for each protein investigated.[234-239]

In contrast to the above reports that the polyphosphates are bound to the amino groups of the proteins, Ferrel and co-workers[865] reported that the highly stable complex between protein and polyphosphate was due to the binding of ortho- and metaphosphoric acid groups to the hydroxyl groups of the amino acids in the proteins, e.g., serine, tyrosine, and threonine, and to a small extent to the phenolic hydroxyl group of tyrosine. The authors reported that the complex, after complete removal of all soluble phosphate and other salts by dialysis, was stable to dilute acid and alkaline hydrolysis.

In seeking the reason for the separation of two polyphosphate fractions when yeast cells are partitioned with trichloroacetic acid, Katchman and VanWazer[235] studied the coprecipitation and formation of complexes between egg albumin and polyphosphates of different chain lengths. The authors reported that the quantity of polyphosphate coprecipitated with the egg albumin upon treating it with trichloroacetic acid increased as the chain length of the polyphosphate increased. Approximately 25% of the added phosphorus was coprecipitated with egg albumin when the precipitating agent was a polyphosphate having an average chain-length of 16 and about 50% of the phosphorus was coprecipitated with egg albumin when it had been treated with a phosphate glass having an average chain-length of 85 to 130. Approximately 80% of the phosphorus of a polyphosphate having a chain length of 230 was coprecipitated with the egg albumin and 100% of the phosphorus of Kurrol's salt with an average chain length of 1600 was coprecipitated with the egg albumin. It was also possible to separate the sodium phosphate glass with an average chain-length of 16 from the Kurrol's salt by means of coprecipitation with the egg albumin in presence of trichloroacetic acid. In each case, the coprecipitates were complexes of the egg albumin with the polyphosphate.

The studies of Braginskaya and El'piner[866] indicated that the formation of protein-polyphosphate and protein-heparin complexes were pH-dependent and that the complexes formed at active sites of the protein which did not inhibit their enzymatic activity. The proteins studied were γ-globulin, lactalbumin, myosin, and polyalanine. Lyons and Siebenthal[238] determined that the differences in the binding of polyphosphates due to variations in chain-length may be due to the number of possible binding sites on the polyphosphate. Thus, the authors speculated that pyrophosphate may have only one possible binding site while tripolyphosphate may have at least two different binding sites. Longer-chain polyphosphates probably have multiple sites for interaction with multiple sites on the proteins and increased opportunity for the binding sites on the two compounds to match each other in space.

The variations in the complexing activity between polyphosphates of various chain-lengths and the different proteins have been useful in a number of food applications. Thus, the polyphosphates can aid in improving dispersion and solubility of proteins, increasing the water-binding and gel-formation of the proteins, improving whipping properties (which appear to be related to increasing the insolubility of the protein), improving the precipitation and insolubilization of the proteins for improved separation, improving their nutritional properties, and purification of the proteins by sequestration of polyvalent metals and unwanted flavor and odor components. These will be discussed in subsequent paragraphs.

Applications in Protein Dispersion

The dispersion of proteins, such as those from milk (casein) or peanuts to prepare a powdered protein food, was accomplished by treating the protein with alkali or alkaline salts, which included sodium and potassium di- or triphosphate (assumed to be TSPP, TKPP, STP, and KTP). The protein dispersion was dried and mixed with various minerals, vitamins, and flavoring agents to obtain a dried protein food.[867] The treatment of the protein with the alkali was reported to produce a dispersible aqueous solution without breakdown of the protein molecule. A number of examples of applications of phosphates to maintain proteins in solution[29] have been reviewed in previous sections including their applications in pasteurized, process cheese, their applications in meat products in which they solubilize proteins and allow them to form protective films about fat globules to improve emulsification, and applications in stabilizing milk proteins to prevent their gelling during storage.

Increasing Protein Water-Holding and Gelling Properties

The ability of polyphosphates to improve the water-holding capacities and, often, to cause the formation of protein gels has been utilized in a number of applications. Hall[868] reported that the addition of a buffering agent, which would maintain the pH between 6.5 and 7.0 in a mixture of gelatin, water, and propylene glycol, would produce a gelling film that was resistant to attack by microorganisms. The film would protect any food product coated with it against spoilage. DSP was among the buffering agents which could be used in this application.

Olcott et al.[865,869-871] reported several methods of preparing a gluten-phosphate complex which had excellent gelling properties. The most satisfactory method involved treating the wet gluten with urea and with 85% phosphoric acid, drying it, heating the dried material for 30 min at 140°C, then neutralizing it with sodium hydroxide, washing, and drying it. The gluten phosphate prepared in this manner was capable of absorbing 200 times its weight of water and would form colorless, tasteless gels that could be used in preparation of food products. Gel Soy®, a soy protein prepared from specially processed defatted, dehulled soy bean flakes[872,873] was reported by Glabe and co-workers[874] to have improved the water- and fat-binding capacities when it was treated with SHMP.

Controlling the pH of protein solutions has long been recognized as important in increasing their water-absorption. Norris and Johnson[875] reported that the water absorption of soy bean protein could be increased if the pH was maintained within the alkaline range of 7.0 to 9.0. A number of alkaline salts could be used for this purpose, including TSP. Pintauro and associates[876] reported that a gelatin food product could be highly improved in water-binding capacity by maintaining a pH of 3.2 to 3.6 with a mixture of MSP, DSP, and citric acid to buffer the gelatin composition within this pH range. A cold water-soluble gelatin composition was reportedly prepared by Wingerd[316] by combining gelatin with lactalbumin phosphate, a compound prepared by precipitating lactalbumin proteins from whey by means of long-chain polyphosphates.

Improving Protein Whipping Properties

The effects of polyphosphates on improving whipping proteins such as egg white have been mentioned previously. The addition of polyphosphates to a number of protein compounds has also demonstrated this improving effect on their whipping properties. Burnett and Gunther[877] prepared a whipping composition from soy bean protein isolated from soy bean flakes by modifying the protein in the flakes with the enzyme, pepsin, in the presence of a peptizing salt such as sodium chloride or sodium phosphate. The salt was added to aid in dissolving any of the unhydrolyzed protein remaining in the preparation. Sevall and Schaeffer[878] prepared protein whipping compositions from soy protein when they combined the protein with SHMP, sodium tetrapolyphosphate, TSPP, or sodium orthophosphates that would stabilize the degraded proteins during whipping in an aqueous system. Patterson[879] prepared whipping compositions similar and equivalent to egg white by combining an alkali metal caseinate, lactose or starch, and a polyphosphate. This composition successfully replaced from 20 to 35% by weight of egg whites and still maintained stable whipped foams in which the egg white was the principal whipping agent.

Gunther[880] produced an entirely new whipping composition by combining gelatin, an enzyme-modified soy protein similar to Gel Soy, and a polyphosphate. The specific polyphosphates mentioned were STP, sodium tetrapolyphosphate, and SHMP. According to Gunther, the mechanisms by which the polyphosphate stabilizes the whipping properties of the soy protein and gelatin combination are not known. However, he speculated that it may be a combination of protein precipitation and their ability to sequester polyvalent metal ions which might interfere. The result of the treatment with the polyphosphate is that the protein was capable of forming thin walls between the air cells which had sufficient stiffness and storage stability to give excellent whipped compositions.

Sutton[881] reported that a combination of a protein material such as nonfat milk solids, soy bean flour, dried egg white solids, dried whole eggs, gelatin, caseinate, and similar proteinaceous substances, combined with the alkyl ester of an aliphatic polycarboxylic acid and a polyphosphate salt would provide highly improved whipping compositions with high foam volume and high foam stability. Among the alkyl esters of aliphatic polycarboxylic acids specified, were those of malonic, succinic, glutaric, tartaric, malic, and citric acids. The polyphosphates that could be used for this purpose were SAPP, TSPP, STP and the longer-chain polyphosphates containing average chain-lengths of four up to several thousand, including SHMP, Graham's salt, TKPP, KTP, and the very long-chain potassium Kurrol's salts. The ammonium salts of the various polyphosphates could also be used for this purpose.

A series of patents was granted to investigators from the research laboratories of the General Foods Corporation for whipping compositions containing proteins that could be precipitated and/or denatured by the presence of polyphosphate compounds in acid medium. These whipping compositions were reported to produce superior whipped dessert products such as chiffons and chiffon-type pie fillings.[882-885] Clausi and co-workers[882] reported that combinations of gelatin, partially hydrolyzed soy protein, pregelatinized starch and polyphosphates, including those with an average chain-length of two through the very long-chain polyphosphates, but preferably SHMP, produced superior whipping compositions with highly stable foams for use in preparation of chiffon desserts and pie fillings. Based on the level of the partially hydrolyzed soy protein, between 1 and 20% phosphate was used with 20% SHMP preferred since it produced the highest foam volume, the shortest whipping time, and the greatest foam stability.

Mancuso and Common[883] prepared foaming compositions for use in preparation of chiffon food products or pie fillings composed of partially degraded soy protein, one of the vegetable gums, gelatin, and a polyphosphate, especially SHMP and sodium polymetaphosphates of longer average-chain-length. Mitchell and Seidel[884] reported that the use of a phosphated gelatin prepared by reacting normal untreated gelatin with a polyphosphoric acid or one of its salts at a pH of less than 4 produced a complex of gelatin-phosphate that could be used to prepare superior chocolate chiffon products. The addition of chocolate to normal chiffon mixes suppressed the foam properties of the mix until the authors discovered the use of the phosphated gelatin. Block[885] eliminated the necessity of preparing a phosphated gelatin used in the preparation of chocolate chiffon desserts by combining the desired phosphate with cocoa powder, gelatin, partially hydrolyzed soy protein, sugar, and a pregelatinized starch. SHMP and Graham's salts were preferred because of the superior properties they imparted to the foaming composition.

Downey[886] prepared a dry mix which could be used to produce marshmallows that did not require the close attention and special processing conditions of conventional marshmallow manufacture and also made it possible to prepare uniform batches of marshmallows in the home kitchen. The composition he described contained a major quantity of dextrose together with small quantities of starch, gelatin, phosphates, and vegetable gums. SHMP was the preferred polyphosphate declared in the patent. Grettie and Tiemstra[887] reported another gelatin-containing marshmallow composition that was reported to have highly improved properties also requiring the addition of polyphosphates.

Applications as Protein Precipitants

The polyphosphates could also be used as protein precipitants useful in coagulating proteins, separating them, and purifying them. Schwartz[888] produced artificial fibers or filaments of casein by first dissolving the casein in an alkaline solution and passing the dissolved casein through a spinnerette which discharged a thin filament of the protein solution into polyphosphoric acid maintained at a pH of approximately 2.5. The polyphosphoric acid solution coagulates filaments which then contain a casein-polyphosphate complex of great stability.

Horvath[889] studied and reported a method for the isolation of vegetable proteins as vegetable protein-polyphosphate complexes for use in numerous food applications. The process involves preparation of a weak solution of the polyphosphate which could be any one of the phosphates having a ratio of alkali metal oxide to P_2O_5 between 0.9:1 and 1.7:1, using it to extract the protein from the finely pulverized protein-containing vegetable material, then precipitating the protein-polyphosphate complex by acidification. A British patent[890] described a similar process for isolating vegetable protein-polyphosphate complexes that then can be precipitated around any food material to form an edible food coating. The process for extraction of the protein from the vegetable material is quite similar to that described by Horvath and the precipitation of the protein-polyphosphate complex as a coating around the food material is performed by acidifying a solution of the complex. Solutions of the complex were readily prepared by dissolving the complex in urea, neutral salts, or alkali.

Rane and Newhouser[891] precipitated proteins without forming a protein-polyphosphate complex by use of the cyclic polyphosphate, sodium tetrametaphosphate. The advantage to this procedure was that the protein could be obtained without the associated phosphate and the procedure was especially applicable for the precipitation of proteins from blood serum, although the authors reported that it could be used for recovery of proteins from animal or vegetable fluids, such as milk, liver extract, corn extract, or industrial protein-bearing wastes. The process involves the addition of the tetrametaphosphate, acidification to precipitate the protein, separation of the protein from the supernatant, and subsequently drying the precipitate or using it directly for the intended purpose. The authors also described the use of this process for fractioning some of the components of blood such as fibrinogen, globulins, and albumins. This involves first lowering the pH to 5 to precipitate fibrinogen and the globulins, then lowering the pH to approximately 4.2 to precipitate albumins, and later lowering the pH to approximately 3.5 to obtain the balance of the blood proteins. Lowering the pH below 3.5 will result in denaturing protein. This process is especially beneficial because it does not denature the proteins in any way.

Nitschmann et al.[892] reported a similar fractionation procedure for blood plasma proteins involving the use of polyphosphoric acid. Again, the proteins were reported to be undenatured and a fractionation of proteins was described. The fractionation procedure described in this publication involved adding increasing increments of the polyphosphoric acid to precipitate the various blood protein fractions.

Keil et al.[893] patented the preparation of a fungus-resistant food coating prepared from gelatin, SHMP, and water acidified to pH values between 2.15 and 3.5 that could be applied by dipping the food product in the solution. The gelatin-polyphosphate composition dried as a transparent film that prevented crystallization of salt and mold growth on the surface of the food products.

Fukamachi and Watanabe[894] reported that sodium citrate and sodium polyphosphates added to soy bean milk resulted in improved formation of curd in the presence of calcium salts.

Waugh[895] isolated casein from skim milk by first increasing the calcium ion concentration to between 0.05 and 0.1 *M*, with 0.06 to 0.08 *M* calcium ion preferred, separating the precipitated casein from the whey, then removing the calcium ion by precipitating it with oxalate, orthophosphate, or carbonate to reduce the calcium ion concentration to less than 0.2% in the final casein compound. The addition of the calcium ion was necessary to increase the size of the calcium caseinate micelles and caused greater precipitation and easier separation of the calcium caseinate curd. After separation by filtration or centrifuging, the calcium curd was washed (to remove additional whey) and resuspended in water. The calcium ion was then precipitated by means of orthophosphate or other calcium precipitating salts or removed in the form of a soluble complex by the addition of one of the polyphosphates, such as STP or SHMP. However, when a soluble calcium phosphate complex has formed, such as when using SHMP, the soluble complex must be removed by dialysis and might complicate the procedure. The resulting casein was reported to be highly soluble and was also compatible with milk. It could be used to increase the protein content of the milk, or to improve the protein and nutritional content of other food products.

The polyphosphates and orthophosphates have also been reported to aid in the purification of gelatin from impure solutions through their ability to precipitate polyvalent metal contaminants from the gelatin composition.[896] The impure gelatin was dissolved in water, phosphoric acid was added, and magnesium chloride was then added to the solution at alkaline pH values, such as pH 9.15. A precipitate of magnesium ammonium phosphate was thus formed which brought down the impurities in the gelatin with it. The filtrate contained highly purified gelatin.

Phosphate Applications in Starch Processing

It is well-known that acids react with the hydroxyl groups in starch. The esters thus formed have been useful in a number of industrial applications and also in some food applications. Phosphoric acid will form esters with starches. There are phosphoric acid ester groups in natural starches, particularly in potato starch.[897-900] Investigations demonstrate that the paste viscosity of the potato starch increases as its phosphoric acid ester content increases. In addition, the potato starch is capable of forming primary and secondary phosphate complexes with metal ions depending upon the amount of phosphate in the starch.

Effects of Phosphates on Starch Properties

Phosphates have been shown to have significant effects on starch viscosities as demonstrated by Nara and co-workers[901] who demonstrated that sodium pyrophosphate had a greater effect in decreasing the swelling of starch granules and depressing viscosity of the starch pastes than various types of monoglycerides and other surface active agents. Similar effects were obtained with flour[902] when the addition of phosphates decreased the degree of swelling of two hard wheat varieties of 70% extraction.

Kuhl[903] reported that H_3PO_4 differed in its gelatinizing effect on starch than did its salts. He found that the degree of gelatinization varied in proportion to the quantity of phosphate salt added and also in proportion to the degree of saturation of the phosphate radical with the metallic ion.

Nutting[904] demonstrated that treatment of potato starch pastes with SHMP decreased the viscosity of the potato starch paste to a greater degree at pH 8.5 than any other treatment tested in his work. Conditions which increased ionization also increased the hydration of the starch, the particle volume, and the viscosity of the paste. All factors which depressed ionization decreased the viscosity of the paste.

Bowen[905] reported that a dextrinized starch for improving bread can be prepared after partially gelatinizing the starch by cooking it, mixing the cooked material with an acidic solution of MCP, calcium sulfate, and ammonium chloride, then further heating the mixture and drying it. Rozenbroek[906] modified potato starch so that it resembled the properties of corn starch in food applications by treating the potato starch with a small quantity of phosphoric acid in a small quantity of water, heating it at 50°C for 10 hr, then neutralizing it with slaked lime. Kunz[907] modified amylose in such a way that it could be used for preparing films and tubes for use as meat casings and other food applications. The amylose was treated with sodium hydroxide in aqueous solution and cast or shaped as desired to form the film or tube. The cast solution was then treated with a coagulating solution prepared with sulfuric acid, sodium sulfate, and a mixture of H_3PO_4 and DSP. The author reported that it was essential that a combination of the acid and salt be used since using acid alone formed weak, unsatisfactory films or tubings. Use of the salts alone did not supply sufficient acidity for coagulation of the amylose. Fredrickson[908] reported that mixing approximately 1% TSP with starch renders it fluidizable for treatment in a fluid bed where it could be heated to dextrinize the starch.

The Starch Phosphates

The most important and best known applications of phosphates and starch are those used in preparing the starch phosphates. The starch phosphates are relatively new, modified starches having very useful properties in food processing. One of the most important properties is the reduction in retrogradation which is believed to be due to the introduction of ionizing groups causing the molecules to repel one another.[909] The resistance to retrogradation is especially important during freezing and thawing of sauces and gravies containing the starch phosphate since the phosphate groups esterified to hydroxyl groups block the cross-bonding between adjacent starch molecules essential for retrogradation of starch. The starch phosphates also have improved clarity and high water-binding capacity as well as reduced gel formation. They are capable of increasing the viscosity of the foods in which they are used without forming thick or highly viscous gels.[910] The processes and properties of the various types of starch phosphates have been thoroughly reviewed by Hamilton and Paschall.[911] There are two types of chemically reacted starch phosphate compounds that can be produced by treating various types of starches with phosphates under specific conditions.

Starch Phosphate Monoesters

One type of starch phosphate is a starch phosphate monoester in which one acidic function of an orthophosphate group forms an ester linkage with one of the hydroxyls in the starch chain. Various degrees of substitution can be obtained and mono- and disodium or potassium salts can be formed at each phosphate linkage. According to Hamilton and Paschall, the reaction which takes place can be represented by the following equation:

Evans[912] formed the monoesters of starch phosphates by treating starch containing no more than about 20% by weight of water with a small quantity of TSP, passing the slurry through a steam-heated cooker during which it is held for about 1 min at temperatures ranging between 160 and 212°F, then drying the slurry on a heated roll to reduce its moisture content to about 4%. The dry pulverized starch could then be used for preparation of instant pudding mixes. Evans also describes the use of MSP, DSP, MKP, and TKP for the same purpose of obtaining starch phosphate monoesters. Kerr and Cleveland[913] prepared starch phosphate monoesters containing the disodium salt of the ester. The process calls for treating the starch with an alkali metal tripolyphosphate in the presence of alkali, as shown in the following reaction:

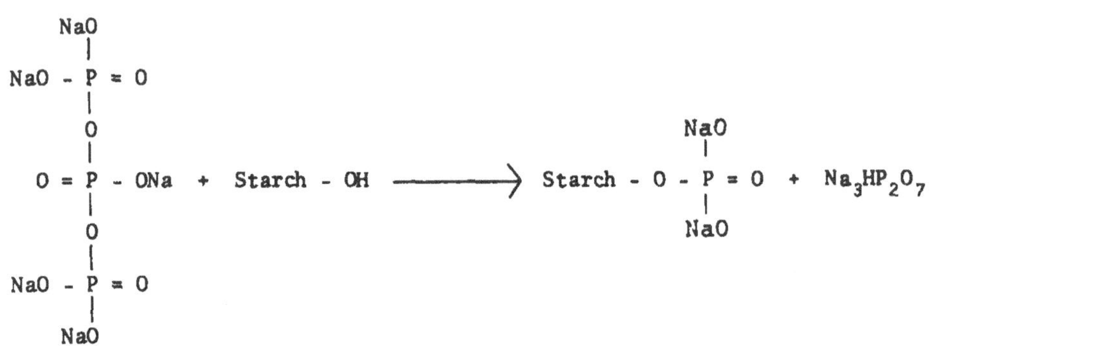

The disodium starch phosphate monoesters were reported to be useful in the preparation of pudding mixes, instant cold milk gelling pudding mixes, pie fillings, instant cream pie filling mixes, powdered salad dressing mixes, and for thickening sauces for vegetables. The authors reported that the addition of as little as 3 phosphate groups per 100 anhydroglucose units in the starch molecule resulted in highly significant changes in the properties of the corn starch which then began to resemble potato starch in many of its properties. Use of the phosphate-modified starch in such foods as soups, canned vegetables, and fruit sauces or fillings resulted in thick-bodied, creamy, relatively clear, thickened sauces, instead of opaque, gel-like stiffened consistencies obtained with the unmodified corn starch. Under suitably adjusted conditions, the authors claimed that the starch phosphate monoesters could also be prepared from the cyclic metaphosphátes, the long-chain polyphosphates, as well as pyrophosphates. The same authors[914] obtained identical starch phosphate monoesters as those from their previous invention by use of the acidic orthophosphate salts that were mixed with the starch to form a slurry containing no more than 20% moisture, heated to a temperature between 160 and 200°C at a pH not above 8.5, then dried. The authors believed that under the conditions of the process, the orthophosphoric acid salts were dehydrated as the moisture decreased and finally polymerized in which form they are capable of reacting with the starch to form the monophosphate ester.

Neukom[915] reported the use of a mixture of MSP, DSP and urea to esterify the starch. The resulting product, containing between 0.2 and 5% and usually between 1 and 5% bound phosphorus based on the weight of the starch, produced starch phosphates having higher viscosities and shorter pastes than those prepared without the urea. The same author later[916] described the preparation of starch phosphate monoesters by soaking the ungelatinized starch in a solution of the alkali metal phosphate, separating the starch from the excess moisture, and heating the treated starch to temperatures between 120 and 175°C for from one to several hours, after which a dry starch phosphate, containing between 1 and 5% phosphorus, based on the weight of the starch, was obtained. The resulting starch phosphates could be used in preparation of various types of gravies and sauces[917] and resulted in gravies and sauces with ideal characteristics.

In a further extension of his work, Neukom[918] reported the process for preparation of starch phosphate monoesters which involves soaking the starch in a solution of any combination of phosphoric acid, MSP, DSP, or TSP or a mixture of phosphoric acid with sodium hydroxide to prepare a solution having a pH between 3.0 and about 10.0. After soaking for a sufficient period of time, the starch is separated from the excess solution and heated to between 120 and 175°C at which time a dry, cold water-soluble starch phosphate monoester has been formed. Various types of starches could be treated in this way to prepare their starch phosphate monoesters. The author reported that, in addition to their ability to swell and dissolve in cold water, they form stable pastes on prolonged standing and also resist bacteriological attack to a greater extent than the parent, untreated starch.

Kodras[919] patented the use of oxygen-containing water-miscible organic liquids to precipitate the alkali metal phosphate-modified starches described by Neukom. Sietsema and Trotter[920] also patented the precipitation and purification of phosphate-modified starches using those reported by Neukom as starting materials. Schierbaum and Boerner[921] prepared the orthophosphate esters of starch by intimately mixing the wet starch with a solution of the alkali metal phosphates at 55°C, drying the mixture, and then heating it for 3 hr at about 140°F. The resulting product was a starch phosphate suitable for making pastes.

Starch Phosphate Diesters

A second type of starch phosphate that can be prepared and is now being used in food applications is the type known as starch phosphate diesters in which an orthophosphate unit is linked by ester groups to two starch molecules. This, therefore, forms a phosphate ester bridge connecting the two starch molecules to each other. These starches are typically highly stable to heat, agitation, and acidity and the starch granules are slow to swell and rupture.[910,911]

There are various methods of obtaining the phosphorus cross-linkage including the use of phosphorus oxychloride, phosphorus pentachloride, and thiophosphoryl chloride.[922] The addition of water-soluble salts of alkali metals, such as sodium chloride, improved and gave better control of the cross-linking reaction.[923] The presence of the salt was thought to retard the hydrolysis of the phosphorus compounds long enough to permit them to penetrate deeper into the granules of the starch and produce a more uniform amount of phosphorylation.

Kerr and Cleveland[924-926] obtained patents for the reaction of starch with the cyclic trimetaphosphate salts in the presence of various alkaline materials which catalyze the cross-linking esterification. An example of the reaction is the preparation of a slurry of the starch with a 2% solution of sodium trimetaphosphate (based on starch) at 50°C for 1 hr at pH 10 to 11. The resulting starch has a high hot-paste viscosity when cooked as in preparations of gravies and sauces. The reaction involved is thought to be:

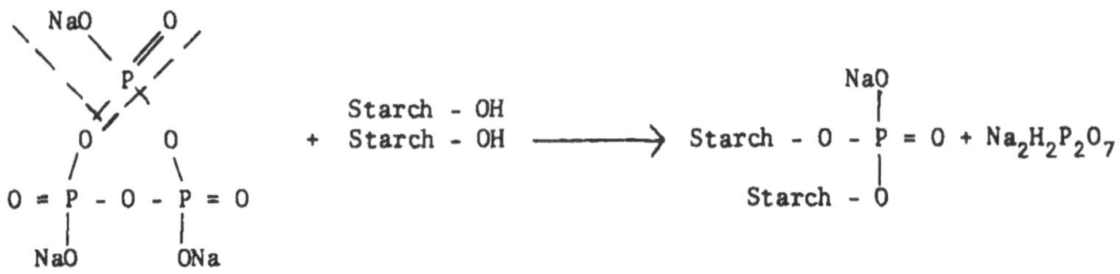

It is thought that small amounts of the mono- and trisubstituted esters occur along with a major proportion of the starch phosphate diesters.[911]

The resulting cross-bonded starches produce exceptionally clear sauces and viscous pastes and are especially desirable for pie fillings. They have excellent freeze-thaw stability, resisting the bleeding of water after as many as 10 freeze-thaw cycles. In contrast to untreated and starch phosphate monoesters, these starches maintain their viscosities even though they are cooked for long periods of time or added to high acidity products such as fruits.[910,911] They are unaffected by long cooking, high temperatures, and low pH values.

Gramera and co-workers[927] analyzed the glucose-phosphate compounds that they were able to isolate upon hydrolysis of this type of starch phosphate. They found that the compounds they isolated indicated that there was 28% phosphate substitution of the C-2 position, 9% of the C-3 position, and 63% of the C-6 position in the glucose units of the starch phosphates.

A cold-water swelling starch was also reportedly prepared by first treating cereal or potato starch for a short period of time in an acidic solution with a pH range of 3.0 to 4.0, then adding TSPP, drying, and gelatinizing the preparation.[928] Vermicelli compositions of improved stability, gloss, and smoother texture were also reported to be prepared by treating the vermicelli with hydrogen peroxide and polyphosphates.[929]

Among the numerous applications for the starch phosphates which have been reported in the literature are their uses as stabilizers for fountain syrups and ice cream toppings, ice cream stabilizers, stabilizers for various types of fat and water emulsions, suspending agents for insoluble solids which could also include fruits in pie fillings, Chinese-type foods in which great thickening and suspending characteristics as well as maximum clarity are desirable, in bakery mixes, as pan greases in combination with fats, as beer foam stabilizers, in instant puddings, in various types of cheese sauces, and numerous other nonedible applications.[910,911, and others] The various starch phosphates have been approved for use in foods providing they contain no more than 0.4% residual phosphate calculated as phosphorus.[930]

Phosphate Applications in Sugar Processing

The term "sugar" is most commonly applied to the disaccharide, sucrose, the major food sweetener in human nutrition. Sucrose is a common constituent of all plant cells. Sugar cane and sugar beets are the most common sources of the sucrose in use as a food.

Sugar refining is the process by which the sugar-bearing juice is extracted from the plant cells of sugar cane or sugar beets, purified, and crystallized. The process involves numerous complicated steps, the purpose of which is rupturing the plant cells containing the sugar, concentrating the sugar while removing impurities which are also extracted from the cells, then crystallizing it. The modern sugar refining processes result in one of the most highly purified chemicals known to man.

Clarification of the Sugar-Bearing Juice

The juices obtained through the rupture of the plant cells (by pressing the stalks in the case of sugar cane and by water extraction of the sliced root of sugar beets) contain numerous soluble nonsugar compounds. These must be removed; the usual practice is to treat the juices with lime, then precipitate the calcium by formation of insoluble calcium phosphates or carbonates, which in turn aid in precipitation of the colloidal nonsugar materials in the solution. Filter aids or Fuller's earth are added if the precipitates are to be removed by filtration. Some processes remove the precipitates by centrifuging the juices. Phosphates are often used to precipitate the calcium in this step. A number of references advocate the use of orthophosphoric acid,[931-944] while some find the addition of phosphoric acid and/or acid phosphates useful.[932,940,941] Bennett studied the effects of calcium and phosphate levels in the sugar clarification step. He reported that the optimum phosphate concentration is 4mM and the optimum calcium ion concentration is about 15mM.[945]

Phosphates other than phosphoric acid can also be used. Various sodium, ammonium, potassium, and calcium orthophosphates have been proposed as additives to the sugar solution along with the lime.[937,946-962] The polyphosphates such as SHMP,[963] superphosphate,[964-966] and pyrophosphate[967] have also been proposed. Barrett et al.[968,969] have proposed a synthetic hydroxyapatite for the clarification step. Most of the phosphates, particularly the alkaline sodium and potassium phosphates, are used to maintain an optimum pH for the precipitation of the lime as proposed by Kubala.[970] The control of pH is critical if all of the lime is to be removed from the juice. According to Kortschak,[971] the calcium phosphate precipitate is not a definite compound. After it precipitates, the calcium phosphate absorbs additional calcium hydroxide from the juice. Jung[972] patented the use of polyphosphoric acid in combination with a dicarboxylic acid for the clarification of sugar juices. Cummings[973] reported that the filtration of sugar liquors is easier when the calcium phosphate is combined with a diatomaceous earth filter aid.

Bleaching the Juice

Another important step in the refining of sugar is that of bleaching the sugar syrup prior to crystallization. This is often done by treating the solution with calcium hypochlorite in the presence of a calcium phosphate to act as a stabilizer and aid in final precipitation of the calcium from the bleached solution.[932,974] Several soluble phosphates have been proposed as a means of precipitating the calcium contributed by the calcium hypochlorite bleach, e.g., MCP,[975] orthophosphoric acid,[976] and a combination of calcium hydroxide with MCP.[977] Andreson[978] advocated the replacement of mineral constituents of raw sugar by the addition of various calcium compounds including calcium phosphates.

Other Applications

The sweetness of beet sugar, reported not to be as sweet as cane sugar,[979] was increased by the addition of calcium sulfate and DCP to the wet beet sugar with subsequent drying in one report. Schongart[980] reported that orthophosphoric acid produced a lower ratio of inversion of sucrose than an equivalent level of hydrogen ion produced by hydrochloric acid. The addition of DSP at levels between 0.01 and 0.05% to clarified sugar cane juice was reported to reduce the hydrolysis of the sugars, and thus reduce caramelization, which leads to dark-colored raw sugar.[981]

The addition of TCP to sugar was reported to reduce caking,[982] provide free-flowing properties,[983] and reduce its hygroscopicity.[984] Mead[985] reported that the addition of pyrophosphoric acid or SAPP to molasses produced a greater amount of clarification and resulted in a clearer, more brilliant molasses.

Food Applications for Organic Phosphate Compounds

Numerous organic phosphate compounds are known in nature. These include the phosphate esters, e.g., those in which phosphate groups are chemically linked through the hydroxyls of organic compounds such as sugars, and the numerous energy-giving compounds containing phosphate esters, such as ATP, the various sugar phosphates, and similar substances. Any biochemical textbook can provide numerous examples of the energy-rich phosphate esters of organic compounds. There also is considerable evidence that phosphate compounds of organic materials occur in which there is a carbon to phosphorus bond.[3]

The methods for preparation of compounds that include the esterification of hydroxyl groups of organic substances such as carbohydrates and polyhydroxylic alcohols, as well as hydroxy-acids, are well known. One of the first patents for the preparation of such compounds for food use was granted as early as 1910.[986] Few food applications for organic phosphates have been developed and fewer have been approved by FDA for commercial use. It is necessary to demonstrate that these are nontoxic compounds, a time-consuming and costly process. As a result, the development costs for these compounds as food additives are very high, and the manufacturer often cannot foresee sufficient profit during the life of the product to return his investment.

Applications as Antioxidants

Epstein and Harris[987] reported that a phosphorylated mono- or diglyceride of a fatty acid could be used as an antioxidant in food applications.

Martin[988] patented the preparation of the orthophosphoric esters of phenolic compounds which he claimed were excellent neutral phosphate ester antioxidants for treatment of animal, vegetable, and marine oils, thus preventing or inhibiting development of rancidity in these compounds.

Nyrop[989] patented the preparation and application of phosphate esters of fermentable sugar compounds such as hexoses, pentoses, and trioses. He reported that these organic phosphate-sugar compounds, when added to organic food materials subject to oxidative deterioration, would stabilize the organic food compounds against rancidity and oxidative deterioration.

Applications as Emulsifying Agents

Harris[990] reported that organic phosphate esters of mono- and diglycerides were able to modify interfaces between aqueous fat systems. Thus, these compounds were reported to be excellent emulsifiers when added to shortenings for use in most of the applications in which shortenings are used, such as for baked products, preparation of icings, and similar applications. Katzman[991] patented the preparation and application of metaphosphoric acid esters of numerous types of high molecular weight alcohols as emulsifying, foaming, and improving agents for baked products, particularly for nonshortening foam cakes such as angel food cakes.

Thurman[992,993] patented the preparation of sodium phosphate esters of phosphatides. He described the preparation of the compounds and their applications as excellent surface active agents for use as emulsifiers and similar applications in foods. Examples claimed emulsifying properties for mayonnaise, margarine, and shortening compositions.

Pader and Gershon[994,995] obtained patents for the application of phosphoric acid esters of mono- and diglycerides as emulsifiers and aerating agents for aerosol toppings which could be dispensed as light foams from pressurized cans having the texture and appearance of whipping cream. The aerosol preparation was reported to be stable to temperatures of $-10°$ to $70°F$.

Applications as Whipping Agents

Katzman[996] reported the preparation of tetrapolyphosphoric acid esters of long-chain alkyl alcohols such as stearyl tetrapolyphosphate. According to the author, these compounds were excellent egg white foaming agents when added to the egg white in small quantities. The author listed 35 examples of organic phosphate esters ranging from the alkyl alcohols through mono- and diglyceride fatty compounds. All were reported to provide a strongly hydrophilic group through the tetrapolyphosphoric acid and a strongly lipophilic group through the alkyl compound.

Thompson[997] reported the methods for preparation and the food applications of an alkyl phosphoryl halide, e.g., octadecyl phosphoryl dichloride, as improvers of air incorporation and foam stability in foamed food products. These compounds were reported to be especially useful when incorporated into baked products requiring a great deal of aeration such as cakes, breads, and similar products.

Cunningham and co-workers[998] patented the application of a triethyl phosphate as a foaming agent for egg whites. The incorporation of 0.001 to 0.1% of the compound, based on the weight of the egg white, would improve the whipping properties of egg whites normally having lowered whipping properties due to pasteurization procedures now required of all such products. Concern over the toxicity of this compound was the reason for studies by Gumbmann et al.[999] who provided a thorough study of the toxicity of the compound in quantities up to 10% of the diet of rats. Feeding and reproductive studies indicated that the compound had no effect at levels that would be encountered if this were used as a food additive in the manner proposed by Cunningham and his co-workers.

Other Applications

Curtin and Gagolski[1000] and Toy[1001] obtained patents for the preparation of sucrose-calcium phosphate compounds of a very complex nature which formed soluble compositions even though a considerable quantity of dicalcium phosphate was detectable in the product. The ability of the sucrose-phosphate complex to form soluble dicalcium phosphate compounds was not understood but was adequately demonstrated. These compounds were shown to increase crispness of the sugar coatings on breakfast cereals[1002,1003] and to inhibit the development of dental caries in the teeth of children who ate food compositions containing the compound.[1004]

A Japanese patent[1005] reports that organic phosphates such as the sugar phosphates and glyceryl phosphates improve the flavor of alcoholic beverages.

Chapter 7

FUTURE TRENDS

As demonstrated by the preceding sections, the food science literature contains much evidence of the usefulness of the phosphates in food processing. However, many of the applications described are not being exploited by the food industry.

For example, the ability of the phosphates, and especially the polyphosphates, to control firmness or tenderness of the tissues of fruits and vegetables is not applied to the extent which the published research indicates. The use of the polyphosphates in increasing the acceptability of meat products, including fresh and frozen meats, poultry, and seafood is well documented, but mostly used only in cured meats.

A major reason that these applications are not exploited is that the regulatory agencies, USDA and FDA, have been very cautious in approving phosphate applications in processed foods. These agencies most frequently cite three reasons. One is fear that the polyphosphates will allow the processor to upgrade lower-grade food products. This has repeatedly been disproved. The phosphates are unable to improve the quality of low-grade foods. They are, however, capable of preventing further lowering of the quality, and in many cases, of improving some of the characteristics which make these foods less acceptable to the consumer.

The second reason often given is that phosphate treatment allows the processor to increase the water content of the foods. This is most often mentioned in regard to meat, poultry, and seafood products. Yet, careful research, especially as reviewed in the sections on poultry and seafoods, has clearly demonstrated that treating these foods with phosphate dips often *reduces* the amount of water absorbed during ice-chilling of the products. In fact, the evidence presented indicates that phosphate treatment seals the surfaces of the food, therefore preventing internal juices from escaping and water or saline solutions from entering.

The third argument given against food applications of phosphates is that their increased use will reach a level which will be toxic to humans. Evidence presented in the section on the toxicology of the food phosphates has demonstrated that they are little, if any, more toxic than common table salt which is now used in most foods. In addition, the effective levels of the phosphates are usually 0.5% or less. This level of phosphate has been consumed in the diets of many animals and has been shown to be nontoxic. Since the phosphates have adverse effects in many foods, it is inconceivable that their use in every practical application could ever result in a level of 0.5% of phosphate in the total human diet.

As the need to conserve every available amount of food increases over the next few years due to the "population explosion," the use of phosphates to aid in making more foods acceptable and in increasing their nutritional value by retaining internal juices should be given further serious consideration. The pressure of the increased need for expansion of the world food supply is certain to result in the increased application of phosphates in food processing.

REFERENCES

1. **Nazario, G.,** *Rev. Inst. Adolfo. Lutz,* 11, 141, 1952.
2. **Sanchelli, V.,** *Phosphates in Agriculture,* Reinhold, New York, 1965, 90.
3. **Quin, L. D.,** *Topics in Phosphorus Chemistry,* 4, 23, 1967.
4. **Harada, K. and Fox, S. W.,** Thermal polycondensation of free amino acids with polyphosphoric acid, in *The Origins of Prebiological Systems,* Fox, S. W., Ed., Academic Press, New York, 1965, 289.
5. **Lipman, F.,** Projecting backwards from the present stage of biosynthesis, in *The Origins of Prebiological Systems,* Fox, S. W., Ed., Academic Press, New York, 1965, 259.
6. **Schramm, G.,** Synthesis of nucleotides and polynucleotides with metaphosphate esters, in *The Origins of Prebiological Systems,* Fox, S. W., Ed., Academic Press, New York, 1965, 299.
7. **Miller, S. L. and Parris, M.,** *Nature,* 204, 1248, 1964.
8. **Ponnamperuma, C., Sagan, C., and Mariner, R.,** *Nature,* 199, 222, 1963.
9. **Lohmann, K.,** *Kondensierte Phosphate in Lebensmitteln,* Symposium, 1957, Mainz, Springer-Verlag, Berlin, 1958, 29.
10. **Harold, F. M.,** *Bacteriol. Rev.,* 30, 772, 1966.
11. **Wiame, J. M.,** *J. Biol. Chem.,* 178, 919, 1949.
12. **Stich, H.,** *Z. Naturforsch.,* 8b, 36, 1953.
13. **Kornberg, S. R.,** *Biochim. Biophys. Acta,* 26, 294, 1957.
14. **Ipata, P. L. and Felicioli, R. A.,** *Boll. Soc. Ital. Biol. Sper.,* 39(2), 85, 1963; *C. A.,* 59, 874g, 1963.
15. **Van Steveninck, J. and Booij, H. L.,** *J. Gen. Phys.,* 48, 43, 1964.
16. **Shiokawa, K. and Yamana, K.,** *Exp. Cell Res.,* 38(1), 180, 1965; *C. A.,* 63, 4722A, 1965.
17. **Miyachi, S.,** *J. Biochem.* (Toyko), 50, 367, 1961; *C. A.,* 56, 9121a, 1962.
18. **Tewari, K. K. and Singh, M.,** *Phytochemistry,* 3(2), 341, 1964; *C. A.,* 61, 963c, 1964.
19. **Grossmann, D. and Lang, K.,** *Biochem. Z.,* 336, 351, 1962.
20. **Griffin, J. B., Davidian, N. M., and Penniall, R.,** *J. Biol. Chem.,* 240, 4427, 1965.
21. **Mattenheimer, H.,** *Kondensierte Phosphate in Lebensmitteln,* Symposium, 1957, Mainz, Springer-Verlag, Berlin, 1958, 45.
22. **Sober, H. A.,** Ed., *Handbook of Biochemistry,* 2nd ed., The Chemical Rubber Co., Cleveland, 1970.
23. **Sherman, H. C.,** *Calcium and Phosphorus in Foods and Nutrition,* Columbia University Press, New York, 1947, 91.
24. **Kiermeier, F. and Mohler, K.,** *Z. Lebensmittel-Unters. Forsch.,* 106, 33, 1957.
25. **Ruf, F.,** *Mitt. Geb. Lebensmittel-Unters. Hyg.,* 48, 451, 1957.
26. **Shettino, O.,** *Farmaco (Pavia) Ed. Pract.,* 20(2), 65, 1965.
27. *Kondensierte Phosphate in Lebensmitteln,* Symposium, 1957, Mainz, Springer-Verlag, Berlin, 1958.
28. **Barackman, R. A. and Bell, R. N.,** *Food Eng.,* 25(6), 68, 108, 1953.
29. The Functions of Phosphates in Food Products, *Food Industry Release No. 1,* Stauffer Chemical Co., New York, 1966.
30. **Van Wazer, J. R.,** *Phosphorus and Its Compounds,* in two volumes. Interscience Publishers, Inc., New York, 1958.
31. Nomenclature and Classification of Inorganic Phosphates, *Technical Service Release No. 1,* Stauffer Chemical Co., New York, 1965.
32. **Van Wazer, J. R.,** Phosphoric acids and phosphates, in *The Encyclopedia of Chemical Technology,* Vol. 15. Kirk, R. E. and Othmer, D. F., Eds., Interscience Encyclopedia, Inc., New York, 1968, 232.
33. **Bell, R. N.,** *Ind. Eng. Chem.,* 40, 1464, 1948.
34. **Huhti, A. L. and Gartaganis, P. A.,** *Can. J. Chem.,* 34, 785, 1956.
35. **Partridge, E. P.,** *Chem. Eng. News,* 27(4), 214, 1949.
36. **Thilo, E.,** *Kondensierte Phosphate in Lebensmitteln,* Symposium, 1957, Mainz, Springer-Verlag, Berlin, 1958, 7.
37. *World Health Organization Technical Report, Series No. 281,* Specifications for the Identity and Purity of Food Additives and Their Toxicological Evaluation: Emulsifiers, Stabilizers, Bleaching and Maturing Agents, World Health Organization, Geneva, 1964. (May be obtained from: National Agency for International Publications, Inc., 317 East 34th Street, New York 10016.)
38. Food Protection Committee, *Food Chemicals Codex,* 1st ed., Publ. No. 1406, National Academy of Sciences - National Research Council, Washington, D.C., 1966.
39. **Brout, H.,** British Patent 883,169, 1961; Diller, I. M. and Brout, H., U.S. Patent 2,851,359, 1958.
40. **Van Wazer, J. R.,** *J. Am. Chem. Soc.,* 72, 647, 1950.
41. **Westman, A. E. R. and Crowther, J.,** *J. Am. Ceram. Soc.,* 37, 420, 1954.
42. **Westman, A. E. R. and Gartaganis, P. A.,** *J. Am. Ceram. Soc.,* 40, 293, 1957.
43. **Bernhart, D. N. and Chess, W. B.,** *Anal. Chem.,* 31, 1026, 1959.
44. **Corbridge, D. E. C.,** *Top. Phosphor. Chem.,* 3, 57, 1967.
45. **Batra, S. C.,** *J. Food Sci.,* 30, 441, 1965.
46. **Van Wazer, J. R. and Campanella, D. A.,** *J. Am. Chem. Soc.,* 72, 655, 1950.
47. **Van Wazer, J. R. and Callis, C. F.,** *Chem. Rev.,* 58, 1011, 1958.
48. **Thilo, E.,** *Angew. Chem.,* 67, 141, 1955.

49. **Irani, R. R. and Callis, C. F.,** *J. Am. Chem. Soc.*, 39, 156, 1962.
50. **Irani, R. R. and Morgenthaler, W. W.,** *J. Am. Chem. Soc.*, 40, 283, 1963.
51. **Kutscher, W.,** *Deut. Lebensm. -Rundsch.*, 57(6), 140, 1961.
52. **Karl-Kroupa, E., Callis, C. F., and Seifter, E.,** *Ind. Eng. Chem.*, 49, 2061, 1957.
53. **Bell, R. N.,** *Ind. Eng. Chem.*, 39, 136, 1947.
54. **Hermann, S.,** *Exp. Med. Surg.*, 5, 160, 1947.
55. **Behrens, B. and Seelkopf, K.,** *Naunyn-Schmiedeberg's Arch. Exp. Pathol. Pharmakol.*, 169, 238, 1932.
56. **Eichler, O.,** *Handbuch der Experimentellen Pharmakologie, Erganzungswerk, Bd. 10,* Berlin, 1950, 363. (Quoted in World Health Organization Technical Report, Series No. 281, Ref. 37.)
57. **McFarlane, D.,** *J. Pathol. Bacteriol.*, 52, 17, 1941.
58. **van Esch, G. J., Vink, H. H., Wit, S. J., and van Genderen, H.,** *Arzneimittel-Forschung,* 7, 172, 1957.
59. **Datta, P. K., Frazer, A. C., Sharratt, M., and Sammons, H. G.,** *J. Sci. Food Agric.*, 13, 556, 1962.
60. **Zipf, K.,** *Arzneimittel-Forschung,* 7, 445, 1957.
61. **Gosselin, R. E. and Megirian, R.,** *J. Pharmacol. Exp. Ther.*, 115, 402, 1955.
62. **Bonting, S. L.,** Ph. D. Thesis, The University of Amsterdam, Amsterdam, Netherlands, 1952.
63. **MacKay, E. M. and Oliver, J.,** *J. Exp. Med.*, 61, 319, 1935.
64. **Hahn, F., Jacobi, H., and Seifen, E.,** *Arzneimittel-Forschung,* 8, 286, 1958.
65. **Dymsza, H. A., Reussner, G., and Thiessen, R.,** *J. Nutr.*, 69, 419, 1959.
66. **Hogan, A. G., Regan, W. O., and House, W. B.,** *J. Nutr.*, 41, 203, 1950.
67. **House, W. B. and Hogan, A. G.,** *J. Nutr.*, 55, 507, 1955.
68. **Hahn, F. and Seifen, E.,** *Arzneimittel-Forschung,* 9, 501, 1959.
69. **Hahn, F.,** *Z. Ernahrungswiss.*, Suppl., 1, 55, 1961.
70. **Hodge, H. C.,** Short-term oral toxicity tests of condensed phosphates in rats and dogs (unpublished mimeographed report), University of Rochester, Rochester, N. Y., 1956.
71. **Hodge, H. C.,** Chronic oral toxicity studies in rats of sodium tripolyphosphate (unpublished mimeographed report), University of Rochester, Rochester, N. Y., 1959.
72. **Hodge, H. C.,** Chronic oral toxicity studies in rats of sodium hexametaphosphate (unpublished mimeographed report), University of Rochester, Rochester, N. Y., 1960.
73. **Hodge, H. C.,** Chronic oral toxicity studies in rats of sodium trimetaphosphate (unpublished mimeographed report), University of Rochester, Rochester, N. Y., 1960.
74. **Gassner, K., Kiekebusch, W., and Lang, K.,** *Biochem. Z.*, 328, 485, 1957.
75. **van Genderen, H.,** *Z. Ernahrungswiss.*, Suppl. 1, 32, 1961.
76. **Lang, K.,** *Kondensierte Phosphate in Lebensmitteln,* Symposium, 1957, Mainz Springer-Verlag, Berlin, 1958, 135.
77. **Borenstein, M. K. and Schwartz, C.,** *J. Nutr.*, 36, 681, 1948.
78. **Schreier, K. and Noeller, H. G.,** *Arch. Exp. Pathol. Pharmakol.*, 227, 199, 1955.
79. **Mitchell, L. C.,** *J. Assoc. Offic. Agr. Chem.*, 41, 185, 1958.
80. **Gosselin, R. E., Rothstein, A., Miller, G. J., and Berke, H. L.,** *J. Pharmacol. Exp. Ther.*, 106, 180, 1952.
81. **Gotte, H.,** *Z. Naturforsch.*, 86, 173, 1953.
82. **Lang, K., Schachinger, L., Karges, O., Blumenberg, F. K., Rossmuller, G., and Schmutte, I.,** *Biochem. Z.*, 327, 118, 1955.
83. **Lehman, A. J.,** *Assoc. Food Drug Officials in U. S. Quart. Bull.*, 24, 45, 1960.
84. **Gosselin, R. E., Tibdall, C. S., Megirian, R., Maynard, E. A., Downs, W. L., and Hodge, H. C.,** *J. Pharmacol. Exp. Ther.*, 108, 117, 1953.
85. Paragraph 121.101(d), *Code of Federal Regulations, Title 21* (Revised as of January 1, 1971), U.S. Government Printing Office, Washington, D. C.
86. Paragraph 318.7 (4), *Code of Federal Regulations, Title 9* (Revised as of January 1, 1971). U.S. Government Printing Office, Washington, D. C.
87. **Navet, P.,** *Tech. Eau (Belg.),* 1947(6), 41; *C. A.*, 41, 6003g, 1947.
88. **Hofer, P.,** *Kondensierte Phosphate in Lebensmitteln,* Symposium, 1957, Mainz , Springer-Verlag, Berlin, 1958, 122.
89. **Adler, H.,** U.S. Patent 2,262,745, 1942.
90. **Corrie, L. M.** (Harry Chin), U.S. Patent 3,231,394, 1966.
91. **Roland, C. T.,** *Milk Plant Mo.*, 31(6), 38, 45, 1942.
92. **Visser, S. A.,** *J. Dairy Sci.*, 45, 710, 1962.
93. **Hamm, R.,** *Z. Lebensmittel-Unters. Forsch.*, 104, 245, 1956.
94. **Rose, D.,** *Dairy Sci. Abstr.*, 31, 171, 1969.
95. **Warning, W. G. and Schille, T. W.,** U.S. Patent 1,865,733, 1932; reissue 18,907, 1933.
96. **Lubeck, W. D.,** U.S. Patent 1,978,040, 1934.
97. **Kemmerer, K. S.,** U.S. Patent 2,829,056, 1958.
98. **Sorgenti, H. A., Nack, H., and Sachsel, G. F.,** U.S. Patent 3,035,918, 1962.
99. **Heyer, W.,** U.S. Patent 2,816,032, 1957.
100. **Bah, S. and McKeen, W. E.,** *Can. J. Microbiol.*, 11, 309, 1965.
101. **Malkov, A. M. and Kormushkina, A. M.,** U.S.S.R. Patent 135,064, 1961.

102. **Kohl, W. F. and Ellinger, R. H.**, Netherlands Patent 68,00059, 1968; South African Patent 67,7804, 1968; Belgian Patent 708,863, 1968; French Patent 1,568,002, 1969; British Patent 1,154,079, 1969.
103. **Kohl, W. F. and Ellinger, R. H.**, Netherlands Patent 68,00060, 1968; Belgian Patent 708,864, 1968; South African Patent 67,7805, 1968; French Patent 1,568,003, 1969.
104. **Libbey, A. G.**, U.S. Patent 2,075,653, 1937.
105. **Peynaud, E.**, *Ind. Agr. Aliment. (Paris)*, 70, 559, 1953.
106. **Freed, E.**, U.S. Patent 2,995,446, 1961.
107. **Skrivanek, M.**, *Chemie* (Prague), 4, 227, 1948.
108. **Ohara, I., Nonomura, H., Kushida, T., and Maruyama, C.**, *J. Ferment. Technol.* (Japan), 34, 431, 1956.
109. **Karlson, W. N.**, U.S. Patent 2,943,939, 1960.
110. 120 Q's and A's about how phosphates are used in foods, *Food Industry Release No. 2.*, Stauffer Chemical Company, New York, 1966.
111. *Ingredients for Food Processors*, Monsanto Co., St. Louis, 1965.
112. **Diller, I. M.** (Assigned to H. Brout), U.S. Patents 2,851,360 and 2,851,361, 1958.
113. **Diller, I. M.** (Assigned to H. Brout), U.S. Patent 2,953,459, 1960.
114. **Common, J. L.**, U.S. Patent 3,023,106, 1962.
115. **DiNardo, A.**, U. S. Patent 3,224,879, 1965.
116. **Stayton, F. J.**, U.S. Patent 2,341,826, 1944.
117. **Hashimoto, K. and Miyazaki, S.**, Japanese Patent 15,550, 1961.
118. **Feldheim, W. and Seidemann, J.**, *Pharmazie*, 14, 12, 1959.
119. **Niwa, S., Katayama, N., and Matsubara, S.**, *Bitamin* (Kyoto), 18(2), 492, 1959; Niwa, S., Katayama, N., Matsubara, S., and Tanaka, H., *Bitamin* (Kyoto), 20(3), 497, 1960; Niwa, S. and Katayama, N., *Bitamin* (Kyoto), 24(2), 73, 1961; *C. A.*, 1002f, 4531g, 1965.
120. **Stahl, J. E. and Ellinger, R. H.**, Use of Phosphates in the Cereal and Baking Industry, presented at the Phosphates in Food Processing Symposium, 1970, University of Guelph, Guelph, Ontario, Canada, to be published.
121. **Kichline, T. P. and Conn, T. F.**, *Bakers Digest*, 44(8), 36, 1970.
122. **Carlin, G. T.**, *Cereal Chem.*, 21, 189, 1944.
123. **Hanssen, E. and Dodt, E.**, *Mikroskopie*, 7, 2, 1952.
124. **Jooste, M. E. and Mackey, A. O.**, *Food Res.*, 17, 185, 1952.
125. **Hunt, F. E. and Green, M. E.**, *Food Technol.*, 9, 241, 1955.
126. **Bradley, W. B. and Tucker, J. W.**, Bakery processes and leavening agents, in *The Encyclopedia of Chemical Technology*, Vol. 3, Kirk, R. E. and Othner, D. F., Eds., The Interscience Encyclopedia, Inc., New York, 1964, 41.
127. **Butterworth, S. W. and Colbeck, W. J.**, *Cereal Chem.*, 15, 475, 1938.
128. **Burhans, M. E. and Clapp, J.**, *Cereal Chem.*, 19, 196, 1942.
129. **Carlin, G. T.**, *Bakers Digest*, 21(8), 22, 1947.
130. **Jongh, G.**, *Cereal Chem.*, 38, 140, 1961.
131. **Ellinger, R. H.**, The Use of Microscopy and Histology in Development of Cereal Products, presented at the Fifth Annual Symposium sponsored by the Central States Section, American Association of Cereal Chemists, St. Louis, Mo., 1964.
132. **Ellinger, R. H.**, The Sodium Aluminum Phosphates: Their Properties and Uses as Leavening Acids, presented at the Sixth Annual Symposium sponsored by the Central States Section, American Association of Cereal Chemists, St. Louis, Mo., 1965.
133. **Barackman, R. A.**, *Cereal Chem.*, 8(5), 423, 1931.
134. **Van Wazer, J. R. and Arvan, P. G.**, *Northwestern Miller, Milling Production Sect. 251*, 3a, Feb., 1954; 5, 17, March, 1954.
135. **Conn, J. F.**, Chemical Leavening, presented at the First Annual Symposium sponsored by the Central States Section, American Association of Cereal Chemists, St. Louis, Mo., 1960.
136. **Parks, J. R., Handleman, A. R., Barnett, J. C., and Wright, F. H.**, *Cereal Chem.*, 37, 503, 1960.
137. Modern leavening with sodium aluminum phosphates, *Food Industry Release No. 4*, Stauffer Chemical Co., New York, 1966.
138. What leavening does for food – what phosphates do in leavening, *Food Industry Release No. 3*, Stauffer Chemical Co., New York, 1965.
139. **Reiman, H. M.**, Properties and Uses of Leavening Agents, *Technical Service Leaflet*, Stauffer Chemical Co., New York, 1965.
140. **Tucker, J. W.**, *Cereal Sci. Today*, 4(4), 91, 1959.
141. **Joslin, R. P.**, *Bakers Digest*, 34(10), 58, 1960.
142. **Joslin, R. P. and Ziemba, J. V.**, *Food Eng.*, 27(9), 59, 184, 1955.
143. Leaven right with Levn-Lite and Pan-O-Lite, *Sodium Aluminum Phosphate Sales Brochure*, The Monsanto Co., St. Louis, Mo.
144. Upgrade your self-rising flours with Stabil-9, *Sales Brochure*, The Monsanto Co., St. Louis, Mo.
145. **Huff, A. E.**, U.S. Patent 2,314,090, 1943; Russihyvili, R., British Patent 608,708, 1948.
146. **Hurka, R. J.**, U.S. Patent 2,366,857, 1945.
147. **McDonald, G. A.**, U.S. Patents 2,550,490 and 2,550,491, 1951.

148. Dyer, J. N., U.S. Patent 2,995,421, 1961.
149. Lauck, R. M. and Tucker, J. W., U.S. Patent 3,041,177, 1962.
150. Vanstrom, R. E., U.S. Patents 3,223,479 and 3,223,480, 1965; Belgian Patents 627,509 and 627,511, 1963.
151. Chemische Fabrik Budenheim Aktiengesellschaft, British Patent 390,743, 1933.
152. Griffith, E. J., U.S. Patent 2,774,672, 1956.
153. Conn, J. F., *Bakers Digest,* 39(4), 66, 70, 1965.
154. Handleman, A. R., Conn, J. F., and Lyons, J. W., *Cereal Chem.,* 38, 294, 1961.
155. Leavening Indicator Chart, Stauffer Chemical Co., New York.
156. Tucker, J. W., U.S. Patent 3,109,738, 1963.
157. Armstrong, L. and Willoughby, L. B., U.S. Patent 2,478,618, 1949.
158. Baker, J. S. and Lindeman, C. G., U.S. Patent 3,297,449, 1967.
159. Erekson, A. B. and Duncan, R. E., U.S. Patent 2,942,988, 1960; Matz, S. A., U.S. Patent 3,397,064, 1968.
160. Barackman, R. A. and Bailey, C. H., *Cereal Chem.,* 4, 400, 1927.
161. Lannuier, G. L., M. S. Thesis, Florida State University, Tallahassee, Fla., 1961; Lannuier, G. L. and Bayfield, E. G., *Bakers Digest,* 35(6), 34, 1961; Bayfield, E. G. and Lannuier, G. L., 36(1), 34, 1962; *Bakers Digest,* 36(6), 57, 1962.
162. Takebayashi, Y., *Nippon Shokuhin Kogyo Gakkcishi,* 10(9), 382, 1963; *C. A.,* 63, 3539d, 1965.
163. Carter, J. E. and Hutchinson, J. B., *Chem. Ind.* (London), 1965(42), 1765.
164. Muhler, J. C., U.S. Patent 3,467,529, 1969; Canadian Patent 823,140, 1969.
165. Schohl, A. T., Mineral Metabolism, Amer. Chem. Soc. Monograph Series No. 82, Reinhold Publishing Corp., New York, 1939.
166. The importance of calcium and phosphorus supplements in foods, *Technical Service Bulletin,* Stauffer Chemical Co., New York.
167. Anon., *Bull. Natl. Res. Coun.,* 40, 38, 1944.
168. Anon., *Food Chem. News,* Sept. 15, 1969, 24.
169. Nizel, A. E., Keating, N., Sundstrom, C., and Harris, R. S., *J. Dent. Res.,* 37, 35, 1958.
170. Buttner, W. and Muhler, J. C., *J. Dent. Res.,* 37, 860, 1958.
171. McClure, F. J. and Muller, A., Jr., *J. Am. Dent. Assoc.,* 58, 36, 1959.
172. Stralfors, A., *J. Dent. Res.,* 43, 1137, 1964.
173. McClure, F. J. and Muller, A., Jr., *J. Dent. Res.,* 38, 776, 1959.
174. McClure, F. J., *J. Nutr.,* 72, 131, 1960.
175. Stookey, G. K., and Muhler, J. C., *J. Dent. Res.,* 45, 856, 1966.
176. Stookey, G. K., Carroll, R. A., and Muhler, J. C., *J. Am. Dent. Assoc.,* 74, 752, 1967.
177. Carroll, R. A., Stookey, G. K., and Muhler, J. C., *J. Am. Dent. Assoc.,* 76, 564, 1968.
178. Luoma, H., Niska, K., and Tustola, L., *Arch. Oral Biol.,* 13(11), 1343, 1968; *Oral Res. Abstr.,* 4, 779, 1969.
179. Averill, H. M. and Bibby, B. G., *J. Dent. Res.,* 43, 1150, 1964.
180. Nizel, A. E. and Harris, R. S., *J. Dent. Res.,* 43, 1123, 1964.
181. Council on Foods and Nutrition, *J. A. M. A.,* 203, 407, 1968.
182. Crosby, W. H., *J. A. M. A.,* 208, 347, 1969.
183. Jacobs, A. and Greenman, D. A., *Br. Med. J.,* 1, 673, 1969.
184. Finch, C. A., *Nutr. Rev.,* 23, 129, 1965.
185. Proposal to Increase the Levels of Iron in Enriched Flour and Bread, submitted to FDA by the Millers National Federation and the American Bakers Association, November 5, 1969; Lampman, H. H., Memorandum of November 17, 1969, Millers' National Federation, Chicago.
186. Nakamura, F. I. and Mitchell, H. H., *J. Nutr.,* 25, 39, 1943.
187. Street, H. R., *J. Nutr.,* 26, 187, 1943.
188. Freeman, S. and Burrill, N. W., *J. Nutr.,* 30, 293, 1945.
189. Steinkamp, R., Dubach, R., and Moore, C. V., *Arch. Intern. Med.,* 95, 181, 1955.
190. Fritz, J. C., *Proc. Med. Nutr. Conf.,* March 20-21, 1969, 1.
191. Fritz, J. C., *Fed. Proc.,* 28, 692, 1969.
192. Iron in flour, in *Reports on Public Health and Medical Subjects, No. 117,* Ministry of Health, London, Her Majesty's Stationery Office, 1968.
193. Hinton, J. J. C., Carter, J. E., and Moran, T., *J. Food Technol.,* 2, 129, 1967.
194. Hinton, J. J. C. and Moran, T., *J. Food Technol.,* 2, 135, 1967.
195. White, H. S. and White, P. L., *Food Nutr. News,* 39(7), 1, 1968.
196. White, H. S., Should Food Iron Be Increased? Presented at Symposium on Nutrition and Food Technology, Chicago, February 12, 1969.
197. Anon., *Cereal Sci. Today,* 15, 121, 1970.
198. Anon., *Fed. Reg.,* April 1, 1970; *Food Eng.,* 42(5), 22, 1970.
199. Anon., *Chem. Eng. News,* 49(6), 52, 1971.
200. Billings, H. J., U.S. Patent 2, 131, 881, 1938.
201. Rao, P. V. S., Ananthachar, T. K., and Desikachar, H. S. R., *Ind. J. Technol.,* 2, 417, 1964; *C. A.,* 62, 13768d, 1965.

202. **Tanaka, M. and Yukami, S.**, U.S. Patent 3, 484, 249, 1969; *Food Sci. Technol. Abstr.*, 2, 4M268, 1970.
203. **Malkov, A. M., Deeva, V. E., and Drizina, T. S.**, *Tr. Leningrad. Tekhnol. Inst. Pishch. Prom.*, 12, 100, 1955; *Zhur. Khim., Biol. Khim.*, No. 11509, 1957.
204. Stauffer Chemical Co., Belgian Patent 634,380, 1963; British Patent 1, 019, 508, 1966.
205. **Payumo, E. M., Broines, P. R., Banzon, E. A., and Torres, M. L.**, *Philippine J. Nutr.*, 22, 216, 1969; *Food Sci. Technol. Abstr.*, 2, 11M1096, 1970.
206. Thames Rice Milling Co., Ltd., British Patent 650, 192, 1951.
207. **Keller, H. M.**, U.S. Patent 2, 909, 431, 1959.
208. **Madrazo, M. G. and Cortina, G. A.**, U.S. Patent 3, 117, 868, 1964.
209. **Sommer, H. M. and Hart, E. B.**, *Wis. Agr. Exp. Sta. Res. Bull.*, 67, 1926.
210. **Vujicic, I., Batra, S. C., and deMan, J. M.**, *J. Agr. Food Chem.*, 15, 403, 1967.
211. **Vujicic, I., deMan, J. M., and Woodrow, I. L.**, *Can. Inst. Food Technol. J.*, 1, 17, 1968.
212. **Puri, B. R., Kalra, K. C., and Toteja, K. K.**, *Ind. J. Dairy Sci.*, 18(4), 126, 1965; *C. A.*, 66, 1652r, 1967.
213. **Tessier, H. and Rose, D.**, *J. Dairy Sci.*, 41, 351, 1958.
214. **Kiermeier, F.**, *Z. Lebensmittel - Unters. Forsch.*, 95, 85, 1952.
215. **Pyne, G. T. and McGann, T. C. A.**, *J. Dairy Res.*, 27, 9, 1960.
216. **Zittle, C. A., Della Monica, E. S., Rudd, R. K., and Custer, J. H.**, *Arch. Biochem. Biophys.*, 76, 342, 1958.
217. **Zittle, C. A.**, *J. Dairy Sci.*, 49, 361, 1966.
218. **Morr, C. V. and Kenkare, D. B.**, *J. Dairy Sci.*, 47, 673, 1964.
219. **Odagiri, S. and Nickerson, T. A.**, *J. Dairy Sci.*, 47, 1306, 1964; *J. Diary Sci.*, 48, 19, 1965.
220. **Ellinger, R.**, and co-workers, unpublished results.
221. **Feagen, J. T., Griffin, A. T., and Lloyd, G. T.**, *J. Dairy Sci.*, 49, 1010, 1966.
222. **Zittle, C. A. and Pepper, L.**, *J. Dairy Sci.*, 41, 1671, 1958.
223. **Rose, D.**, *J. Dairy Sci.*, 44, 430, 1961.
224. **Rose, D.**, *Dairy Sci. Abst.*, 25, 45, 1963.
225. **Rose, D.**, *J. Dairy Sci.*, 45, 1305, 1962.
226. **Kenkare, D. B., Morr, C. V., and Gould, I. A.**, *J. Dairy Sci.*, 47, 947, 1964.
227. **Edmondson, L. F., Landgrebe, A. R., Sadler, A. M., and Walter, H. E.**, *J. Dairy Sci.*, 48, 1597, 1965.
228. **Tessier, H., and Rose, D.**, *J. Dairy Sci.*, 44, 1238, 1961.
229. **Puri, B. R., Arora, K., and Toteja, K. K.**, *Ind. J. Dairy Sci.*, 22, 85, 1969; *Ind. J. Dairy Sci.*, 22, 155, 1969; *Food Sci. Technol. Abstr.*, 2, 11P1459, 11P1460, 1970.
230. **Dusek, B., Semjan, S., and Kazimir, L.**, *18th Intern. Dairy Congr.*, (Sydney), 1E, 92, 1970; *Food Sci. Technol. Abstr.*, 2, 11P1539, 1970.
231. **Fox, K. K., Harper, M. K., Holsinger, V. H., and Pallansch, M. J.**, *J. Dairy Sci.*, 47, 179, 1964.
232. **Vujicic, I. and deMan, J. M.**, *J. Inst. Can. Technol. Aliment.*, 1, 171, 1968.
233. **Lauck, R. M. and Tucker, J. W.**, *Cereal Sci. Today*, 7(9), 314, 1962.
234. **Briggs, D. R.**, *J. Biol. Chem.*, 134, 261, 1940.
235. **Katchman, B. J. and Van Wazer, J. R.**, *Biochim. Biophys. Acta*, 14, 445, 1954.
236. **Perlmann, G. E.**, *J. Biol. Chem.*, 137, 707, 1941.
237. **Horvath, A. A.**, *Ind. Eng. Chem., Anal. Ed.*, 18, 229, 1946.
238. **Lyons, J. W. and Siebenthal, C. D.**, *Biochim. Biophys. Acta*, 26, 174, 1966.
239. **Leach, A. A.**, *GGRA Literature Review, Series D-8*, Gelatin and Glue Research Assoc., Birmingham, Eng.
240. **Leviton, A.**, *J. Dairy Sci.*, 47, 670, 1964.
241. **Herreid, E. O. and Wilson, H. K.**, *Mfr. Milk Prod. J.*, 54, 14, 1963.
242. **Morr, C. V.**, *J. Dairy Sci.*, 50, 1038, 1967.
243. **Hall R. E.**, U.S. Patent 2,064,110, 1936. **Schwartz, C.**, U.S. Patent 2,135,054, 1938.
244. **Kawanishi, G.**, *Nippon Shokuhin Kogyo Gakkaishi*, 10(4), 123, 1963; *C. A.*, 63, 1147g, 1965.
245. **Wouters, O. J.**, U.S. Patent 2,319,362, 1943.
246. **Jackson, J. M.**, U.S. Patent 2,396,265, 1946.
247. **Roland, C. T.**, U.S. Patent 2,665,212, 1954.
248. **Malkov, A. M., Trippel, A. I., and Kirichkova, G. A.**, *Sb. Tr. Leningr. Inst. Sov. Torgovli*, No. 23, 112, 1964.
249. **North, C. E.**, U.S. Patent 2,550,288, 1951.
250. **Meyer, A. and Weck, G.**, U.S. Patent 2,977,232, 1961.
251. **Corbin, E. A. and Long, J. E.**, U.S. Patent 3,340,066, 1967.
252. **Rothwell, J.**, *J. Soc. Dairy Tech.*, 21, 136, 1968.
253. **Wilson, H. K. and Herreid, E. O.**, *J. Dairy Sci.*, 52, 1229, 1969.
254. Paragraph 18.520 (a)(1)(i), *Code of Federal Regulations, Title 21*, U.S. Government Printing Office, Washington, D.C., 1971.
255. **Edmondson, L. F.**, *J. Dairy Sci.*, 42, 910, 1959.
256. **Martynova, K.**, *Moloch. Promy.*, 23(11), 17, 1962.
257. **Shtal'berg, S. and Radaeva, I.**, *Proc. 17th Int. Dairy Cong.*, Munich, 5, 153, 1966.

258. **Muller, L. L.,** *Aust. J. Dairy Technol., Suppl., Aust. Soc. Dairy Technol., Tech. Publ.,* No. 12, 25, 1963; *C. A.,* 60, 13791e, 1964.
259. **Seehafer, M. E.,** The Development and Manufacture of Sterilized Milk Concentrate, *FAO Agricultural Series No. 72,* 1967.
260. **Leviton, A. and Pallansch, M. J.,** *J. Dairy Sci.,* 45, 1045, 1962.
261. **Leviton, A., Anderson, H. A., Vettel, H. E., and Vestal, J. H.,** *J. Dairy Sci.,* 46, 310, 1963.
262. **Wilson, H. K., Vetter, J. L., Sasago, K., and Herreid, E. O.,** *J. Dairy Sci.,* 46, 1038, 1963.
263. **Leviton, A. and Pallansch, M. J.,** (U.S.A., per U.S.D.A.), U.S. Patent 3,119,702, 1964.
264. **Herreid, E. O. and Wilson, H. K.,** *Mfr. Milk Prod. J.,* 54(1), 14, 1963.
265. **Swanson, A. M. and Calbert, H. E.,** *Food Tech.,* 17, 721, 1963.
266. **Magnino, P. J., Jr.,** Ph.D. Thesis, University of Wisconsin, 1968. University Microfilms, Inc., No. 68-5, 332, Ann Arbor, Mich.
267. **Bixby, H. H. and Swanson, A. M.,** *J. Dairy Sci.,* 52, 892, 1969.
268. **Kadan, R. S. and Leeder, J. G.,** *J. Dairy Sci.,* 49, 709, 1966.
269. **Stewart, A. P.,** U.S. Patent 3,348,955, 1967.
270. **Bolt, M. G. and Kastelic, J.,** *J. Dairy Sci.,* 51, 693, 1968.
271. **Calbert, H. E. and Jackson, H. C.,** *J. Dairy Sci.,* 43, 846, 1960.
272. *A New Marketing Opportunity – 2.5 Billion Quarts of Beverage Quality Canned Milk – As a Minimum,* United States Steel Corp., Pittsburg, Pa., 1965.
273. Continental Can Company and United States Steel Corporation, *Summary Report: One Week Family Test of Commercially Produced Aseptically Canned Sterilized Milk Concentrate,* United States Steel Corp., Pittsburg, Pa., 1965.
274. **Alwes, M. L. and Swanson, A. M.,** *J. Dairy Sci.,* 52, 891, 1969.
275. **Nakanishi, T. and Itoh, T.,** *Agr. Biol. Chem.,* 33, 1270, 1969; *J. Agr. Chem. Soc. Jap.,* 43, 725, 1969; *Food Sci. Technol. Abstr.,* 2, 2P215, 3P385, 1970.
276. **Tamate, R. and Ohtaka, F.,** *Jap. J. Dairy Sci.,* 18(2), A37, 1969; *Food Sci. Technol. Abstr.,* 2, 6P773, 1970.
277. **Hoff, J. E., Sunyach, J., Procter, B. E., and Goldblith, S. A.,** *Food Technol.,* 14, 24, 1960; *Food Technol.,* 14, 27, 1960.
278. **Leviton, A., Vestal, J. H., Vettel, H. E., and Webb, B. H.,** *Proc. 17th Int. Dairy Cong., Munich, E.,* 133, 1966.
279. **Johnson, C. E. and Winder, W. C.,** *Quick Frozen Foods,* 31(10), 43, 1969.
280. **Kennedy, M. H. and Castagna, M. P.,** U.S. Patent 2,607,692, 1952; Canadian Patent 501,696, 1954; British Patent 709,931, 1954.
281. **Breivik, O. N., Slupatchuk, W., Carbonell, R. J., and Weiss, G.,** U.S. Patent 3,231,391, 1966.
282. **Clausi, A. S.,** U.S. Patent 2,801,924, 1957; British Patent 741,076, 1955; Canadian Patent 565,651, 1958.
283. **Charie, H. J. and Savage, J. P.,** U.S. Patent 2,927,861, 1960.
284. **Moirano, A. L.,** U.S. Patent 3,443,968, 1969; Canadian Patent 847,287, 1970.
285. **Pfrengle, O. and Zboralski, U.,** German Patent 1,129,813, 1962.
286. **Partridge, E. P.,** U.S. Patent 2,582,353, 1952.
287. **McGowan, J.,** *Food Prod. Dev.,* 4(5), 16, 1970.
288. **Roland, C. T.,** U.S. Patent 2,665,212, 1954.
289. **Freund, E. H. and Danes, E. N., Jr.,** U.S. Patent 2,957,770, 1960.
290. **Salzberg, H. K.,** U.S. Patent 2,181,003, 1939; British Patent 503,841, 1939.
291. **Wouters, O. J.,** U.S. Patent 2,319,362, 1943.
292. **Curry, C. C.,** U.S. Patent 2,341,425, 1944.
293. **Lindewald, T. W. and Kimstad, S. G.,** U.S. Patent 2,468,677, 1949.
294. **Lewis, M. A., Marcelli, V., and Watts, B. M.,** *Food Technol.,* 7, 261, 1953.
295. **Kempf, C. A. and Blanchard, E. L.,** U.S. Patent 2,645,579, 1953.
296. **Kumetat, K.,** U.S. Patent 2,739,898, 1956.
297. **Hayes. J. F.. Dunkerley, J., and Muller, L. L.,** *Aust. J. Dairy Technol.,* 24(2), 69, 1969; *Food Sci. Technol. Abstr.,* 1, 12P1299, 1969.
298. **Prodanski, P.,** *Moloch. Promy.,* 31(3), 43, 1970; *Food Sci. Technol. Abstr.,* 2, 7P916, 1970.
299. **Parsons, C. H. and Scott, E. C.,** U.S. Patent 2,279,205, 1942.
300. **Frazeur, D. R.,** *Ice Cream Field,* 73(3), 25, 1959.
301. **Weinstein, B.,** U.S. Patent 2,856,289, 1958.
302. **Keeney, P. G.,** *J. Dairy Sci.,* 45, 430, 1962.
303. **Anon.,** *Fed. Reg.,* 32, 12750, 1967.
304. **Wallander, J. F.,** *Diss. Abstr. Intern., Sect. B., Sci. Eng.,* 29, (12, Part 1), 4523, 1969; *Food Sci. Technol. Abstr.,* 2, 1P68, 1969.
305. **Amundson, C. H.,** *Am. Dairy Rev.,* 29(7), 22, 1967.
306. **Zboralski, U.,** German Patent 1,008,220, 1957.
307. **Stein, W.,** German Patent 1,097,380, 1962.
308. **Ward, P. J., Johnson, J. D., and Robertson, R. G.,** U.S. Patent 3,061,442, 1962.

309. **Arena, J. M.,** *Nutr. Today*, 5(4), 2, 1970.
310. **Gordon, W. G.,** U.S. Patent 2,377,624, 1945.
311. **McKee, J. I. and Tucker, J. W.,** U.S. Patent 3,269,843, 1966; British Patent 1,045,860, 1967; French Patent 1,415, 969, 1965.
312. **Wingerd, W. H.,** Canadian Patent 790,580, 1968.
313. **Hartman, G. H., Jr.,** Ph.D. Thesis, University of Wisconsin, 1966.
314. **Hartman, G. H., Jr. and Swanson, A. M.,** *J. Dairy Sci.*, 49, 697, 1966.
315. Protolac®-Sodium, Technical Data Sheet 402, and Protolac®-Calcium, Technical Data Sheet, The Borden Co., New York.
316. **Wingerd, W. H.,** U.S. Patent 3,332,782, 1967.
317. **Wingerd, W. H.,** U.S. Patent 3,356,507, 1967.
318. **Ellinger, R. H.,** U.S. Patent 3,563,761, 1971; British Patent 1,145,638, 1969; French Patent 1,549,831, 1968; Netherlands Patent 67,16873, 1968; Belgian Patent 708,289, 1968; Mexican Patent 99,510, 1969.
319. **Broadhead, S. A.,** British Patent 1,162,735, 1969; Belgian Patent 708,279, 1968; Netherlands Patent 67,17398, 1968; French Patent 1,549,751, 1968; Mexican Patent 99,256, 1967.
320. **Endo, N.,** Japanese Patent 8531,'62, 1962; Japanese Patent 9271, '62, 1962; *C. A.*, 58, 10666c, 1963.
321. **Ellinger, R. H. and Schwartz, M. G.,** U.S. Patent 3,615,661 and 3,615,662, 1971; French Patent 2,012,982, 1970; Belgian Patent 736,123, 1970; also French Patent 2,012,983, 1970; Belgian Patent 736,124, 1970.
322. **Holland, R. F. and Bandler, D. K.,** *Cornell Food Plant Tech. Serv. Lett.*, 1(4), 2, 1967.
323. **McCormick, R. D. and Beck, K. M.,** *Food Prod. Dev.*, 2(1), 26, 1968.
324. **Moede, H. H.,** *Dairy Sit.*, DS-320, 30, May, 1968, Economic Research Service, U.S. Department of Agriculture, Washington, D.C.
325. **Kozin, N. I. and Rodionova, I. F.,** *Izv. Vyssh. Ucheb. Zaved. Pishch. Tekhnol.*, 1964(5), 51-53; *C. A.*, 62, 9691d, 1965.
326. **Sabharwal, K. and Vakaleris, D. G.,** *J. Dairy Sci.*, 52, 891, 1969.
327. Nutritionally Balanced Non-Dairy Bases, *Food Prod. Dev.*, 2(5), 99, 1968.
328. **Kempf, C. A. and Blanchard, E. L.,** U.S. Patent 2,645,579, 1953.
329. **Brochner, H. S.,** U.S. Patent 3,241,975, 1966.
330. The Use of Carrageenan in Liquid Coffee Whiteners, *Tech. Bull. No. 255*, Marine Colloids, Inc., Springfield, N. J., 1966.
331. **Miller, D. E.,** *J. Dairy Sci.*, 51, 1330, 1968.
332. **Knightly, W. H.,** *Food Technol.*, 23, 171, 1969.
333. **Anon.,** *Food Eng.*, 38, 135, 1966.
334. Paramount C in Imitation Coffee Cream, Durkee Famous Foods, Chicago, Ill., 1963.
335. Coffee Whitener, DS-214-C, Durkee Famous Foods, Chicago, Ill.
336. Coffee Whitener, DS-237-C, Durkee Famous Foods, Chicago, Ill.
337. Test Formulas for Coffee Whiteners, Stauffer Chemical Co., New York, 1968.
338. Powdered Coffee Whitener, Stauffer Chemical Co., New York, 1964.
339. **Miller, D. E. and Ziemba, J. V.,** *Food Eng.*, 38, 97, Aug., 1966.
340. **Mason, R. D. and Justesen, A. C.,** U.S. Patent 2,407,027, 1946.
341. **Noznick, P. P. and Tatter, C. W.,** U.S. Patent 3,098,748, 1963.
342. **Nash, N. H. and Cheselka, G.,** *Bakers Digest*, 40, 58, June, 1966.
343. **Thalheimer, W. G.,** *Food Eng.*, 40(5), 112, 1968.
344. **Knightly, W. H.,** *Food Technol.*, 22, 731, 1968.
345. Guidelines to the Formulation of Whipped Toppings, Atlas Chemical Industries, Inc., Wilmington, Del., 1968.
346. Chip Dip Base, Crest Foods Co., Inc., Ashton, Ill.
347. **Little, L. L.,** U.S. Patent 3,391,002, 1968.
348. Frozen Dessert Base, Non-Dairy, Crest Foods Co., Inc., Ashton, Ill.
349. Frozen Dessert (Containing No Milk Products), LDS-288-D, Durkee Famous Foods, Chicago, Ill.
350. **Ryberg, J. R.,** *Food Prod. Dev.*, 2(5), 60, 1968.
351. **Webster, D. P.,** U.S. Patent 3,310,406, 1967.
352. **Terada, K. and Yoshida, K.,** Japanese Patent 9261, 1962; *C. A.*, 58, 11900d, 1963.
353. Paragraph 19.530(b)(5), *Code of Federal Regulations, Title 21*, U.S. Government Printing Office, Washington, D.C., 1971.
354. **Bristol, D. C. and Martin, J. H.,** *J. Dairy Sci.*, 53, 1381, 1970.
355. Continuous Cottage Cheese Process, *Form 744-8046*, CP Division, St. Regis, Chicago, Ill.
356. **Kielsmeier, E. W.,** U.S. Patent 2,851,363, 1958.
357. **Vorob'ev, A. I.,** *Proc. 16th Intern. Dairy Congr.*, Copenhagen, 1962, Sect. B, 576, *Dairy Sci. Abstr.*, 24, 2778, 1962.
358. **Anon.,** *Food Eng.*, 40, 15, May, 1968.
359. **Olson, N. F.,** *Food Prod. Dev.*, 2(5), 90, 1968.
360. **Batra, S. C. and deMan, J. M.,** *J. Dairy Sci.*, 47, 673, 1964.
361. **deMan, J. M. and Batra, S. C.,** *J. Dairy Sci.*, 47, 954, 1964.
362. **Odagiri, S. and Nickerson, T. A.,** *J. Dairy Sci.*, 47, 673, 1964.

363. deMan, J. M., *Proc. 17th Intern. Dairy Congr.*, 2, 365, 1966; *C. A.*, 67, 2210a, 1967.
364. Samuelsson, E. G., *Nord. Mejeri-Tidsskrift*, 35(1), 9, 1969; *Food Sci. Technol. Abstr.*, 1, 5P497, 1969.
365. Shew, D. I., *Nature*, 164, 492, 1949.
366. Potter, N. N. and Nelson, F. E., *J. Bacteriol.*, 64, 113, 1952.
367. Babel, F. J., *J. Dairy Sci.*, 41, 697, 1958.
368. Hargrove, R. E., *J. Dairy Sci.*, 42, 906, 1959.
369. Olson, H. C., *J. Dairy Sci.*, 43, 439, 1960.
370. Hargrove, R. E., McDonough, F. E., and Tittsler, R. P., *J. Dairy Sci.*, 44, 1799, 1961.
371. Galesloot, T. E. and Hassing, F., *Versl. Ned. Inst. Zuivelonerzoek*, 1962(70), 117.
372. Kadis, V. W. and Babel, F. J., *J. Dairy Sci.*, 45, 486, 1962.
373. Hargrove, R. E., U.S. Patent 3,041,248, 1962.
374. Waters, H. L. H., U.S. Patent 1,910,195, 1933.
375. Schulz, M. E., British Patent 692,055, 1953.
376. Conochie, J. and Sutherland, B. J., *Aust. J. Dairy Technol.*, 16, 227, 1961.
377. Kraft, J. L., U.S. Patent 1,186,524, 1916.
378. Sommer, H. H. and Templeton, H. L., *Dairy Ind.*, 5, 185, 218, 265, 292, 316, 1940.
379. Van Slyke, L. L. and Price, W. V., *Cheese*, Orange Judd Publishing Co., New York, 1952, 355.
380. Irvine, D. M. and Price, W. V., *Process Cheese Abstr.*, Department of Dairy and Food Industries, University of Wisconsin, Madison, Wis., 1955.
381. Garstin, G. H., U.S. Patent 1,368,624, 1921.
382. Ney, K. H. and Garg, O. P., *Fette, Seifen, Anstrichm.*, 72, 279, 1970; *Food Sci. Technol. Abstr.*, 2, 11P1489, 1970.
383. Rohlfs, H. A., Koch, W., Scheurer, (Benckiser-Knapsack G.m.b.H.), British Patent 1,180,716, 1970; *Food Sci. Technol. Abstr.*, 2, 7P864, 1970.
384. Stauffer Chemical Co., British Patent 1,189,003, 1970.
385. Lauck, R. M., Vanstrom, R. E., and Tucker, J. W., U.S. Patent 3,097,949, 1963.
386. Kichline, T. P., Stahlheber, N. E., and Vetter, J. L., U.S. Patent 3,337,347, 1967.
387. Paragraph 19,750(c), *Code of Federal Regulations, Title 21*, U.S. Government Printing Office, Washington, D. C.,1971.
388. Bolanowski, J. P. and Ziemba, J. V., *Food Eng.*, 38(11), 86, 1966.
389. Templeton, H. L. and Sommer, H. H., *J. Dairy Sci.*, 13, 203, 1930.
390. Templeton, H. L. and Sommer, H. H., *J. Dairy Sci.*, 15, 29, 1932.
391. Templeton, H. L. and Sommer, H. H., *Wis. Agr. Exp. Sta. Res. Bull.*, 428, 43, 1934.
392. Templeton, H. L. and Sommer, H. H., *J. Dairy Sci.*, 19, 561, 1936.
393. Templeton, H. L. and Sommer, H. H., *J. Dairy Sci.*, 20, 231, 1937.
394. Templeton, H. L., Sommer, H. H., and Stewart, J. K., *Wis. Agr. Exp. Sta. Res. Bull.*, 439, 66, 1937.
395. Templeton, H. L., *J. Dairy Sci.*, 20, 470, 1937.
396. Templeton, H. L., Sommer, H. H., and Stewart, J. K., *Natl. Butt. Cheese J.*, 29(2), 16, 1938; *C. A.*, 32, 17967, 1938.
397. Palmer, H. J. and Sly, W. H., *J. Soc. Chem. Ind.*, 63, 363, 1944.
398. Holtstorff, A. F., Mularz, V., and Traisman, E., *J. Dairy Sci.*, 34, 486, 1951.
399. Kiermeier, F. and Mohler, K., *Z. Lebensmittel.-Unters. Forsch.*, 112, 175, 1960.
400. Kiermeier, F., *Z. Lebensmittel.-Unters. Forsch.*, 118, 128, 1962.
401. Roesler, H., *Milchwissenschaft*, 21(2), 104, 1966.
402. Scharpf, L. G., Jr. and Kichline, T. P., *J. Agr. Food Chem.*, 15, 787, 1967.
403. Rank, B. and Siebenlist, E., *Deut. Molkerei-Ztg.*, 62, 1036, 1941; *C. A.*, 37, 60455, 1943.
404. Blanchard, J. F., *Food Ind.*, 21, 51, 1949; *C. A.*, 43, 3534c, 1949.
405. Ovchinnikov, A. and Alyamovskii, *Moloch. Prom.*, 13(10), 21, 1952; *C. A.*, 47, 1860h, 1953.
406. Morris, H. A., Manning, P. B., and Jenness, R., *J. Dairy Sci.*, 52, 900, 1969.
407. Sommer, H. H. and Templeton, H. L., *Dairy Ind.*, 6, 14, 1941.
408. Kozin, N. I. and Starodubtsev., N. V., *Sb. Nauch. Rab. Mosk. Inst. Narod. Khoz. 1959*, No. 16, 202.
409. Hayter, T. C., Olson, N. F., and Price, W. V., *J. Dairy Sci.*, 52, 900, 1969.
410. Pikarr, H. R., U.S. Patent 3,211,659, 1965.
411. Chin, R. G. L. and Redfern, S., Canadian Patent 723,609, 1965; U.S. Patent 3,383,221, 1968.
412. Anon., *Fed. Reg.*, 31, 4677, 1966; 33, 8225, 1968.
413. Hall, G. O., U.S. Patent 2,445,879, 1948.
414. Bellamy, A. J., British Patent 570,268, 1945.
415. Tongur, V. S., *J. Appl. Chem.*, 20, 1191, 1947.
416. Sourby, J. C., Kohl, W. F., and Ellinger, R. H., U.S. Patent 3,561,980, 1971; Belgian Patent 726,348, 1969; Netherlands Patent 69-00026, 1969; New Zealand Patent 155,019, 1969; British Patent 1,219,519, 1971; French Patent 1,600,264, 1970.
417. Kohl, W. F. and Ellinger, R. H., U.S. Patent 3,549,388, 1970; Belgian Patent 734,812, 1969; South African Patent 69/3784, 1969; Argentine Patent 176,511, 1970.

418. Kothe, H. J., Canadian Patent 503,484, 1954.
419. Finucane, T. P. and Mitchell, W. A., U.S. Patent 2,671,730, 1954.
420. Finucane, T. P., U.S. Patent 2,916,379, 1959.
421. Chang, P. K., Ph. D. Thesis, University of Wisconsin, Madison, 1969.
422. Chang, P. K., Powrie, W. D., and Fennema, O., *Food Technol.*, 24, 63, 1970.
423. Kohl, W. F., Ellinger, R. H., and Sourby, J. C., A New Process for Pasteurizing Egg Whites, presented before the 30th Annual Meeting of the Institute of Food Technologists, San Francisco, Calif., May 26, 1970.
424. Kohl, W. F., Sourby, J. C., and Ellinger, R. H., U.S. Patent 3,520,700, 1970; Netherlands Patent 69,00027, 1969; Belgian Patent 726,347, 1969; New Zealand Patent 154,947, 1968.
425. Rousseau, P. M., *Oleagineux,* 10, 183, 1955.
426. Lever Bros. Co., British Patent 661,703, 1951.
427. Sullivan, F. E., U.S. Patent 2,702,813, 1955; British Patent 736,885, 1952.
428. Hayes, L. P. and Wolff, H., U.S. Patent 2,881,195, 1959.
429. Rini, S. J., U.S. Patent 2,507,184, 1950.
430. Appleton, L. F., Canadian Patent 330,967, 1933.
431. Eckey, E. W., U.S. Patent 1,993,152, 1935.
432. Johnson, J. Y., British Patent 337,336, 1932.
433. Colomb, P., *Lack-U. Farben-Chem.*, 5(9/12), 14, 1951.
434. Taussky, I., U.S. Patent, 2,654,766, 1953.
435. Braae, B. and Nyman, M., *Nord. Symp. Harskning Fedstoffer,* 2, 231, Elsinore, Denmark, 1957.
436. Bergman, L. O., German Patent 1,135,600, 1962; Belgian Patent 595,217, 1961.
437. James, E. M., U.S. Patent 2,115,668, 1938.
438. Hempel, H., U.S. Patent 2,226,036, 1941.
439. Clayton, B., U.S. Patent 2,292,822, 1942.
440. Mitani, M., U.S. Patent 3,008,972, 1961.
441. Coith, H. S. and Votaw, V. M., U.S. Patents 2,061,121 and 2,061,122, 1936.
442. Irmen, P., British Patent 503,607, 1939.
443. Beck, T. M. and Klein, G. I., U.S. Patent 2,214,520, 1940.
444. Roy, A. C., *J. Sci. Ind. Res.*, (India), 13B, 376, 1954.
445. Julian, P. L. and Iveson, H. T., U.S. Patent 2,752,378, 1956.
446. Morris, S. G., Myers, J. S., Jr., Kip, M. L., and Riemenschneider, R. W., *J. Am. Oil Chem. Soc.*, 27, 105, 1950.
447. Dron, D. W. and Lindsey, F. A., U.S. Patent 2,650,931, 1953.
448. Eckey, E. W., U.S. Patent 1,982,907, 1934.
449. Lever Bros. and Unilever, Ltd., British Patent 577,879, 1946.
450. Lever Bros. and Unilever, Ltd., British Patent 612,169, 1948.
451. Kukn, K., British Patent 742,233, 1955; German Patent 1,012,014, 1957.
452. Tsuchiya, T., *J. Nippon Oil Technol. Soc.*, 1(3), 24, 1948; 2 (2/3), 1, 1949.
453. Koyama, R., Japanese Patent 2584(52), 1952.
454. Loury, M., *Rev. Franc. Corps Gras,* 2, 15, 1955.
455. Skau, E. L., U.S. Patent 2,684,377, 1954.
456. Van Akkeren, L. A. and Ast, H. J., U.S. Patent 2,878,274, 1959.
457. Morris, C. E. and Khym, F. P., U.S. Patent 2,602,807, 1952.
458. Merker, D. R., U.S. Patent 2,783,260, 1957.
459. Merker, D. R., U.S. Patent 2,862,941, 1958.
460. Richardson, A. S. and Eckey, E. W., U.S. Patent 2,132,437, 1938.
461. Feuge, R. O. and Gros, A. T., *J. Am. Oil Chem. Soc.*, 27, 117, 1950.
462. Van Akkeren, L. A., U.S. Patent 3,284,213, 1966.
463. Richardson, A. S., Vibrans, F. C., and Andrews, J. T. R., U.S. Patent 1,993,181, 1935.
464. Musher, S., U.S. Patents 2,314,364, and 2,314,365, 1943.
465. Calkins, V. P., *J. Am. Chem. Soc.*, 69, 384, 1947.
466. Kraybill, H. R., Dugan, L. R., Jr., Beadle, B. W., Vibrans, F. C., et al., *J. Am. Oil Chem. Soc.*, 26, 449, 1949.
467. Kraybill, H. R. and Beadle, B. W., U.S. Patent 2,451,748, 1948.
468. Beadle, B. W. and Kraybill, H. R., U.S. Patent 2,648,608, 1953.
469. Kring, P., *Dan. Tidsskr. Farm.*, 24, 211, 1950; abstracted in *J. Am. Oil Chem. Soc.*, 28 (3), 123, 1951.
470. Morris, S. G., Gordon, C. F., Brenner, N., Meyers, J. S., Jr., Riemenschneider, R. W., and Ault, W. C., *J. Am. Oil Chem. Soc.*, 29, 441, 1952.
471. Lips, H. J., *Food Can.*, 12(6), 9, 1952.
472. Privett, O. S. and Quackenbush, F. W., *J. Am. Oil Chem. Soc.*, 31, 225, 1954; 31, 321, 1954.
473. Hall, L. A., U.S. Patent 2,511,803, 1950.
474. Magoffin, J. E., U.S. Patent 2,607,746, 1952; British Patent 679,192, 1952; Canadian Patent 511,767, 1955.
475. Abe, Y., Japanese Patent 5323('53), 1953.
476. Gleim, W. K. T., U.S. Patent 2,694,645, 1954.

477. **Sumiki, Y. and Tamura, S.,** Japanese Patent 8521 ('54), 1954.
478. **Bentz, R. W.,** U.S. Patent 2,772,243, 1956.
479. **Mattil, K. F. and Sims, R. J.,** U.S. Patent 2,759,829, 1956.
480. **Brickman, A. W., Conquest, V., Madden, F. J., Filbey, E. T., and Oleson, W. B.,** British Patents 754,388 and 754,389, 1956.
481. **Thompson, R. B.,** U.S. Patent 2,746,871, 1956.
482. **Griffith, C. L. and Sair, L.,** Canadian Patent 531,497, 1956.
483. **Kuhrt, N. H.,** U.S. Patent 2,665,991, 1954; Canadian Patent 540,553, 1957; German Patent 1,042,363, 1958.
484. **Kuhrt, N.,** U.S. Patent 2,681,281. 2,681,283, 1954; U.S. Patents 2,701,769, 2,701,770, 1955; British Patent 731,177, 1955.
485. **Watts, B. M., Lehmann, B., and Goodrich, F.,** *J. Am. Oil Chem. Soc.*, 26, 481, 1949.
486. **Lea, C. H.,** *Rancidity in Edible Fats,* Chemical Publishing Co., New York, 1950.
487. **Watts, B.,** *J. Am. Oil Chem. Soc.*, 27, 48, 1950.
488. **Lehmann, B. T. and Watts, B. M.,** *J. Am. Oil Chem. Soc.*, 28, 475, 1951.
489. **Watts, B. M. and Moss, H. V.,** U.S. Patent 2,629,664, 1953.
490. **Lehmann, B. T. and Watts, B. M.,** U.S. Patent 2,707,154, 1955; Canadian Patent 538,129, 1957.
491. **Ozawa, T. and Ota, A.,** *Shokuhin Eiseigaku Zasshi,* 6(1), 10, 1965; C. A., 62, 16877g, 1965.
492. **Holman, G. W.,** U.S. Patent 2,871,130, 1959; German Patent 1,236,361, 1967.
493. **Mahon, J. H. and Chapman, R. A.,** *J. Am. Oil Chem. Soc.*, 31, 108, 1954.
494. **van Hees, G.m.b.H.,** German Patent 919,779, 1954.
495. Modernas Aplicaciones de la Refrigeracion Industrial, S. A., Spanish Patent 217,490, 1955.
496. **Mitchell, J. H.,** U.S. Patent 2,511,119, 1950.
497. **Vincent, J. F.,** U.S. Patent 2,511,136, 1950.
498. **Langendorf, H. and Lang, K.,** *Z. Lebensmittel – Unters. Forsch.*, 115, 400, 1961.
499. **Benk, E.,** *Flussiges Obst.*, 36, 468, 1969; *Food Sci. Technol. Abstr.*, 2, 5H511, 1970.
500. **Diemair, W. and Pfeifer, K.,** *Z. Lebensmittel – Unters. Forsch.*, 117, 209, 1962.
501. **Laucks, I. F., Banks, H. P., and Rippey, H. F.,** U.S. Patent 1,732,816, 1929.
502. **Bates, H. R.,** U.S. Patent 1,774,310, 1930.
503. **Rippey, H. F.,** U.S. Patent 1,935,599, 1933.
504. **Kalmar, A. F.,** U.S. Patent 2,374,209, 1945.
505. **Kalmar, A. F.,** U.S. Patent 2,374,210, 1945.
506. **Smith, J. D. and Krause, H. J.,** U.S. Patent 2,700,613, 1955.
507. **Brandt, K.,** German Patent 1,417,432, 1968.
508. **Post, F. J., Coblentz, W. S., Chou, T. W., and Salunhke, D. K.,** *Appl. Microbiol.*, 16, 138, 1968.
509. **Kohl, W. F. and Ellinger, R. H.,** Polyphosphates as Preservatives for Apple Cider, paper read before the 30th Annual Meeting of the Institute of Food Technologists, San Francisco, Calif., May 26, 1970.
510. **Deobald, H. J., McLemore, T. A., Bertoniere, N. R., and Martinez, J. A.,** *Food Technol.*, 18, 1970, 1964.
511. **Maggi, A. G.,** Netherlands Patent Appl. 6,600,754, 1966; *C. A.*, 66, 1740c, 1967.
512. **Rhee, K. S. and Watts, B. M.,** *J. Food Sci.*, 31, 669, 1966.
513. **Molsberry, C. C.,** U.S. Patent 3,342,610, 1967.
514. **Hanada, S., Yokoo, Y., Suzuki, E., Yamaguchi, N., and Yoshida, T.,** *Eiyo To Shokurya,* 11, 306, 1959; *C. A.*, 58, 2097f, 1963.
515. **Feldheim, W. and Seidemann, J.,** *Fruchtsaft-Ind.*, 7, 166, 1962; *C. A.*, 57, 14250c, 1962.
516. **Hall, G. O.,** U.S. Patent 2,478,266, 1949.
517. **Nelson, R. F. and Finkle, B. J.,** *Phytochemistry,* 3, 321, 1964.
518. **Bolin, H. R., Nury, F. S., and Finkle, B. J.,** *Bakers Dig.*, 38(3), 46, 1964.
519. **Fleischman, D. L.,** *Dissertation Abstracts International, Section B, The Sciences and Engineering,* 30(2), 696, 1969.
520. **Michels, P.,** German Patent 1,082,792, 1962; *C. A.*, 56, 3872i, 1962.
521. **Smith, O.,** *Potatoes: Production, Storing, Processing,* Avi Publishing Co., Westport, Conn., 1968.
522. **Juul, F.,** Thesis, *I Kommission Hos Jul. Gjellerups Forlag,* Copenhagen, Denmark, 1949; Abstracted in *Am. Potato J.*, 27, 32, 1950; *Biol. Abstr.*, 24, 3610, 11938, 1950.
523. Stauffer Chemicals for Modern Potato Processing, *Food Industry Release No. 5.*, Stauffer Chemical Co., New York.
524. **Smith, O.,** *Am. Potato J.*, 35, 573, 1958.
525. **Smith, O. and Davis, C. O.,** *Am. Potato J.*, 39, 45, 1962.
526. **Talburt, W. F. and Smith, O.,** *Potato Processing,* Avi Publishing Co., Westport, Conn., 1967.
527. **Talley, E. A., Zaehringer, M. V., Reeve, R. M., Hyde, R. B., Dinkel, D. H., Heisler, E. G., and Pressey, R.,** *Am. Potato J.*, 46, 302, 1969.
528. **Hoover, M. W.,** *Food Technol.*, 17, 636, 1963.
529. **Kertesz, Z. I.,** *Canner,* 88(7), 26, 1939.
530. **Baker, G. L. and Woodmansee, C. W.,** *Fruit Prod. J.*, 23(6), 164, 1944.
531. **Baker, G. L. and Gilligan, G. M.,** *Food Packer,* 27(6), 56, 1946.
532. **Baker, G. L. and Murray, W. G.,** *Food Res.*, 12, 129, 1947.

533. McCready, R. M., Shepherd, A. D., and Maclay, W. D., *Fruit Prod. J.*, 27, 36, 1947.
534. Woodmansee, C. W. and Baker, G. L., *Del. Agr. Exp. Sta. Bull. No. 305*, 5, 1954.
535. Waller, G. R. and Baker, G. L., *J. Agr. Food Chem.*, 1, 1213, 1953.
536. Woodmansee, C. W., *Del. Agr. Exp. Sta. Circular No. 5.*, University of Delaware, Newark, Del., 1967.
537. Olsen, A. G. and Fehlberg, E. R., U.S. Patent 2,334,281, 1943.
538. Leo, H. T. and Taylor, C. C., U.S. Patents 2,483,548, and 2,483,549, 1949.
539. Leo, H. T., Taylor, C. C., and Lindsey, J. W., U.S. Patent 2,483,550, 1949.
540. Pedersen, K., U.S. Patent 2,540,545, 1951.
541. Briccoli-Bati, M., Italian Patent 472,159, 1952.
542. Avoset Co., British Patent 748,402, 1956.
543. Leo, H. T. and Taylor, C. C., U.S. Patent 2,824,007, 1958.
544. Baker, G. L. and Gilligan, G. M., *Fruit Prod. J.*, 26(9), 260, 1947.
545. Morse, R. E., *Food Packer*, 33(6), 30, 1952.
546. Peters, G. L., Brown, H. D., Gould, W. A., and Davis, R. B., *Food Technol.*, 8, 220, 1954.
547. Wagner, J. R. and Miers, J. C., *Food Technol.*, 21, 920, 1967; Wagner, J. R., Miers, J. C., Sanshuck, D. W., and Becker, R., *Food Technol.*, 22, 1484, 1968.
548. Smith, O. and Davis, C. O., *Am. Potato J.*, 40, 67, 1963.
549. Cole, M. S., U.S. Patent 3,219,464, 1965.
550. Kintner, J. A. and Tweedy, E., *Food Technol.*, 21, 865, 1967.
551. Nielsen, J. P., Campbell, H., and Boggs, M., *West. Can. Pack.*, 35(7), 49, June, 1943.
552. Holmquist, J. W., Schmidt, C. F., and Guest, A. E., *Can. Trade*, 70(40), 7, 1948.
553. Neubert, A. M. and Carter, G. H., *Food Technol.*, 8, 518, 1954.
554. Anon., *Food Eng.*, 27, 188, Nov., 1955 .
555. Blakemore, S. M., U.S. Patent 3,307,954, 1967.
556. Loconti, J. O. and Kertesz, Z. I., *Food Res.*, 6, 499, 1941.
557. Powers, J. J. and Esselen, W. B., Jr., *Fruit Prod. J.*, 25(7), 200, 1946.
558. Van Buren, J. P., *Food Technol.*, 22, 790, 1968.
559. Anon., *Code of Federal Regulations, Title 21*, Parts 27, 51, 53, U.S. Government Printing Office, Washington, D. C., 1971.
560. Anon., *Food Chem. News*, 18, Sept. 15, 1969 .
561. Dakin, J. C., *Nature*, 199, 383, 1963.
562. Takeda Chemical Industries, Ltd., Japanese Patent 8619/70, 1970; *Food Sci. Technol. Abstr.*, 2, 11T414, 1970.
563. Peschardt, W. J. and Hume, S. R., West German Patent Application 1,280,646, 1968; *Food Sci. Technol. Abst.*, 1(5), 5J415, 1969.
564. Glicksman, M., *Adv. Food Res.*, 11, 110, 1962.
565. Sergeeva, Z. I., *Tr. Vses. Nauch.-Issled. Inst. Konditersk. Prom.*, No. 15, 150, 1960; *C. A.*, 57, 384le, 1962.
566. Lucas, H. J., U.S. Patent 2,097,228, 1937.
567. Steiner, A. B., U.S. Patent 2,441,729, 1948.
568. Steiner, A. B., U.S. Patent 2,485,934, 1949.
569. Roland, C. T., U.S. Patent 2,665,211, 1954.
570. Gibsen, K. F., U.S. Patent 2,808,337, 1957.
571. Gibsen, K. F., U.S. Patent 2,918,375, 1959.
572. Rocks, J. K., U.S. Patent 2,925,343, 1960.
573. Hunter, A. R. and Rocks, J. K., U.S. Patent 2,949,366, 1960.
574. Merton, R. R. and McDowell, R. H., U.S. Patent 2,930,701, 1960; British Patent 828,350, 1960.
575. Hunter, A. R., U.S. Patent 2,987,400, 1961.
576. Miller, A. and Rocks, J. K., U.S. Patent 3,266,906, 1966; Canadian Patent 811,682, 1969.
577. Andrew, T. R. and MacLeod, W. C., *Food Prod. Dev.*, 4(5), 99, 1970.
578. Stoloff, L. S., U.S. Patent 2,801,923, 1957.
579. Stoloff, L. S., U.S. Patent 2,864,706, 1958.
580. Zabik, M. E. and Aldrich, P. J., *J. Food Sci.*, 30, 111, 1965.
581. Weinstein, B., U.S. Patent 2,856,289, 1958.
582. Green, J., Schuller, E. J., Rickert, J. A., and Borders, B., U.S. Patent 2,992,925, 1961.
583. Marcus, J. K., U.S. Patent 2,605,229, 1952.
584. Grau, R., *Kondensierte Phosphate in Lebensmitteln*, Symposium, 1957, Mainz, Springer-Verlag, Berlin, 1958, 89.
585. Hamm, R., *Adv. Food Res.*, 10, 355, 1960.
586. Saffle, R. L., *Adv. Food Res.*, 16, 105, 1968.
587. Karmas, E., *Fresh Meat Processing*, Noyes Data Corp., Park Ridge, N. J., 1970.
588. American Meat Institute Foundation, *The Science of Meat and Meat Products*, W. H. Freeman & Co., San Francisco, 1960.
589. McLoughlin, J. V., *Food Mfr.*, 44(1), 36, 1969.
590. Kotter, L., *Die Fleischwirtschaft*, 9, 697, 1957.

591. Kotter, L., *Die Fleischwirtschaft,* 13, 186, 1961.
592. Grau, R., Hamm, R., and Baumann, A., *Biochem. Z.*, 325, 1, 1953.
593. Bendall, J. R., *J. Sci. Food Agr.*, 5, 468, 1954.
594. Fukazawa, T., Hashimoto, Y., and Yasui, T., *J. Food Sci.*, 26, 541, 1961.
595. Fukazawa, T., Hashimoto, Y., and Yasui, T., *J. Food Sci.*, 26, 550, 1961.
596. Yasui, T., Sakanishi, M., and Hashimoto, Y., *J. Agr. Food Chem.*, 12, 392, 1964.
597. Yasui, T., Fukazawa, T., Takahashi, K., Sakanishi, M., and Hashimoto, Y., *J. Agr. Food Chem.*, 12, 399, 1964.
598. Klepacka, M., *Przem. Spozyw.*, 19(8), 501, 1965; *C. A.*, 65, 17596a, 1966.
599. Grau, R., Hamm, R., and Baumann, A., *Naturwissenschaften,* 40, 535, 1953.
600. Morse, R. E., *Food Eng.*, 27(10), 84, 1955.
601. Epstein, F., U.S. Patent 2,876,115, 1959.
602. Kotter, L., *Zur Wirkung Kondensierter Phosphate Und Anderer Salze Auf Tierisches Eiweiss,* M. & H. Schaper, Hannover, Germany, 1960.
603. Hellendoorn, E. W., *Food Technol.*, 16(9), 119, 1962.
604. Inklaar, P. A., *J. Food Sci.*, 32, 525, 1967.
605. Baldwin, T. T. and deMan, J. M., *J. Inst. Can. Technol. Aliment.*, 1, 164, 1968.
606. Turner, E. W. and Olson, F. C., U.S. Patent 2,874,060, 1959.
607. Saffle, R. L. and Galbreath, J. W., *Food Technol.*, 18, 1943, 1964.
608. Acton, J. C. and Saffle, R. L., quoted in Saffle, R. L., *Adv. Food Res.*, 16, 105, 1968.
609. Kotter, L., *Kondensierte Phosphate in Lebensmitteln,* Symposium, 1957, Mainz , Springer-Verlag, Berlin, 1958, 99.
610. Hansen, L. J., *Food Technol.*, 14, 565, 1960.
611. Swift, C. E., Lockett, C., and Fryar, A. J., *Food Technol.*, 15, 468, 1961.
612. Hegarty, G. R., Bratzler, L. J., and Pearson, A. M., *J. Food Sci.*, 28, 663, 1963.
613. Carpenter, J. A. and Saffle, R. L., *Food Technol.*, 19, 1567, 1965.
614. Trautman, J. C., *Food Technol.*, 18, 1065, 1964.
615. Swift, C. E., *Proc. Meat Ind. Res. Conf.*, Kolari, O. E. and Aunan, W. J., Eds., American Meat Institute Foundation, Chicago, 1965, 78.
616. Ellinger, R. H. and Harkey, C. N., unpublished work, 1960.
617. Mahon, J. H., *Proc. 30th Res. Conf., Amer. Meat Inst. Found.*, No. 64, 59, 1961.
618. Sherman, P., *Food Technol.*, 15, 79, 1961.
619. Depner, M., *Fleischwirtschaft,* 1, 252, 1949.
620. Hamm, R., *Naturwissenschaften,* 42, 394, 1955.
621. Hamm, R., *Fleischwirtschaft,* 7, 196, 1955.
622. Hamm, R., *Fleischwirtschaft,* 8, 266, 1956.
623. Hamm, R., *Fleischwirtschaft,* 8, 340, 1956.
624. Hamm, R., *Z. Lebensmittel-Unters. Forsch.*, 106, 281, 1957.
625. Hamm, R., *Z. Lebensmittel-Unters. Forsch.*, 107, 428, 1958.
626. Hamm, R. and Grau, R., *Z. Lebensmittel-Unters. Forsch.*, 108, 280, 1958..
627. Hamm, R., *Natl. Prov.*, 140 (15), 78, 1959.
628. Hamm, R., *Z. Lebensmittel-Unters. Forsch.*, 110, 95, 1959.
629. Swift, C. E. and Berman, M. D., *Food Technol.*, 13, 365, 1959.
630. Wierbicki, E., Kunkle, L. E., and Deathrage, F. E., *Fleischwirtschaft,* 15, 404, 1963.
631. Swift, C. E. and Ellis, R., *Food Technol.*, 10, 546, 1956.
632. Wierbicki, E., Tiede, M. G., and Burrell, R. C., paper presented at the 19th Annual Meeting, Institute of Food Technologists, Philadelphia, May 17-21, 1959.
633. Wismer-Pedersen, J., *J. Food Sci.*, 30, 85, 1965.
634. Wismer-Pedersen, J., *J. Food Sci.*, 30, 91, 1965.
635. Popp, H. and Muhlbrecht, F. N., *J. Food Sci.*, 10, 399, 1958.
636. Watts, B. M., *Adv. Food Res.*, 5, 1, 1954.
637. Fox, J. B., *J. Food Agr. Chem.*, 14, 207, 1966.
638. Hall, G. O., U.S. Patent 2,513,094, 1950.
639. Brissey, G. E., U.S. Patent 2,596,067, 1952.
640. Chang, I. and Watts, B. M., *Food Technol.*, 3, 332, 1949.
641. Bendall, J. R., *Nature* (London), 172, 586, 1953.
642. Fukazawa, T., Hashimoto, Y., and Yasui, T., *J. Food Sci.*, 26, 331, 1961.
643. Giacino, C., U.S. Patent 3,394,015, 1968.
644. Keil, H. L., Hagen, R. F., and Flaws, R. W., U.S. Patent 2,953,462, 1960.
645. Kamstra, L. D. and Saffle, R. L., *Food Technol.*, 13, 652, 1959.
646. Carpenter, J. A., Saffle, R. L., and Kamstra, L. D., *Food Technol.*, 15, 197, 1961.
647. Voegeli, M. M. and Gorsica, H. J., U.S. Patent 3,154,421, 1964.
648. Voegeli, M. M. and Gorsica, H. J., U.S. Patent 3,154,423, 1964.
649. Rust, M. E., Ph.D. Thesis, Kansas State University, 1963.

650. Allen, H. E., U.S. Patent 2,140,781, 1938.
651. Mullins, A. M., Kelley, G. G., and Brady, D. E., *Food Technol.*, 12, 227, 1958.
652. Hopkins, E. W. and Zimont, L. J., U.S. Patent 2,999,019, 1961.
653. Williams, B. E., U.S. Patent 3,006,768, 1961.
654. Komarik, S. L., U.S. Patent 3,147,123, 1964.
655. Brown, W. L., Denny, D., and Schmucker, M. L., U.S. Patent 3,207,608, 1965.
656. Delaney, W., U.S. Patent 3,188,213, 1965.
657. Huffman, D. L., Palmer, A. Z., Carpenter, J. W., and Shirley, R. L., *Quart. J. Fla. Acad. Sci.*, 28, 106, 1965; *J. Anim. Sci.*, 28, 443, 1969.
658. Miller, E. M. and Harrison, D. L., *Food Technol.*, 19, 94, 1965.
659. Hopkins, E. W. and Zimont, L. J., U.S. Patent 3,049,428, 1962.
660. Unilever, N. V., Netherlands Patent Application 6,506,834, 1965.
661. Brendl, J., Klein, S., and Kocianova, M., *Sb. Vys. Sk. Chem.-Technol. Praze, Potrav.*, 14, 49, 1967; *C. A.*, 67, 115889x, 1967.
662. Tims, M. J. and Watts, B. M., *Food Technol.*, 12, 240, 1958.
663. Tarladgis, B. G., Younathan, M. T., and Watts, B. M., *Food Technol.*, 13, 635, 1959.
664. Chang, P., Younathan, M. T., and Watts, B. M., *Food Technol.*, 15, 168, 1961.
665. Ramsey, M. B. and Watts, B. M., *Food Technol.*, 17, 1056, 1963.
666. Zipser, M. W., Kwon, T., and Watts, B. M., *J. Agr. Food Chem.*, 12, 105, 1964.
667. Watts, B. M., Proc. Flavor Chem. Symp., Campbell Soup Co., Camden, N. J., 1961, 83.
668. Watts, B. M., Meat products, in *Lipids and Their Oxidation,* Academic Press, New York, 1962, Chap. 11.
669. Savich, A. L. and Jansen, C. E., U.S. Patent 2,830,907, 1958.
670. Ziegler, J. A., U.S. Patent 3,073,700, 1963.
671. Jacoby, D. and Berhold, R. J., U.S. Patent 3,469,995, 1969.
672. Bickel, W., U.S. Patent 3,029,150, 1962.
673. Fetty, H. J., U.S. Patent 3,216,827, 1965.
674. Sair, L. and Cook, W. H., *Can. J. Res.,* D16, 255, 1938.
675. Sherman, P., *Food Technol.*, 16, 91, 1962.
676. Zwanenberg's Fabrieken N. V., Netherlands Patent 62, 273, 1948.
677. Anon., *Food Eng.*, 25, 214, Feb., 1953.
678. Wilson, G. D., Proc. Res. Report to Industry, March 24, 1954, American Meat Institute Foundation, Chicago, 1954, 32.
679. Mahon, P., Hogue, D., Leeking, P., Lim, E., and Fenton, F., *Food Technol.,* 10, 265, 1956.
680. Fenton, F., Sheffy, B. E., Naumann, H. D., Wellington, G. H., Hogue, D., and Mahon, P., *Food Technol.*, 10, 272, 1956.
681. Wilson, G. D., *Natl. Prov.*, 131, 112, 1954.
682. Watts, B. M., Proc. 9th Res. Conf., American Meat Institute, Chicago, 1957, 61.
683. Swift, C. E. and Ellis, R., *Food Technol.,* 11, 450, 1957.
684. Wasserman, M., U.S. Patent 2,812,261, 1957.
685. Gisske, W., *Fleischwirtschaft,* 10, 21, 1958.
686. Schoch, W. S., U.S. Patent 2,824,809, 1958.
687. Suri, B. R., *Fleischwirtschaft*, 13, 403, 1961.
688. Bailey, M. E., Frame, R. W., and Naumann, H. D., *J. Agr. Food Chem.*, 12, 89, 1964.
689. Siedlecki, E., *Gospodarka Miesna,* 16(1), 7, 1964; *C. A.*, 62, 1009h, 1965.
690. Lauck, R. M. and Tucker, J. W., U.S. Patent 3,118,777, 1964.
691. Savic, I., Suvakov, M., Nikolic, J., and Korolija, S., *Technol. Mesa*, 6(10), 274, 1965; *C. A* , 64, 14868h, 1966.
692. Rahelic, S., Rede, R., and Jovicic, V., *Technol. Mesa*, 7(9), 252, 1966; *C. A.*, 67, 36667z, 1967.
693. Sair, L. and Komarik, S. L., U.S. Patent 3,391,007, 1968.
694. Swift, C. E., Weir, C. E., and Hankins, O. G., *Food Technol.*, 8, 339, 1954.
695. Szent-Gyorgyi, A., Proc. 4th Research Conference, American Meat Institute, Chicago, 1952, 22.
696. Szent-Gyorgyi, A. G., *Proteins of the Myofibril in Structure and Function of Muscle,* Vol. 2, Bourne, G. H., Ed., Academic Press, New York, 1960, 1.
697. Schleich, H. and Arnold, R. S., U.S. Patent 3,037,870, 1962.
698. Maas, R. H. and Olson, F. C., U.S. Patent 3,075,843, 1963.
699. Leeking, P., Mahon, P., Hogue, D., Lim, E., and Fenton, F., *Food Technol.,* 10, 274, 1956.
700. Frank, S. S., M. S. Thesis, Cornell University, Ithaca, N. Y., 1955.
701. Swift, C. E. and Ellis, R., *Food Technol.*, 10, 546, 1956.
702. Volovinskaya, V. P., Rubashkina, S. S., Solov'ev, V. I., and Dyklop, V. K., *Tr. Vses. Nauch.-Issled. Inst. Myasnoi Prom.,* 1958(8), 78, 1958; *C. A.*, 55, 18822a, 1960.
703. Volovinskaya, V. P., Rubashkina, S. S., Poletaev, T., Kel'man, B., and Merkulova, V., *Myasnaya Ind. S.S.S.R.*, 30(4), 48, 1959; *C. A.*, 55, 9712f, 1961.

704. **Rupp, V. R., Brockmann, M. C., and Nickolson, L. W.,** U.S. Patent 2,937,094, 1960; Canadian Patent 594,779, 1960.
705. **Zeller, M.,** *Deut. Lebensm.-Rundsch.*, 57, 205, 1961.
706. **Grau, R.,** *Fleischwirtschaft*, 13, 183, 1961.
707. **Mahon, J. H. and Schneider, C.,** Canadian Patent 788,939, 1968; Netherlands Appl. 6,600,579, 1966.
708. **Rongey, E. H. and Bratzler, L. J.,** *Food Technol.*, 20, 1228, 1966.
709. **Keller, H.,** *Fleischwirtschaft,* 7, 15, 1955.
710. **Sherman, P.,** *Food Technol.*, 15, 87, 1961.
711. **Pfaff, W.,** British Patent 881,397,1961.
712. **Zipser, M. W. and Watts, B. M.,** *J. Agr. Food Chem.*, 15, 80, 1967.
713. **Evans, J. B.,** *Proc. Res. Report to Industry,* American Meat Institute Foundation, Chicago, 1954, 12.
714. **Bickel, W.,** U.S. Patent 2,735,776, 1956.
715. **Wollmann, E.,** *Fleischwirtschaft,* 11, 291, 1959.
716. **Peat, M. R.,** U.S. Patent 2,925,344, 1960.
717. **Gorsica, H. J.,** U.S. Patent 3,008,832, 1961.
718. **Hall, L. A.,** U.S. Patent 2,145,417, 1939.
719. **Hall, L. A.,** U. S. Patent 2,688,771, 1954.
720. **Hall, L. A. and Kalchbrenner, W. S.,** U.S. Patents 2,770,548 and 2,770,551, 1956.
721. **Hall, L. A.,** U.S. Patent 2,770,549, 1956.
722. **Hall, L. A. and Griffith, C. L.,** U.S. Patent 2,770,550, 1956.
723. Chemische Fabrik Budenheim A. G., British Patent 754,406, 1956.
724. **Huber, H. and Vogt, K.,** U.S. Patent 2,852,392, 1958; British Patent 696,617, 1953.
725. **Barnett, H. W.,** U.S. Patent 2,903,366, 1959.
726. **Barth, W.,** German Patent 1,041,338, 1961.
727. **Miyake, M. and Noda, H.,** Japanese Patent 9134, 1961, *C. A.*, 56, 6437a, 1962.
728. Chemische Fabrik, Joh. A. Benckiser G.M.B.H. British Patent 897,682, 1962.
729. **Buchholz, K.,** U.S. Patent 3,032,421, 1962.
730. **Oliver, R. J. and Voegeli, M. M.,** U.S. Patent 3,028,246, 1962.
731. **Pfrengle, O.,** German Patent 1,181,030, 1964.
732. **Flesch, P. and Bauer, G.,** French Patent 1,446,678, 1966.
733. **Haynes, R. D.,** U.S. Patent 2,868,654, 1959.
734. **Bickel, W.,** German Patent 1,074,380, 1960.
735. Chemische Fabrik Budenheim, A. G., British Patent 831,132, 1960.
736. **Elder, E. V., Jr.,** U.S. Patent 3,104,978, 1963.
737. **Sair, L. and Donahoo, W. P.,** U.S. Patent 3,139,347, 1964.
738. **Sair, L.,** U.S. Patent 3,231,392, 1966.
739. **Mahon, J. H.,** U.S. Patent 3,401,046, 1968.
740. **Murphy, L. W. and Jones, R. L.,** British Patent 862, 109, 1961.
741. **Beerend, R. F. and Needle, H. C.,** U.S. Patent 3,192,055, 1965.
742. **Takeda Chemical Industries, Ltd.,** French Patent 1,386,675, 1965.
743. **deFremery, D. and Pool, M. F.,** *Food Res.*, 25, 73, 1960.
744. **deFremery, D.,** *Proc. Meat Tenderness Symposium,* Campbell Soup Co., Camden, N. J., 1963, 99.
745. **deFremery, D.,** *J. Agr. Food Chem.*, 14, 214, 1966.
746. **Wells, G. H., May, K. N., and Powers, J. J.,** *Food Technol.* 16, 137, 1962.
747. **Klose, A. A., Campbell, A. A., and Hanson, H. L.,** *Poult. Sci.*, 42, 743, 1963.
748. **Mahon, J. H.,** *Poult. Proc. Market.*, 68(8), 16, 1962.
749. **Mahon, J. H.,** *Food Eng.*, 34(11), 108, 1962.
750. **Mahon, J. H.,** U.S. Patent 3,104,170, 1963; Belgian Patent 614,742, 1962; Canadian Patent 691,480 1963; Netherlands Patent Appl. 6,613,628, 1967.
751. **Froning, G. W.,** *Poult. Sci.*, 44, 1104, 1965.
752. **Froning, G. W.,** *Poult. Sci.*, 45, 185, 1966.
753. **Dodge, J. W. and Stadelman, W. J.,** *Food Technol.*, 13, 81, 1959.
754. **May, K. N., Helmer, R. L., and Saffle, R. L.,** *Poult. Sci.*, 41, 1665, 1962.
755. **Klose, A. A., Campbell, A. A., and Hanson, H. L.,** *Poult. Sci.*, 41, 1655, 1962.
756. **Spencer, J. V. and Smith, L. E.,** *Poult. Sci.*, 41, 1685, 1962.
757. **Baker, R. C. and Darfler, J.,** *Poult. Sci.*, 47, 1590, 1968.
758. **Schwall, D. V., Rogers, A. B., and Corbin, D.,** U.S. Patent 3,399,063, 1968.
759. **Schwall, D. V. and Rogers, A. B.,** U.S. Patent 3,285,753, 1966; British Patent 1,198,670, 1970.
760. **Schwall, D. V. and Rogers, A. B.,** U.S. Patent 3,413,127, 1968.
761. **Schnell, P. G., Vadehra, D. V., and Baker, R. C.,** *J. Can. Inst. Food Technol.*, 3,(2), 44, 1970.
762. **Vadehra, D. V. and Baker, R. C.,** *Food Technol.*, 24, 766, 1970.
763. **Schlamb, K. F.,** U.S. Patent 3,499,767, 1970.

764. Mountney, G. J. and Arganosa, F. C., *Poultr. Sci.*, 41, 1668, 1962.
765. Schermerhorn, E. P. and Stadelman, W. J., *Poult. Sci.*, 41, 1680, 1962.
766. Monk, J. A., Mountney, G. J., and Prudent, I., *Food Technol.*, 18, 226, 1964.
767. Schermerhorn, E. P. and Stadelman, W. J., *Food Technol.*, 18, 101, 1964.
768. Mahon, J. H., U.S. Patent 3,462,278, 1969; Netherlands Patent Appl. 6,613,627, 1967.
769. Montemayor, E., Vadehra, D. V., and Baker, R. C., quoted in Vadehra, D. V. and Baker, R. C., *Food Technol.*, 24, 766, 1970.
770. Hudspeth, J. P. and May, K. N., *Food Technol.*, 21, 1141, 1967.
771. Marion, W. W. and Forsythe, R. H., *Poult. Sci.*, 41, 1663, 1962; *J. Food Sci.*, 29, 530, 1964.
772. Thomson, J. E., *Food Technol.*, 18, 1805, 1964.
773. Jimenez, A. V., Fernandez, R. P., and Girauta, M. E., *Symp. Substances Entrangeres Aliments, 6, Madrid, 1960*, 1961, 530; *C. A.*, 59, 3256e, 1963.
774. Melnick, D., Perry, G. A., and Lawrence, R. L., U.S. Patent 3,065,084, 1962.
775. Kraft, A. A. and Ayres, J. C., *Appl. Microbiol.*, 9, 549, 1961; *J. Food Sci.*, 29, 218, 1964.
776. Tomiyama, T., Kitahara, K., Shiraishi, E., and Imaizumi, K., *Intern. Symp. Food Microbiol., 4th Gotborg, Swed.*, 1964, 261.
777. Elliott, R. P., Straka, R. P., and Garibaldi, J. A., *Appl. Microbiol.*, 12, 517, 1964.
778. Linko, R. R. and Nikkila, O. E., *J. Food Sci.*, 26, 606, 1961.
779. Kuusi, T., Nikkila, O. E., and Kytokangas, R., *Valtion Tek. Tutkimuslaitos, Tiedotus, Sarja IV, No. 78*, 1, 1965, 1; *C A.*, 64, 20525a, 1966.
780. Nikkila, O. E., Kuusi, T., and Kytokangas, R., *J. Good Sci.*, 32, 686, 1967.
781. Scheurer, P. G., *J. Food Sci.*, 33, 504, 1968.
782. Love, R. M. and Abel, G., *J. Food Technol.*, 1, 323, 1966.
783. Akiba, M., Motohiro, T., and Tanikawa, E., *J. Food Technol.*, 2, 69, 1967.
784. Spinelli, J., Pelroy, G., and Miyauchi, D., *Fish. Ind. Res.*, 4, 37, 1968.
785. Ravesi, E. M. and Anderson, M. L., *Fish. Ind. Res.*, 5, 175, 1969.
786. Kuusi, T. and Kytokangas, R., *Maataloustieteellinen Aikak.*, 42(1), 30, 1970; *Food Sci. Technol. Abstr.*, 2, 11R378, 1970.
787. Meyer, A., U.S. Patent 2,735,777, 1956.
788. Yamaga, M., Morioka, K., and Kawakami, T., *Reito*, 36, 904, 1961; *C. A.*, 57, 7758d, 1962.
789. Sen, D. P. and Lahiry, N. L., *Food Technol.*, 18, 1611, 1964.
790. Miyauchi, D., Spinelli, J., Pelroy, G., and Steinberg, M. A., *Isotop. Radiat. Technol.*, 5(2), 136, 1967; *Commer. Fish. Abstr.*, 22(3), 7, 1969.
791. Sen, D. P., Visweswarish, K., and Lahiry, N. L., *Food Sci.* (Mysore), 10, 144, 1960; *C. A.*, 56, 1812a, 1962.
792. Garnatz, G., Volle, N. H., and Deatherage, F. E., U.S. Patent 2,488,184, 1949.
793 MacCallum, W. A., Chalker, D. A., Lauder, J. T., Odense, P. H., and Idler, D. R., *J. Fish. Res. Board Can.*, 21, 1397, 1964; *C. A.*, 62, 8310, 1965.
794. Tanikawa, E., *Adv. Food Res.*, 12, 367, 1963.
795. Okamura, K., Matsuda, T., and Yokoyama, M., *Bull. Jap., Soc. Sci. Fish.*, 24, 545, 1958; *C. A.*, 53, 14376e, 1959.
796. Okamura, K., Matsuda, T., and Yokoyama, M., *Nippon Suisangaku Kaishi*, 24, 821, 1958; *C. A.*, 53, 15411g, 1959.
797. Okamura, K., Matsuda, T., and Yokoyama, M., *Nippon Suisan Gakkaishi*, 24, 978, 986, 1958; *C. A.*, 54, 15750a, 1960.
798. Okamura, K., *Nippon Suisan Gakkaishi*, 26, 600, 1960; *C. A.*, 55, 14749b, 1961.
799. Okamura, K., *Nippon Suisan Gakkaishi*, 27, 48, 1961.
800. Okamura, K., *Nippon Suisan Gakkaishi*, 27, 52, 1961.
801. Okamura, K., *Nippon Suisan Gakkaishi*, 27, 58, 1961.
802. Miyake, M., Hayashi, K., and Noda, H., *Mie Kenritsu Daigaku Suisan Gakubu Kiyo*, 6(1), 39, 1963; *C. A.*, 63, 1965.
803. Akiyama, N., *Hokkaido-Ritsu Eisei Kenkyu.*, No. 12, 113, 1961; *C. A.*, 62, 15348c, 1965.
804. Anon., *Commer. Fish.*, 7(1), 19, 1968; *Commer. Fish. Abstr.*, 21(11), 13, 1968.
805. Seagran, H. L., *Food Res.*, 23, 143, 1958.
806. Albright & Wilson (Mfg.), Ltd., British Patent 916,208, 1961.
807. Mahon, J. H., U.S. Patent 3,036,923, 1962.
808. Anon., *Food Eng.*, 34(9), 106, 1962.
809. Tanikawa, E., Akiba, M., and Shitamori, A., *Food Technol.*, 17, 1425, 1963.
810. Mahon, J. H. and Schneider, C. G., *Food Technol.*, 18, 117, 1964.
811. Boyd, J. W. and Southcott, B. A., *J Fish. Res. Board Can.*, 22, 53, 1965.
812. Swartz, W. E., Canadian Patent 847,280, 1970.
813. Sutton, A. H., Paper presented before the FAO Technical Conference on the Freezing and Irradiation of Fish, Madrid, Sept. 4-8, 1967. Reviewed in Anon., *Food Mfg.*, 43(1), 26, 1968.
814. Sutton, A. H. and Ogilvie, J. M., *J. Fish. Res. Board Can.*, 25, 1475, 1968.
815. Miyauchi, D. T., Spinelli, J., Pelroy, G., and Stoll, N., *AEC* Accession No. 29583, Rept. No. TID-22515. 1965, U.S. Government Printing Office, Washington, D. C.; *Nucl. Sci. Abstr.*, 20, 3600, 1966.
816. Spinelli, J. and Weig, D., *Canner/Packer*, 137(12), 28, 1968; *Commer. Fish. Abstr.*, 22(4), 1, 1968.
817. Anon., *Quick Frozen Foods*, 31, 146, April, 1969.

818. Anon., *Quick Frozen Foods*, 31, 150, April, 1969.
819. Anon., *Food Eng.*, 40(2), 11, 1968.
820. Barnett, H. J., Nelson, R. W., and Dassow, J. A., *Fish. Ind. Res.*, 5, 103, 1969.
821. Dyer, W. J., Brockerhoff, H., Hoyle, R. J., and Fraser, D. I., *J. Fish. Res. Board Can.*, 21, 101, 1964.
822. Love, R. M., *Rep. Progr. Appl. Chem.*, 50, 292, 1965.
823. Anon., *Torry Res. Sta. Ann. Rept.*, 1964, 13; quoted in Love, R. M., *Rep. Progr. Appl. Chem.*, 50, 292, 1965.
824. Okamura, K., *Nippon Suisan Gakkaishi*, 26, 60, 1960; *C. A.*, 55, 6720c, 1961.
825. Okamura, K., Kodama, M., Sonada, H., and Yamada, J., *Nippon Suisan Gakkaishi*, 32, 80, 1966; *C. A.*, 67, 10497q, 1967.
826. Tanikawa, E., Motohiro, T., and Akiba, M., Paper presented before the FAO Technical Conference on the Freezing and Irradiation of Fish, Madrid, Sept. 4-8, 1967. Reviewed in Anon., *Food Mfg.*, 43(1), 26, 1968.
827. Ueoka, Y., Oka, H., Suemitso, H., and Sugimoto, T., *Nippon Shokuhin Kogyo Gakkaishi*, 13, 475, 1966; *C. A.*, 67, 2233k, 1967.
828. Takei, M. and Takahashi, T., *Tokaiku Suisan Kenkyusho Kenkyu Hokoku*, No. 30, 105, 1961; *C. A.*, 57, 14248g, 1962.
829. Ichisugi, T., Kuroda, K., and Ikawa, K., *Hokusuishi Geppo*, 22, 273, 319, 1965; *C. A.*, 63, 10576c, 1965.
830. Mathen, C., *Fish Technol.*, 5(2), 104, 1968; *Food Sci. Technol. Abstr.*, 2, 5R167, 1970.
831. Jones, R., *Fish. Ind. Res.*, 5, 83, 1969.
832. Sjostrom, L. B., U.S. Patent 2,446,889, 1948.
833. Tarr, H. L. A., Gardner, L. J., and Ingram, P., *J. Food Sci.*, 34, 637, 1969.
834. Zipser, M. W. and Watts, B. M., *Food Technol.*, 15, 318, 1961.
835. Greig, R. A., Emerson, J. A., and Fliehman, G. W., *Fish. Ind. Res.*, 5, 2, 1969.
836. Hempel, H., U.S. Patent 2,397,547, 1946.
837. Aktieselskabet Protan and Helberud, O., Norwegian Patent 86,681, 1955; *C. A.*, 50, 7348a, 1950.
838. Uchiyama, H. and Amano, K., *Bull. Jap. Soc. Sci. Fish.*, 25(7-9), 531, 1959.
839. Ozawa, T., Nagaoka, S., and Aragaki, M., *Shokuhin Eiseigaku Zasshi*, 4, 332, 339, 1963; *C. A.*, 60, 13788h, 1964.
840. Levin, R. E., *J. Milk Food Technol.*, 30, 277, 1967.
841. McFee, E. P. and Peters, J. A., U.S. Patent 2,554,625, 1951.
842. Kreidl, E. L. and McFee, E. P., U. S. Patent 2,555,236, 1951.
843. McFee, E. P. and Swaine, R. L., *Food Eng.*, 25(12), 67, 1953.
844. Yamada, H., *Nippon Suisan Chuo Kenkyusho Hokoku*, 9, 78, 1961; *C. A.*, 58, 11895f, 1963.
845. Kelch, F. and Buhlmann, X., *Fleischwirtschaft*, 10, 325, 1958.
846. Pacheco, G. and Echaniz, J. S., *Brasil-Med.*, 67(2), 43, 1953.
847. Pacheco, G. and Dias, V. M., *Mem. Inst. Oswaldo Cruz*, 52(2), 405, 1954.
848. Heiligman, F., Desrosier, N. W., and Broumand, H., *Food Res.*, 21, 63, 1956.
849. Williams, O. B. and Hennessee, A. D., *Food Res.*, 21, 112, 1956.
850. Wollmann, E., Veterinary Medical Dissertation, Munich, Germany, 1957; abstracted in *Fleischwirtschaft* 11(4), 291, 1959.
851. Post, F. J., Krishnamurty, G. B., and Flanagan, M. D., *Appl. Microbiol.*, 11, 430, 1963.
852. Gould, G. W., *Fourth Int. Symp. Food Microbiol.*, 1964, Swedish Inst. Food Preserv. Res., Goteborg, Sweden, 17; also *Chem. Ind.*, 1967, 234.
853. Irani, R. R. and Callis, C. F., *J. Am. Chem. Soc.*, 39, 156, 1962.
854. Irani, R. R. and Morgenthaler, W. W., *J. Am. Chem. Soc.*, 40, 283, 1963.
855. Nowakowska-Waszczuk, A., *Acta Microbiol. Polon.*, 14, 207, 1965; *C. A.*, 63, 7367b, 1965.
856. Gray, G. W. and Wilkinson, S. G., *J. Gen. Microbiol.*, 39, 385, 1965.
857. Hamilton-Miller, J. M. T., *Biochem. Biophys. Res. Commun.*, 20, 688, 1965.
858. Brown, M. R. W. and Richards, R. M. E., *Nature*, 207, 1391, 1965.
859. Burkard, G., Weil, J. H., and Ebel, J. P., *Bull. Soc. Chim. Biol.*, 47, 561, 1965; *C. A.*, 63, 10218f, 1965.
860. Tomiyama, T., Kitahara, K., Shiraishi, E., and Imaizumi, K., *Fourth Int. Symp. Food Microbiol.*, 1964, Swedish Inst. Food Preserv. Res., Goteborg, Sweden, 261.
861. Kooistra, J. A. and Troller, J. A., U.S. Patent 3,404,987, 1968.
862. Vishniac, W., *Arch. Biochem.*, 26, 167, 1950.
863. Brock, R. D., *J. Aust. Inst. Agr. Sci.*, 18, 41, 1952.
864. Fuld, E., *Beitr. Chem. Physiol. Path.*, 2, 155, 1902; quoted in Horvath, A. A., *Ind. Eng. Chem. Anal. Ed.*, 18, 229, 1946.
865. Ferrel, R. E., Olcott, H. S., and Fraenkel-Conrat, H., *J. Am. Chem. Soc.*, 70, 2101, 1948.
866. Braginskaya, F. I. and El'piner, I. E., *Biofizika*, 8, 34, 1963; *C. A.*, 58, 12792c, 1963.
867. Council of Scientific and Industrial Research, Indian Patent, 45,580, 1953; *C. A.*, 47, 6576c, 1953.
868. Hall, L. A., U.S. Patent 2,477,742, 1949.
869. Ferrel, R. E. and Olcott, H. S., U.S. Patent 2,522,504, 1950.
870. Mohammad, A., Mecham, D. K., and Olcott, H. S., *J. Agr. Food Chem.*, 2, 136, 1954.

871. Knight, J. W., *The Chemistry of Wheat Starch and Gluten and Their Conversion Products*, Leonard Hill, London, 1965, 92.
872. Beckel, A. C., DeVoss, L. I., Belter, P. A., and Smith, A. K., U.S. Patent 2,444,241, 1948.
873. Beckel, A. C., Belter, P. A., and Smith, A. K., U.S. Patent 2,445,931, 1948.
874. Glabe, E. F., Goldman, P. F., Anderson, P. W., Finn, L. A., and Smith, A. K., *Food Technol.*, 10, 51, 1956.
875. Norris, F. A. and Johnson, D. C., U.S. Patent 3,155,524, 1964.
876. Pintauro, N. D., Reynolds, J. M., and Newman, K. R., U.S. Patent 3,147,125, 1964.
877. Burnett, R. S. and Gunther, J. K., U.S. Patent 2,489,173, 1949.
878. Sevall, H. E. and Schaeffer, R. P., U.S. Patent 2,588,419, 1952.
879. Patterson, B. A., U.S. Patent 2,716,606, 1955.
880. Gunther, R. C., U.S. Patent 2,844,468, 1958; Canadian Patent 565,025, 1958.
881. Sutton, W. J. L., U.S. Patent 2,929,715, 1960.
882. Clausi, A. S., Common, J. L., and Horti, H. M., U.S. Patent 2,954,299, 1960; Canadian Patent 600,649, 1960.
883. Mancuso, J. J. and Common, J. L., U.S. Patent 2,965,493, 1960.
884. Mitchell, W. A. and Seidel, W. C., U.S. Patent 2,968,565, 1961.
885. Block, H. W., U.S. Patent 2,983,617, 1961.
886. Downey, H. A., U.S. Patent 3,018,183, 1962.
887. Grettie, D. P. and Tiemstra, P. J., U.S. Patent 3,490,920, 1970.
888. Schwartz, C., U.S. Patent 2,215,137, 1940.
889. Horvath, A. A., U.S. Patent 2,429,579, 1947.
890. Albright & Wilson, Ltd., British Patent 600,933, 1948.
891. Rane, L. and Newhouser, L. R., U.S. Patent 2,726,235, 1955.
892. Nitschmann, H., Rickli, E., and Kistler, P., *Helv. Chim. Acta*, 42, 2198, 1959; *C. A.*, 54, 12227i, 1959.
893. Keil, H. L., Hagen, R. F., and Flawe, R. W., German Patent 1,111,917, 1961; German Patent 1,133,228, 1962.
894. Fukamachi, C. and Watanabe, T., *Nippon Shokuhin Kogyo Gakkaishi*, 9, 19, 1962; *C. A.*, 59, 13264a, 1963.
895. Waugh, D. F., U.S. Patent 2,744,891, 1956.
896. Consolazio, G. A. and Moses, R. E., U.S. Patent 3,153,030, 1964.
897. Wegner, H., *Starke*, 9, 196, 1957.
898. Winkler, S., *Starke*, 12, 34, 1960.
899. Palasinski, M. and Bussek, J., *Rocz. Technol. Chem. Zywnosci*, 10, 47, 1964; *C. A.*, 61, 9658c, 1964.
900. Palasinski, M., *Zesz. Nauk. Wyzszej Szkoly Rolniczej Krakowie, Rolnictwo*, No. 11(21), 65, 1964; *C. A.*, 62, 15342e, 1965.
901. Nara, S., Maeda, I., and Tsujino, T., *Nippon Nogei Kagaku Kaishi*, 38, 356, 1964; *C. A.*, 63, 5881c, 1965.
902. Knyaginichev, M. I. and Komarov, V. I., *Izv. Vyssh. Ucheb. Zaved., Pishch. Tekhnol.*, 1966(1), 41; *C. A.*, 64, 13286h, 1966.
903. Kuhl, H., *Muhlenlab*, 7, 95, 1937; *C. A.*, 31, 8979, 1937.
904. Nutting, G. C., *J. Colloid Sci.*, 7, 128, 1952.
905. Bowen, W. S., U.S. Patent 2,185,368, 1940.
906. Rozenbroek, M. D., Netherlands Patent 59,843, 1947.
907. Kunz, W. B., U.S. Patent 3,030,667, 1962.
908. Fredrickson, R. E. C., U.S. Patent 3,003,894, 1961.
909. Osman, E. M., *Starch: Chemistry and Technology*, Whistler, R. L. and Paschall, E. F., Eds., Academic Press, New York, 1967, 177.
910. Whistler, R. L., *Cereal Science Today*, 16(2), 54, 1971.
911. Hamilton, R. M. and Paschall, E. F., *Starch: Chemistry and Technology*, Whistler, R. L. and Paschall, E. F., Eds., Academic Press, New York, 1967, 351.
912. Evans, J. W., U.S. Patent 2,806,026, 1957.
913. Kerr, R. W. and Cleveland, F. C., Jr., U.S. Patent 2,884,413, 1959.
914. Kerr, R. W. and Cleveland, F. C., Jr., U.S. Patent 2,961,440, 1960.
915. Neukom, H., U.S. Patent 2,824,870, 1958.
916. Neukom, H., U.S. Patent 2,865,762, 1958.
917. Ferrara, L. W., U.S. Patent 2,865,763, 1958.
918. Neukom, H., U.S. Patent 2,884,412, 1959.
919. Kodras, R., U.S. Patent 2,971,954, 1961.
920. Sietsema, J. W. and Trotter, W. C., U.S. Patent 2,993,041, 1961.
921. Schierbaum, F. and Boerner, O., East German Patent 36,806, 1965; *C. A.*, 63, 3150f, 1965.
922. Felton, G. F. and Schopmeyer, H. H., U.S. Patent 2,328,537, 1943.
923. Wetzstein, H. L. and Lyon, P., U.S. Patent 2,754,232, 1956.
924. Kerr, R. W. and Cleveland, F. C., Jr., U.S. Patent 2,801,242, 1957.
925. Kerr, R. W. and Cleveland, F. C., Jr., U.S. Patent 2,852,393, 1958.
926. Kerr, R. W. and Cleveland, F. C., Jr., U.S. Patent 2,938,901, 1960.
927. Gramera, R. E., Heerema, J., and Parrish, F. W., *Cereal Chem.*, 43, 104, 1966.

928. Deutsche Maizena Werke G.m.b.H., German Patent 1,109,500, 1961.
929. **Ogawa, H.**, Japanese Patent 5,017/70, 1970; Chiyoda Kagako Kogyosho Co., Ltd., Japanese Patent 5018/70, 1970; *Food Sci. Technol. Abstr.*, 2, 9M798, 9M799, 1970.
930. **Anon.**, *Fed. Reg.*, 26, 188, 1961; *Fed. Reg.*, 29, 14403, 1964; *Fed. Reg.*, 29, 15256, 1964; *Fed. Reg.*, 32, 8359, 1967.
931. **Baikow, V. E.**, *Sugar*, 51, 39, 1956; *J. Sci. Food Agr.*, 7, ii, 1956.
932. **Taussig, C. W. and Roland, A. C.**, U.S. Patent 2,249,920, 1941; British Patent 525,014, 1940.
933. **Cummins, A. B.**, *Ind. Eng. Chem.*, 34, 398, 1942.
934. **Wright, A.**, *Ind. Eng. Chem.*, 34, 425, 1942.
935. **Boyd, R.**, Canadian Patent 422,170, 1944.
936. **Bliss, L. R.**, *Int. Sugar J.*, 50, 205, 1948; *C. A.*, 42, 8502, 1948.
937. **Cook, H. A. and Kilby, M.**, *Repts. Hawaiian Sugar Technol.*, 6, 195, 1948; *C. A.*, 44, 7076, 1950.
938. **Bojorquez, M. O., Jasso, J. deJ. U., and Acosta, J.**, *Bol. Azucar. (Mex.)*, 2, 23, July, 1951; *Sugar Ind. Abstr.*, 13, 182, 1951.
939. **Black, R. F. and Zemanek, L.**, *Sugar Ind. Tech.*, 1952, 9; *C. A.*, 46, 9331, 1952.
940. **Ishigaki, H.**, *Sci. Rep. Hyogo Univ. Agr. Ser. Agr. Chem.*, 1, 15, 1953.
941. N. V. Algemeene Norit Mfg., German Patent 1,008,219, 1957.
942. **Zemanek, L. A.**, *Publ. Tech. Papers Proc. Ann. Meeting Sugar Ind. Tech.*, 18, Sect. VI, 9, 1959.
943. **Delfos, J. J.**, U.S. Patent 2,964,428, 1960.
944. **Payet, P. R.**, French Patent 1,230,842, 1960.
945. **Bennett, M. C. and Ragnauth, J. M.**, *Int. Sugar J.*, 62, 13, 41, 1960.
946. **de Sornay, P.**, *Bull. Soc. Chim. Maurice*, 6, 41, 1915; *C. A.*, 10, 1109, 1916.
947. **Schulmann, V.**, *Listy Cukrov.*, 56, 501, 1938; *C. A.*, 32, 8813, 1938.
948. **Klopfer, V.**, U.S. Patent 2,243,381, 1941.
949. **Anon.**, *Food Ind.*, 14(2), 104, 1942.
950. **Alvarex, A. S.**, *Rev. Ind. Agr. Tucuman*, 32, 127, 1942.
951. **Bollaert, A. R. and Halvorsen, G. G.**, U.S. Patent 2,470,332, 1949.
952. **Golovin, P. V. and Gerasimenko, A. A.**, U.S.S.R. Patent 116,101, 1958.
953. **Golovnyak, Y. D., Kartashov, A. K., and Kurilenko, O. K.**, *Izv. Vyssh. Ucheb. Zaved., Pishch. Tekhnol.*, 1962(5), 78; *C. A.*, 60, 4321f, 1964.
954. **Skorobogat'ko, N. I.**, *Sakh. Prom.*, 37(5), 41, 1963; *C. A.*, 59, 6599g, 1963.
955. **Muehlpforte, H.**, *Zuckererzeugung*, 7(4), 106, 1963; *C. A.*, 60, 10909g, 1964.
956. **Onanchenko, L. I.**, *Sakh. Prom.*, 40(2), 11, 1966; *C. A.*, 64, 14399f, 1966.
957. **Katana, A. I.**, *Sakh. Prom.*, 40(12), 12, 1966; *C. A.*, 66, 77201m, 1967.
958. **Golubeva, A. D., Kartashov, A. K., Limanskaya, A. Y., and Verchenko, L. M.**, *Sakh. Prom.*, 43(10), 55, 1969; *Food Sci. Technol. Abstr.*, 2, 6L394, 1970.
959. **Bird, M.**, *Facts About Sugar*, 23, 139, 1928; *Planter and Sugar Mfr.*, 80, 143, 1928.
960. **Ruping, H.**, German Patent 617,706, 1935.
961. **Spengler, O. and Todt, F.**, *Z. Wirtschaftsgruppe Zuckerind.*, 91, 87, 1941.
962. **Loewy, K.**, *Rev. Brasil, Quim.*, 15, 105, 1943; *C. A.*, 37, 6921, 1943.
963. **Gururaja, J. S. and Ray, P. K.**, *Indian Sugar*, 7, 127, 1946.
964. **Andres, P.**, *Z. Wirtschaftsgruppe Zuckerind.*, 90, 228, 1940; *C. A.*, 35, 4987, 1941.
965. **Neumann, M. W.**, *Arch. Suikerind, Nederland en Ned.-Indie*, 1, 51, 1940; *C. A.*, 34, 4604, 1940.
966. **Saha, J. M. and Jain, N. S.**, *Sugar*, 49(3), 38, 1954; *C. A.*, 48, 5533, 1954.
967. Chemische Werke Albert and Sueddeutsche Zucker A.-G., Belgian Patent 611,251, 1962; *C. A.*, 57, 14042h, 1962.
968. **Barrett, E. P., Brown, J. M., and Olech, S. M.**, *Ind. Eng. Chem.*, 43, 639, 1951.
969. **Barrett, E. P.**, *Sugar Ind. Tech.*, 1952, 9; *C. A.*, 46, 9331, 1952.
970. **Kubala, S.**, *Cukoripar*, 5, 229, 1952; *C. A.*, 47, 11775, 1953.
971. **Kortschak, H. P.**, *Hawaiian Plant. Rec.*, 46, 105, 1942.
972. **Jung, E. V.**, U.S. Patent 3,347,705, 1967.
973. **Cummings, A. B.**, Canadian Patent 421,985, 1944.
974. **Fowler, A. P.**, *Proc. Cuban Sugar Tech. Assoc.*, 10, 52, 1936.
975. **Sanchez, P. and Ehrhart, E. N.**, U.S. Patent 2,216,753, 1940.
976. **Isbell, H. S.**, U.S. Patent 2,421,380, 1947.
977. **Vincent, G. P.**, U.S. Patent 2,430,262, 1947.
978. **Andresen, V. V. J.**, U.S. Patent 1,732,492, 1929.
979. **Yasuo, I.**, *Sugar J.*, 18(5), 29, 1955.
980. **Schongart, H. A.**, *Z. Zuckerind.*, 9(1), 10, 1959.
981. **Gupta, S. C. and Ramaiah, N. A.**, *Ind. Sugar*, 14, 719, 1965; *C. A.*, 63, 18435f, 1965.
982. **Moss, H. V.**, *Canner*, 79(6), 18, 1934; *C. A.*, 28, 7054, 1934.
983. **White, J. R. and Dunn, J. A.**, U.S. Patent 2,225,894, 1940.
984. **Nelson, T. J.**, *Food Technol.*, 3, 347, 1949.
985. **Mead, M. W.**, U.S. Patent 2,075,127, 1937.

986. Byk, H., German Patent 247,809, 1910.
987. Epstein, A. K. and Harris, B. R., U.S. Patent 2,075,807, 1937.
988. Martin, G. D., U.S. Patent 2,247,280, 1941.
989. Nyrop, J. E., British Patent 640,241, 1950.
990. Harris, B. R., U.S. Patent 2,177,983, 1939.
991. Katzman, M., U.S. Patent 2,243,868, 1941.
992. Thurman, B. H., U.S. Patent 2,271,409, 1942.
993. Thurman, B. H., U.S. Patent 2,272,616, 1942.
994. Pader, M. and Gershon, S. D., U.S. Patent 3,224,883, 1965; *C. A.*, 67, 2295g, 1967.
995. Pader, M. and Gershon, S. D., U.S. Patent 3,224,884, 1967.
996. Katzman, M. B., U.S. Patent 2,176,078, 1939.
997. Thompson, J. E., U.S. Patent 3,458,323, 1969.
998. Cunningham, F. E., Kline, L., and Lineweaver, H., U. S. Patent 3,328,175, 1967.
999. Gumbmann, M. R., Gagne, W. E., and Williams, S. N., *Toxicol. Appl. Pharmacol.*, 12, 360, 1968.
1000. Curtin, J. H. and Gagolski, J., U.S. Patent 3,375,168, 1968.
1001. Toy, A. D. F., U.S. Patent 3,428,624, 1969.
1002. Stauffer Chemical Co., Netherlands Patent Appl. 6,608,960, 1966; British Patent 1,139,684, 1969.
1003. Tucker, J. W. and Toy, A. D. F., Canadian Patent 825,087, 1969.
1004. Colonial Sugar Refining Co., Ltd., Japanese Patent 21,383/69, 1969.
1005. Kyowa Hakko Kogyo Co., Ltd., Japanese Patent 13,486/70, 1970; *Food Sci. Technol. Abstr.*, 3, 1H58, 1971.

INDEX

A

Agar, 94
  in confectionary, 94
  effect of phosphates on, 94
Alcoholic beverages, phosphates in processing of, 33
Algin, 94-96
  effect of phosphates on, 94-96
Anticaking agents, phosphates as, 31,123
Antioxidants, 46,54,79,82-87,107,121-122,124,130, 143-145,161-162
  phosphates as, 46,54,79,82-87,107,121-122,130,143-145,161-162
  mechanism of action of phosphates as, 83,84
  synergism with chelates, 54,82-89,107,133
Ascorbic acid, effects of phosphates on, 35,83-85,87-89
ATP (adenosine triphosphate), 1,98,100-101,106,117, 124-125,161

B

Bacteria, action of preservatives, phosphates, 33,80,85, 122-123,130-131,145-152
Baked goods, phosphates in, 35-51
Baking powders, 40
  acidic phosphates in, 40
  double-acting, 40
  single-acting, 40
Beer, phosphates in processing of, 31,33
Beverages, 33-35
  carbonated, phosphates in processing of, 31,34
  metal complexing in, 33
  mineral supplementation in nutritional, 33
  phosphates in processing of, 33-34
Bread, phosphates in dough conditioners for, 44-45
Butter, phosphates in processing of, 53-54
Buttermilk, phosphates in processing of, 54

C

Cake donut mixes, leavening systems for, 43-44
Cake mix, leavening systems for, 41-42
Carbonated beverages, acidification, 34
  complexing metals in, 34
  flow improvement of powders, 34
  phosphates in processing of, 34
  water treatment for, 31
Carrageenan, 96
  gels, effect of phosphates on, 96
  in milk products, 96
Casein, phosphate interactions with, 31,52-63,69-76
  calcium-phosphate-casein micelle, effects of phosphates on, 51-53
Cereal products, 35-51
  effects of phosphates on cooking time of, 50-51
  leavening systems in, 35-45
  mineral enrichment of, 47-50
  phosphates in processing of, 35-51
Cheese and cheese products, 31,69-76
  cottage, 69
  natural, 69-72
  phosphates in processing of, 31,69-76
  pasteurized, processed cheese, 31,72-76
  preservation of, by phosphates, 70
Cheese cake mix, phosphates in processing of, 59
Chelating agents, phosphates as, 1-2,24,31-34,51-53,60, 62, 82-85,88-93,94-96,98-105,124,131,146,149,151-152
Chemical preservatives, phosphates as, 1-2,33,46-47,72-74,79-80,85-86,131,145-147,148-152
Cloud and haze in beer and wine, prevention of, 33
Coffee, instant, phosphates in processing of, 34
Complexing agents, phosphates as, 1,24,31,33-34,52-54, 61-62,70,82-85,89-105,124,131,147,149,151-153
Condensed milk, phosphates in processing of, 54
Cream, phosphates in processing of, 54

D

Dairy products, 51-76
  cheese and cheese products, 69-76
    cake mix, 59
    cottage, 69
    natural, 69-72
    pasteurized, processed, 72-76
  coagulation in, 69-74
  condensed milk, 54
  cream, 54
    imitation, 68
  evaporated milk, 54
  frozen desserts, 62-63,68-69
    imitation, 68-69
  ice cream, 62-63
  imitation, 65-69
    cheese products, 69
    cream, 68
    frozen desserts, 68-69
    milk, 66
    sour cream products, 68
    whipping cream, 68
  margarine, 69
  milk products, 54-63
    condensed, 54
    evaporated, 54
    gels, 57-60
    non-fat, dry, 60-62
    puddings, 57-60
    sterile, concentrated, 55-57
  puddings, milk, 57-60
  skim milk, dry, 60-62
Dental caries, phosphates in prevention of, 48,162-163
Dessert gels, phosphates in processing of, 57-60,89-90, 94-97,155-157
Dough improvers, phosphates in, 46
  refrigerated, phosphate leavening acids in, 44-46
    SAPP flavor in, 44-46
Drupex process, 80
Dry mix products, 40-44
  for baking, 40-44
  biscuit, 43
  cake, 41-42

cake donut, 44-45
leavening systems for, 40-44
pancake, 42-43
phosphates in processing of, 40-44
self-rising corn meal, 43
self-rising flour, 43
waffle, 42-43

E

EDTA, comparison with phosphates, 52,70,85,89,93-94, 102-105,114,142-143
Egg products, phosphates in processing of, 76-80
color stabilization, 77-80
egg white, 76-80
foam improvers, 77,79-80,162,163
inhibition of fat oxidation, 177
pasteurization processes, improved, 79-80
shell cleaning, 76-77
whole egg processing, 79-80
Emulsion, 85,162
phosphates in processing of, 85,162
stabilization, 31,63,66,72-76,85,120-121,124-125, 127,162
Enzyme, hydrolysis of polyphosphates, 1,100
inhibition by phosphates, 46,79,86-87,143-144
rennet, phosphates to control activity, 52,70
synthesis of polyphosphates, 1

F

Fats and oils, 80-85
antioxidants for, phosphates as, 82-85
extraction, phosphates in, 80
processing of, phosphates in, 80-82
O/W emulsions, 85
Ferric orthophosphates, 34,49-50
Ferric pyrophosphate, 34
Fish and fish products, 131-147
binding, increasing, 134-135
biochemistry of, phosphate action in, 131-133
color preservation in, 133
fat oxidation, prevention of, 133-147
flavor improvement, 143-145
microbiological preservation, 145-147
moisture retention, increasing, 135-143
phosphates in processing of, 131-147
rate of phosphate incorporation, 131-133
struvite, prevention of formation, 147
tenderness, increasing, 133-134
thawing-drip, reduction of, 135,138-143
Foods, enhancement of nutritive value, 47-50
preservation of, 1,33,46-47,79-80,86,122,130-131, 145-152
Food and Drug Administration, 14,27,50,63,76-77,93
Flour, phosphated, 35-36
Frozen desserts, 62-63,68-69
Fruits and vegetables, 85-94
phosphates in processing of, 85-94
preservation of, by phosphates, 85-86
sequestrants in, 85-94
Fruit juices, phosphates in processing of, 85-94

G

Gelatin, polyphosphate interaction with, 97,154-157
Gelatin desserts, phosphates in processing of, 154-157
Gel inducing agents, 57-60,89-91,94-97,149-160
Gluten, polyphosphate interaction with, 154
Gluten gel, 154
Gums, 94-97
Gums, individual, agar, 94-96
algin, 94-96
carageenan, 96
guar, 96-97
locust bean, 96-97
phosphates in processing and use of, 94-97
Guar gum, 96-97
phosphate interaction with, 96-97

H

Hydrogenation of fats, phosphates in, 82

I

Ice cream, agglomeration of fat particles, 62-63
phosphates in processing of, 62-63
Imitation, cheese products, 69
cream, 68-69
dairy products, 65-69
frozen desserts, 69
milk, 66
sour cream products, 68
whipping cream, 68
Iron enrichment, 48-50

L

Lactose, phosphates in processing of, 63-64
Leavening, 35-45
acids, 35-45
controlling properties of, 35-36,41-43
disagreeable flavors of, 38,40,44
dough reaction rates of, 35
measuring characteristics of, 35,42-43
neutralizing values of, 35,37
phosphates as, 35-45
properties required, 35
rates of reaction of, 35
description of, 35

M

Macaroni, shortening cooking time, phosphates in, 50
Margarine, 69
phosphates in processing of, 69
Meat products, 97-124
binding of, 106,110,113,117-118,123-124
emulsification of, 120-122
flavor improvement in, 107,111,114-117,121-122, 123-124
maintaining reddish color of, 105-106,108,111-112, 114-117,123-124
microbiological preservation of, 108,122-123

moisture retention, improvement of, 98-107,110-111, 113-114,120,123-124
phosphates in processing of, 97-124
preservatives for, 108,122
proteins, effect of phosphates on, 98-105,108-111, 113
rancidity, prevention of, 108,111,121-122
tenderizing, 105-106,110,112-113,117
Mechanism of microbial action, 148-149
of phosphates, 148-149
Metals ions, chelating, complexing and sequestering, 1, 24,31-34,51-53,59-62,70,98-105,146-153
Milk products, see dairy products
Milk puddings, phosphates in, 57-60
Mineral enrichment, 47-50
Molds, action of preservatives on, phosphates, 33,86,152
Mono- and di-glycerides, 83

O

Oils, extraction of, phosphates in, 80
phosphates in processing of, 80-85
Organic phosphates as antioxidants, 161-162
in food processing, 161-162
preparation, 161-162
as whipping agents, 160

P

Peanut butter, 85
Pectins, 89-91
gels, phosphates in 89-91
pH, phosphates in adjustment of, 31-33,35-45,53-54,69, 76-85,87-89,105-106,108-110,114-117,122-123
Phosphates, 1-165
acceptable daily intake, 24-25
applications in foods, 2,14,15,33,162-163
beverages, 33-35
cereal products, 35-51
dairy products, 51-76
egg products, 76-80
fats and oils, 80-85
food protein processing, 152-157
fruit and vegetable products, 85-94
gums and gels, 94-97
meat processing, 97-131
microbiological inhibitors, 148-152
poultry products, 124-131
seafood products, 131-148
starch processing, 157-160
sugar processing, 160-161
cyclic polyphosphates, 1,3,11,14-15,18,20,23-24,25-28,31,57,65,72,107,159-160
hydrolysis, 17
in metal chelation, 31
nomenclature, classification and structure, 3,14
occurrence in living organisms, 1
regulatory status, 14,24-25
toxicity, 14,19-24
flavors contributed, 38-40,45,75-76
functions in foods, 1-2,7,15-16,24,31-54,57-72,76-80,89-91,94-112,114-124,131-162
in acidification or lowering pH, 31,33-45,53-54, 69,76,87,90-91,108,131-133
in alkalization or raising pH, 31,72-76,114,122
in buffering or pH stabilization, 1,31,33,35,41-42, 85,87-89,105-106,108-110,114-117,122
in chelating, complexing and sequestering metal ions, 1,15,17,24,31-34,51-53,60-62,70,98-103,105,146,149,151-153
in contro of enzymatic activity, 46,47,52,70,87, 96,144
in direct chemical reactions with food constituents, 18,43,106,157-159
in dispersion of food constituents, 32,52,64,65,69, 79,98-105,131-133
in emulsion stabilization, 31,63,65,68,169,72-76, 80,120,121,127
in flavor stabilization, 45,84,107,111,121,122, 127,130,143-145
in food gel preparation, 57-62,89-91,94-97
in increasing hydration, moisture retention and water holding capacity of food constituents, 17,31, 32,97,101,103-105,110,111,113,114,118-120,126-127,135,138,139,140-143,154
in microbiological inhibition and control, 1,33,46, 47,129-130,152,147-148
in mineral enrichment or supplementation, 6,7,31, 35,47-50,
in preservation of natural food colors, 50,86-87, 105,108,114-117,120-122,124,133
in prevention of caking of dry foods, 105-107,122, 123
in prevention of oxidation of food constituents, 46,82-87,108,111,122,143-147,161,162
in protein precipitation and extraction, 52,53,63-66,98-100,131-133,155-157
in reactions with other polyelectrolytes, 1,15,17, 32,52,53,57,66,68-69,76,78-80,94-97,98-105,111, 114,117-121,124-127,131-145,152-157
hydrolysis of polymeric, 17,56-57,75,100-101
ionization characteristics, 15
nomenclature, classification and structure, 3-14,15
occurrence in living organisms, 1
organic, in food processing, 161-163
in origin of life on earth, 1
regulatory status, 26,27-29,51,62,76-78,93-94,105
stability in foods, 17,56-57,75
straight chain polyphosphates,
anions of, 10,15-17
average chain length, 10-14
composition, 3,10-14
hydrolysis, 15-18
ionization, 15
occurrence in living organisms, 1
regulatory status, 27-29,76,105
stability, 17
structure of chains, 15,17
toxicity, 19-24
synthesis by living organisms, 1
titration characteristics, 15
toxicity, 14,19-25
Phosphoric acid, 3,9,10,14-15,19,22,33,35,53-56,69-72, 107,130-131

use in foods, 33-34,53-54,69,72,76-91,107,130-131
Phosphorus, 1,24,46-48,85-86
acceptable daily intake, 24-25
enrichment in foods, 47-48
in foods, 1,24-25,85-86
in plants, 1,85-86
requirements in living organisms, 1
Potato products, 86,88-89,91
after-cooking darkening, prevention of, 88-89
improving texture, 91
peeled, preservation of, 86
phosphates in processing of, 86,88-89,91
Poultry products, 124-131
binding of, 125-126
color preservation, 124
fat emulsification in, 127
flavor improvement, 127-130
microbiological preservation of, 130-131
moisture retention, improvement of, 126-127
phosphates in processing of, 124-131
rancidity in, prevention of, 124,130
tenderness in, increasing, 124-125
Preservatives, chemical, phosphates as, 1-2,33,46-47,70-72,80,85-86,130-131,145-152
Processed meats, phosphates in, 114-124
Proteins, phosphates in processing of, 1,17,33,50,52-53,57-60,62-69,72-80,94-107,113-117,123-145,152-157
Puddings, 57-60
phosphates in processing of, 57-60

R

Rennet, 52,70
effects of phosphates on activity of, 52,70

S

Seafood products, see fish and fish products
Sequestrants, phosphates as, 1-2,17,24,31,33-35,51-54,61-63,68,82-85,98-105,123-124,131,147-153
Skim milk, phosphates in processing of, 60-62
Sodium benzoate, synergism with polyphosphates, 86,108-110
Sodium caseinate, 65-69
Sorbates, synergism with polyphosphates, 86,108-110
Sour cream products, imitation, 68
Starch phosphates, processing of, 157-160
phosphates in processing of, 12,31,157-160
puddings, instant, 57-60
Sterile concentrated milk, 55-57
Struvite, phosphates in prevention of, 147
Sugar, 160-161
processing, phosphates in, 160-161
Sweet potatoes, preventing after-cooking discoloration, 85
Synergists for antioxidants, 54,82-85,86-87,107,133

T

Tomato products, phosphates in processing of, 90-91

V

Vegetables and fruits, phosphates in processing of, 85-94
Viruses, phosphates as inhibitors of, 71,152

W

Water treatment, phosphates in, 31
Whey products, phosphates in processing of, 50-51,63-65
proteins, effects of phosphates on, 50-51,64-65
Whipped cream and whipped toppings, 68
phosphates in processing of, 68
Wines, phosphates in processing of, 33
preservatives for, 33

Y

Yeast, phosphates as inhibitors of growth of, 33,84,86,152
phosphates as promoters of growth of, 51

"SOYBEANS AS A FOOD SOURCE"

Authors: W. J. WOLF, B.S., Ph.D., Head of Meal Products Investigation in the Oilseed Crops Laboratory, Northern Marketing and Nutrition Research Division, Peoria, Illinois.

J. C. COWAN, A.B., Ph.D., Chief of the Oilseed Crops Laboratory, Northern Marketing and Nutrition Research Division, Peoria, Illinois.

Approx. 80 pp., 7½ x 10½, 1971. (Reprinted from CRC Critical Reviews™ in Food Technology.)

Edited by: Thomas E. Furia, Technical Development Manager, CIBA-GEIGY Corp.

Animal proteins are expensive and in shortest supply where needed. Because of this plant proteins have received a great deal of attention. Lists of potential sources of unconventional food proteins rank soybeans at or near the top. They have a long history as a foodstuff in the Orient and supply protein, calories and essential fatty acids. Soybeans are especially attractive as a protein source when compared to animal proteins on a cost basis. An important advantage of soybeans over other protein materials is their availability in large quantities. A variety of food-grade soy proteins are commercially available whereas other proteins often found in the potential category are still floundering with economic, technical and legal problems.

This review summarizes the conversion of soybeans to food. Those newly concerned with soy proteins as a food should find this monograph very helpful. Emphasis is given to the protein because of the present high level of interest.

CAT. NO. 112 (Add 50¢ Outside U.S.A. — U. S. Currency Only) **Each $11.50**

"THE USE OF FUNGI AS FOOD AND IN FOOD PROCESSING"

Author: WILLIAM D. GRAY, A.B., Ph.D., Department of Botany, Southern Illinois University, Carbondale, Illinois.

114 pages, 7½ x 10½. Hard cover. Well documented with 348 references. (Reprinted from Critical Reviews™ in Food Technology.) 1970.

Edited by: Thomas E. Furia, Technical Development Manager, CIBA-GEIGY Corp.

This review is an attempt to describe how filamentous fungi have been used in the past and are used in the present in various areas of the world in connection with man's eternal attempt to satisfy his urge for food. Attention has been given to the potential which such fungi seem to possess for making large contributions to the world protein supply.

CONTENTS

INTRODUCTION. DIRECT USE OF FUNGI AS FOOD — *mushrooms as foods; cultivation of truffles, common mushroom* and the Padi-Straw mushroom. **FUNGI AS ANIMAL FOOD. USE OF FUNGI IN FOOD PROCESSING** — fungi in the processing of cheese, Miso, Shoyu, Hamanatto, Tempeh, Aug-klah, Ontjom, Sufu, Meifauza, Ketjap, Katsuobushi and other fungus-fermented foods. **POTENTIAL ADDITIONAL USES OF FUNGI AS FOOD. MYCELIA OF FLESHY FUNGI. MOLD TYPE OF MYCELIUM MYXOMYCETE PLASMODIA. TOXIC SUBSTANCES IN FUNGI** — Stachybotryotoxicosis, Aspergillustoxicosis. Moldy Corn Toxicosis, Facial Eczema in Ruminants, Alimentary Toxic Aleukia (ATA), Estrogenic Condition in Swine, Toxic Moldy Rice, Aflatoxin and Other Toxics. **SUMMATION.**

CAT. NO. 104 (Add 50¢ Outside U.S.A. — U. S. Currency Only) **Each $12.75**

"CRC HANDBOOK OF FOOD ADDITIVES"

Edited by THOMAS E. FURIA, Technical Development Manager, CIBA-GEIGY Corp. 16 specialist contributors.

771 pp., 7½ x 10½. 1968.

Properties and uses of direct food additives are described by well-qualified specialists. Intended for the practicing technologist, the book will also be of distinct value as a classroom or laboratory reference. The text is written in an uncomplicated fashion to provide rapid reference, and to create a better interdisciplinary understanding of the subject. To further aid the user, each chapter is followed by an extensive list of references, typically extending to well over 100 entries. The book is concluded by a 184-page section which summarizes in concise form the U. S. Government regulations regarding the use of over 2,000 frequently employed direct food additives.

CAT. NO. 541 (Add 50¢ Outside U.S.A. — U. S. Currency Only) **Each $28.50**

"CRC FENAROLI'S HANDBOOK OF FLAVOR INGREDIENTS"

Translated and Edited by: THOMAS E. FURIA and NICOLO BELLANCA, of the CIBA-GEIGY Corp., Ardsley, New York. From the Italian Language Works of Prof. Giovanni Fenaroli, Director, Center for Studies of Aromatic Substances, University of Milano, Milano, Italy. 803 pp., 7½ x 10½. 1971.

The chemical armament of the flavor industry contains by far the most numerous constituents of all food ingredients. In the U. S. A., it constitutes no less than 1,200 ingredients of which 200 or so are well characterized products of natural origin while approximately 1,000 are precisely defined synthetics. The task of selecting and blending those ingredients capable of imparting a desired flavor characteristic to complex substrates (often times having flavor backgrounds of their own) and at the same time being mindful of toxicological, regulatory and economic considerations is indeed formidable.

While detailed treatises on various aspects of flavor technology have appeared from time-to-time, this is the first effort to include this pertinent information in one volume. In an effort to present as broad an information base as possible into a logical sequence capable of future expansion, the information is presented in four interrelated parts: Flavor Ingredients, Natural Flavors, Synthetic Flavors, and The Use of Flavor Ingredients in Food.

CAT. NO. 531 (Add 50¢ Outside U.S.A. — U. S. Currency Only) **Each $35.00**